| DATE | | | |
|---|---|---|---|
| | | | |
| | | | |
| | | | |
| | | | |
| | | | |
| | | | |
| | | | |
| | | | |
| | | | |
| | | | |
| | | | |
| | | | |

# CONTROLLING VOLATILE EMISSIONS
# AT HAZARDOUS WASTE SITES

# CONTROLLING VOLATILE EMISSIONS AT HAZARDOUS WASTE SITES

by

**John R. Ehrenfeld, Joo Hooi Ong**

Arthur D. Little, Incorporated
Cambridge, Massachusetts

**William Farino, Peter Spawn**
**Michael Jasinski, Brian Murphy**

GCA Corporation
Bedford, Massachusetts

**Douglas Dixon, Edwin Rissmann**

Versar, Incorporated
Springfield, Virginia

**NOYES PUBLICATIONS**
Park Ridge, New Jersey, U.S.A.

Published in the United States of America by
Noyes Publications
Mill Road, Park Ridge, New Jersey 07656

10 9 8 7 6 5 4 3 2 1

Library of Congress Cataloging-in-Publication Data
Main entry under title:

Controlling volatile emissions at hazardous waste
    sites.

    (Pollution technology review, ISSN 0090-516X ; no.
126)
    Bibliography: p.
    Includes index.
    1. Hazardous wastes. 2. Organic compounds--
Environmental aspects. I. Ehrenfeld, John. II. Series.
TD811.5.C655  1986     628.5'3     85-25951
ISBN 0-8155-1063-2

# Foreword

An evaluation of emission control technologies for hazardous waste treatment, storage, and disposal facilities (TSDFs) is presented in this book. At present, numerous chemical substances and chemical waste streams are "listed" as hazardous. Volatile emissions from TSDFs represent a major source of organic contaminants entering the atmosphere. Previous studies estimate that volatile emissions from TSDFs range from 1.6 to 5 million metric tons/year in the U.S.A. The need to control these volatile emissions is obvious. This book should be extremely useful to those concerned with controlling volatile hazardous emissions.

The book is divided into three parts. Part A evaluates controls for volatile emissions arising from the treatment, storage, and disposal of hazardous wastes. For each principal type of hazardous waste management facility, sources of atmospheric emissions are described and controls representing different approaches are examined and compared. The evaluation is based on actual data and on theoretical models, where data are lacking, where the control technologies have been borrowed from other types of applications, or are novel concepts.

Mathematical models describing the release rate of volatile air emissions from hazardous waste TSDFs are compiled and reviewed in Part B. Mathematical modeling techniques which predict volatile air emissions release rates from landfills, landfarms, surface impoundments, storage tanks, wastewater treatment processes, and drum handling and storage facilities are assessed. Existing field test validation efforts are also reviewed.

Part C represents (1) physical-chemical properties of Resource Conservation and Recovery Act (RCRA) wastes related to volatility and (2) a waste categorization scheme based on the volatility of RCRA wastes from hazardous waste TSDFs. Physical-chemical properties of RCRA wastes covered include molecular weight, boiling point, vapor pressure, solubility, Henry's Law constant, diffusion coefficients, and mass transfer coefficients. RCRA wastes are categorized by vapor pressure, aqueous volatility, and relative soil volatility. The physical-chemical properties include basic input parameters for air emission modeling.

The information in the book is from:

> *Evaluation of Emission Controls for Hazardous Waste Treatment, Storage, and Disposal Facilities,* prepared by John R. Ehrenfeld and Joo Hooi Ong of Arthur D. Little, Incorporated for the U.S. Environmental Protection Agency, November 1984.

> *Evaluation and Selection of Models for Estimating Air Emissions from Hazardous Waste Treatment, Storage, and Disposal Facilities,* prepared by William Farino, Peter Spawn, Michael Jasinski, and Brian Murphy of GCA Corporation for the U.S. Environmental Protection Agency, December 1984.

> *Physical-Chemical Properties and Categorization of RCRA Wastes According to Volatility,* prepared by Douglas Dixon and Edwin Rissmann of Versar, Incorporated for the U.S. Environmental Protection Agency, February 1985.

The table of contents is organized in such a way as to serve as a subject index and provides easy access to the information contained in the book.

> Advanced composition and production methods developed by Noyes Publications are employed to bring this durably bound book to you in a minimum of time. Special techniques are used to close the gap between "manuscript" and "completed book." In order to keep the price of the book to a reasonable level, it has been partially reproduced by photo-offset directly from the original reports and the cost saving passed on to the reader. Due to this method of publishing, certain portions of the book may be less legible than desired.

## NOTICE

> These reports have been reviewed by the Office of Air Quality Planning and Standards, U.S. Environmental Protection Agency, and approved for publication as received from Arthur D. Little, Incorporated, GCA/Technical Division of GCA Corporation, and Versar Incorporated. Approval does not signify that the contents necessarily reflect the views and policies of the U.S. Environmental Protection Agency or the Publisher, nor does mention of trade names or commercial products constitute endorsement or recommendation for use.

# Contents and Subject Index

PART A
EVALUATION OF EMISSION CONTROL TECHNOLOGIES

I.  SUMMARY AND CONCLUSIONS . . . . . . . . . . . . . . . . . . . . . . . . . .2
    Reference . . . . . . . . . . . . . . . . . . . . . . . . . . . . . . . . . . . . .6

II. INTRODUCTION . . . . . . . . . . . . . . . . . . . . . . . . . . . . . . . . . .7
    A. Objective . . . . . . . . . . . . . . . . . . . . . . . . . . . . . . . . . . . .7
    B. Framework for the Analysis . . . . . . . . . . . . . . . . . . . . . . . .7
    C. Hazardous Waste Treatment, Storage and Disposal Facility—
       Definitions and Descriptions . . . . . . . . . . . . . . . . . . . . . . .11
       1. Landfills . . . . . . . . . . . . . . . . . . . . . . . . . . . . . . . . .11
       2. Land Treatment . . . . . . . . . . . . . . . . . . . . . . . . . . . .12
       3. Surface Impoundments . . . . . . . . . . . . . . . . . . . . . . . .12
       4. Storage and Treatment Tanks . . . . . . . . . . . . . . . . . . .13
    D. Controls/Definitions and Description . . . . . . . . . . . . . . . .13
       1. Pretreatment Controls . . . . . . . . . . . . . . . . . . . . . . . .13
       2. Design and Operating Practices . . . . . . . . . . . . . . . . . .14
       3. In-Situ Controls . . . . . . . . . . . . . . . . . . . . . . . . . . . .14
       4. Post-Treatment Techniques . . . . . . . . . . . . . . . . . . . .14
       References . . . . . . . . . . . . . . . . . . . . . . . . . . . . . . . . .16

III. CONTROLS FOR SURFACE IMPOUNDMENTS . . . . . . . . . . . .17
    A. Surface Impoundment Description . . . . . . . . . . . . . . . . . .17
       1. Definition . . . . . . . . . . . . . . . . . . . . . . . . . . . . . . . .17
       2. Types, Construction and Uses . . . . . . . . . . . . . . . . . . .17
       3. Operation . . . . . . . . . . . . . . . . . . . . . . . . . . . . . . . .18
          3.1 Active . . . . . . . . . . . . . . . . . . . . . . . . . . . . . . . .18
          3.2 Closure . . . . . . . . . . . . . . . . . . . . . . . . . . . . . . .20

3.3 Post-Closure . . . . . . . . . . . . . . . . . . . . . . . 23
**B. Emission Sources and Models** . . . . . . . . . . . . . . . . . . . 23
   1. General Description of Factors Affecting Emissions . . . . . . . . . . 23
   Active . . . . . . . . . . . . . . . . . . . . . . . . . . . . 23
   Closure . . . . . . . . . . . . . . . . . . . . . . . . . . . 24
   Post-Closure . . . . . . . . . . . . . . . . . . . . . . . . . 24
   2. Emission Models . . . . . . . . . . . . . . . . . . . . . . . 25
   Gas Film, Nonaerated . . . . . . . . . . . . . . . . . . . . . 27
   Liquid Film, Nonaerated . . . . . . . . . . . . . . . . . . . . 28
   Gas-Film, Aerated . . . . . . . . . . . . . . . . . . . . . . . 29
   Liquid Film, Aerated . . . . . . . . . . . . . . . . . . . . . . 30
   Floating Immiscible Organic Layer . . . . . . . . . . . . . . . . 32
   3. Parameters that Control Emissions . . . . . . . . . . . . . . . . 32
**C. Potential Controls** . . . . . . . . . . . . . . . . . . . . . . . . 33
   1. Introduction . . . . . . . . . . . . . . . . . . . . . . . . . 33
   2. Pretreatment . . . . . . . . . . . . . . . . . . . . . . . . . 34
   3. Design and Operating Practices . . . . . . . . . . . . . . . . . 34
      3.1 Design . . . . . . . . . . . . . . . . . . . . . . . . . 35
         3.1.1 Surface Area Minimization . . . . . . . . . . . . . . 35
         3.1.2 Freeboard Depth . . . . . . . . . . . . . . . . . . . 35
         3.1.3 Inflow/Outflow Drainage Pipe Locations . . . . . . . . 36
      3.2 Operating Practices—Active . . . . . . . . . . . . . . . . 36
         3.2.1 Temperature of Influent . . . . . . . . . . . . . . . 36
         3.2.2 Dredging, Draining, and Cleaning Frequency . . . . . . 37
         3.2.3 Handling of Sediments and Sludge . . . . . . . . . . . 37
      3.3 Operating Practices—Closure . . . . . . . . . . . . . . . . 37
         3.3.1 Dewatering . . . . . . . . . . . . . . . . . . . . . 37
         3.3.2 Proper Consolidation . . . . . . . . . . . . . . . . . 38
         3.3.3 Fugitive Dust Abatement . . . . . . . . . . . . . . . 38
   4. In-Situ Controls . . . . . . . . . . . . . . . . . . . . . . . 39
      4.1 Rafts . . . . . . . . . . . . . . . . . . . . . . . . . . 39
      4.2 Barriers . . . . . . . . . . . . . . . . . . . . . . . . . 41
      4.3 Shades . . . . . . . . . . . . . . . . . . . . . . . . . 41
      4.4 Synthetic Covers . . . . . . . . . . . . . . . . . . . . . 43
      4.5 Floating Spheres . . . . . . . . . . . . . . . . . . . . . 44
      4.6 Surfactant Layers . . . . . . . . . . . . . . . . . . . . . 46
   5. Post-Treatment . . . . . . . . . . . . . . . . . . . . . . . . 48
      5.1 Gaseous Carbon Adsorption . . . . . . . . . . . . . . . . 49
      5.2 Afterburners . . . . . . . . . . . . . . . . . . . . . . . 50
**D. Effectiveness of Surface Impoundment Controls** . . . . . . . . . . 52
   1. Methodology . . . . . . . . . . . . . . . . . . . . . . . . . 52
      1.1 Selection of Parameter Values . . . . . . . . . . . . . . . 52
      1.2 Calculation of Mass Transfer Coefficients and Emissions . . . . 52
      1.3 Emission Reduction and Efficiencies of Controls . . . . . . . 55
      1.4 Cost-Effectiveness of Controls . . . . . . . . . . . . . . . 55
   2. Parameters . . . . . . . . . . . . . . . . . . . . . . . . . . 56
   3. Mass Transfer Coefficients and Emissions . . . . . . . . . . . . . 58
   4. Emissions Reduction and Efficiencies . . . . . . . . . . . . . . . 58

Rafts. . . . . . . . . . . . . . . . . . . . . . . . . . . . . . . . . . . . . . .66
Barriers . . . . . . . . . . . . . . . . . . . . . . . . . . . . . . . . . . . . .71
Shades. . . . . . . . . . . . . . . . . . . . . . . . . . . . . . . . . . . . . .71
Floating Spheres . . . . . . . . . . . . . . . . . . . . . . . . . . . . . . .71
Post-Treatment. . . . . . . . . . . . . . . . . . . . . . . . . . . . . . . .72
  5. Costs. . . . . . . . . . . . . . . . . . . . . . . . . . . . . . . . . . . . . .72
Pretreatment . . . . . . . . . . . . . . . . . . . . . . . . . . . . . . . . .72
In-Situ Controls . . . . . . . . . . . . . . . . . . . . . . . . . . . . . . .72
    Rafts. . . . . . . . . . . . . . . . . . . . . . . . . . . . . . . . . . . . .72
    Barriers . . . . . . . . . . . . . . . . . . . . . . . . . . . . . . . . . . .77
    Shades. . . . . . . . . . . . . . . . . . . . . . . . . . . . . . . . . . . .77
    Spheres . . . . . . . . . . . . . . . . . . . . . . . . . . . . . . . . . . .77
Post-Treatment. . . . . . . . . . . . . . . . . . . . . . . . . . . . . . . .78
  6. Cost-Effectiveness of Controls . . . . . . . . . . . . . . . . . . . . .80
 E. Summary. . . . . . . . . . . . . . . . . . . . . . . . . . . . . . . . . . . . .85
 F. References. . . . . . . . . . . . . . . . . . . . . . . . . . . . . . . . . . . .86

IV. CONTROLS FOR TANKS . . . . . . . . . . . . . . . . . . . . . . . . . .89
 A. Tank Description. . . . . . . . . . . . . . . . . . . . . . . . . . . . . . . .89
  1. Definition . . . . . . . . . . . . . . . . . . . . . . . . . . . . . . . . . . .89
  2. Types, Construction and Uses . . . . . . . . . . . . . . . . . . . . .89
  3. Operation . . . . . . . . . . . . . . . . . . . . . . . . . . . . . . . . . .90
 B. Emission Sources and Models. . . . . . . . . . . . . . . . . . . . . . .90
  1. General Description of Factors Affecting Emissions . . . . . . . . .90
  2. Emission Models . . . . . . . . . . . . . . . . . . . . . . . . . . . . . .91
  3. Parameters that Control Emissions . . . . . . . . . . . . . . . . . . .91
 C. Potential Controls . . . . . . . . . . . . . . . . . . . . . . . . . . . . . .91
  1. Summary of Applicable Controls . . . . . . . . . . . . . . . . . . . .91
  2. Pretreatment . . . . . . . . . . . . . . . . . . . . . . . . . . . . . . . .91
  3. Design and Operating Practices. . . . . . . . . . . . . . . . . . . . .94
  4. In-Situ Controls . . . . . . . . . . . . . . . . . . . . . . . . . . . . . .94
    4.1 Fixed Roofs. . . . . . . . . . . . . . . . . . . . . . . . . . . . . . .94
    4.2 Floating Roofs . . . . . . . . . . . . . . . . . . . . . . . . . . . . .95
    4.3 Rafts. . . . . . . . . . . . . . . . . . . . . . . . . . . . . . . . . . .96
    4.4 Floating Spheres . . . . . . . . . . . . . . . . . . . . . . . . . . . .96
  5. Post-Treatment. . . . . . . . . . . . . . . . . . . . . . . . . . . . . . .96
 D. Effectiveness of Tank Controls. . . . . . . . . . . . . . . . . . . . . . .97
  1. Methodology . . . . . . . . . . . . . . . . . . . . . . . . . . . . . . . .97
    1.1 Selection of Parameter Values . . . . . . . . . . . . . . . . . . . .97
    1.2 Calculation of Mass Transfer Coefficients and Emissions . . . . .97
    1.3 Emission Reduction and Efficiencies of Controls. . . . . . . . . .98
    1.4 Cost-Effectiveness of Controls . . . . . . . . . . . . . . . . . . . .98
  2. Parameters. . . . . . . . . . . . . . . . . . . . . . . . . . . . . . . . . .99
  3. Emissions. . . . . . . . . . . . . . . . . . . . . . . . . . . . . . . . . . .99
  4. Emissions Reduction and Efficiencies . . . . . . . . . . . . . . . . .99
In-Situ. . . . . . . . . . . . . . . . . . . . . . . . . . . . . . . . . . . . . .99
Post-Treatment. . . . . . . . . . . . . . . . . . . . . . . . . . . . . . . .103
  5. Costs. . . . . . . . . . . . . . . . . . . . . . . . . . . . . . . . . . . . .103

Pretreatment . . . . . . . . . . . . . . . . . . . . . . . . . . . . . . 103
In-Situ Controls . . . . . . . . . . . . . . . . . . . . . . . . . . . . 103
Post-Treatment. . . . . . . . . . . . . . . . . . . . . . . . . . . . . 103
6. Cost-Effectiveness of Controls . . . . . . . . . . . . . . . . . . . 105
E. Summary. . . . . . . . . . . . . . . . . . . . . . . . . . . . . . . . . 110
References . . . . . . . . . . . . . . . . . . . . . . . . . . . . . . . . . 112

V.  LANDFILLS . . . . . . . . . . . . . . . . . . . . . . . . . . . . . . . . 114
A. Landfill Description. . . . . . . . . . . . . . . . . . . . . . . . . . 114
B. Emission Sources and Models. . . . . . . . . . . . . . . . . . . . 115
1. General . . . . . . . . . . . . . . . . . . . . . . . . . . . . . . . 115
2. Emission Models . . . . . . . . . . . . . . . . . . . . . . . . . . 116
Covers. . . . . . . . . . . . . . . . . . . . . . . . . . . . . . . . . 116
Bulk Liquids. . . . . . . . . . . . . . . . . . . . . . . . . . . . . 119
Bulk Solids. . . . . . . . . . . . . . . . . . . . . . . . . . . . . . 119
3. Controlling Parameters. . . . . . . . . . . . . . . . . . . . . . 120
Covers. . . . . . . . . . . . . . . . . . . . . . . . . . . . . . . . . 120
Working Faces . . . . . . . . . . . . . . . . . . . . . . . . . . . 120
C. Controls . . . . . . . . . . . . . . . . . . . . . . . . . . . . . . . . . 120
1. Introduction. . . . . . . . . . . . . . . . . . . . . . . . . . . . 120
2. Pretreatment . . . . . . . . . . . . . . . . . . . . . . . . . . . . 121
3. Design and Operating Practices. . . . . . . . . . . . . . . . . 121
3.1 Handling Bulk Liquids or Solids. . . . . . . . . . . . . . 121
3.2 Temporary Cover Practices . . . . . . . . . . . . . . . . 121
3.3 Final or Permanent Covers . . . . . . . . . . . . . . . . 127
4. In-Situ Controls . . . . . . . . . . . . . . . . . . . . . . . . . . 129
5. Post-Treatment. . . . . . . . . . . . . . . . . . . . . . . . . . . 129
D. Effectiveness . . . . . . . . . . . . . . . . . . . . . . . . . . . . . . 130
1. Introduction. . . . . . . . . . . . . . . . . . . . . . . . . . . . 130
2. Effectiveness . . . . . . . . . . . . . . . . . . . . . . . . . . . . 130
2.1 Covers. . . . . . . . . . . . . . . . . . . . . . . . . . . . . 130
2.2 Post-Treatment. . . . . . . . . . . . . . . . . . . . . . . 136
3. Costs. . . . . . . . . . . . . . . . . . . . . . . . . . . . . . . . . 138
3.1 Pretreatment . . . . . . . . . . . . . . . . . . . . . . . . 138
3.2 Design and Operating Practice . . . . . . . . . . . . . 138
3.3 In-Situ Controls . . . . . . . . . . . . . . . . . . . . . . 138
3.4 Post-Treatment. . . . . . . . . . . . . . . . . . . . . . . 138
4. Cost-Effectiveness . . . . . . . . . . . . . . . . . . . . . . . . 141
E. Summary. . . . . . . . . . . . . . . . . . . . . . . . . . . . . . . . . 142
F. References . . . . . . . . . . . . . . . . . . . . . . . . . . . . . . . . 143

VI. LAND TREATMENT FACILITIES . . . . . . . . . . . . . . . . . . 144
A. Description . . . . . . . . . . . . . . . . . . . . . . . . . . . . . . . 144
B. Emission Sources and Models. . . . . . . . . . . . . . . . . . . . 145
1. General . . . . . . . . . . . . . . . . . . . . . . . . . . . . . . . 145
2. Emission Models . . . . . . . . . . . . . . . . . . . . . . . . . . 145
2.1 Surface Emissions . . . . . . . . . . . . . . . . . . . . . 145
2.2 Emissions from Incorporated Wastes. . . . . . . . . . 146

3. Controlling Parameters. . . . . . . . . . . . . . . . . . . . . . . . . . . 148
   3.1 Liquid Pools. . . . . . . . . . . . . . . . . . . . . . . . . . . . 148
   3.2 Incorporated Wastes. . . . . . . . . . . . . . . . . . . . . . . 148
C. Controls . . . . . . . . . . . . . . . . . . . . . . . . . . . . . . . . . . 149
1. Introduction. . . . . . . . . . . . . . . . . . . . . . . . . . . . . . . 149
2. Pretreatment . . . . . . . . . . . . . . . . . . . . . . . . . . . . . . 149
3. Operating Practices . . . . . . . . . . . . . . . . . . . . . . . . . . 149
4. In-Situ Controls . . . . . . . . . . . . . . . . . . . . . . . . . . . . 150
5. Post-Treatment. . . . . . . . . . . . . . . . . . . . . . . . . . . . . 150
D. Effectiveness . . . . . . . . . . . . . . . . . . . . . . . . . . . . . . . 151
1. Introduction. . . . . . . . . . . . . . . . . . . . . . . . . . . . . . . 151
2. Emissions Reduction and Effectiveness . . . . . . . . . . . . . . . 151
3. Costs. . . . . . . . . . . . . . . . . . . . . . . . . . . . . . . . . . . 155
   3.1 Pretreatment . . . . . . . . . . . . . . . . . . . . . . . . . . . 155
   3.2 Operating Practices . . . . . . . . . . . . . . . . . . . . . . . 155
   3.3 Post-Treatment. . . . . . . . . . . . . . . . . . . . . . . . . . 155
4. Cost Effectiveness. . . . . . . . . . . . . . . . . . . . . . . . . . . 155
E. References. . . . . . . . . . . . . . . . . . . . . . . . . . . . . . . . . 156

## PART B
## EVALUATION OF MODELS FOR ESTIMATING AIR EMISSIONS

ACKNOWLEDGMENT. . . . . . . . . . . . . . . . . . . . . . . . . . . . . . 158

1. INTRODUCTION. . . . . . . . . . . . . . . . . . . . . . . . . . . . . 159
Background and Purpose. . . . . . . . . . . . . . . . . . . . . . . . . . . 159
Project Scope and Technical Approach . . . . . . . . . . . . . . . . . . 159
Technical Introduction. . . . . . . . . . . . . . . . . . . . . . . . . . . . 161
   Background . . . . . . . . . . . . . . . . . . . . . . . . . . . . . . . . 161
   General Discussion of Available Models . . . . . . . . . . . . . . . . 164

2. SUMMARY AND CONCLUSIONS. . . . . . . . . . . . . . . . . . . . 168
Introduction. . . . . . . . . . . . . . . . . . . . . . . . . . . . . . . . . . 168
Availability and Selection of Mass Transfer Coefficients (K-Values). . . . 169
   Soil-Phase Mass Transfer Coefficient ($k_{soil}$, $k_s$) . . . . . . . . . . . . . . 169
   Liquid-Phase Mass Transfer Coefficient ($k_{liquid}$, $k_L$) . . . . . . . . . . 174
      Molecular Diffusion—$k_L$ for No Wind Conditions . . . . . . . . . . 174
      Turbulent Diffusion from Wind and Flow Effects . . . . . . . . . . 174
      Turbulent Diffusion from Mechanical Mixing . . . . . . . . . . . . 180
   Gas-Phase Mass Transfer Coefficients ($k_{gas}$, $k_G$) . . . . . . . . . . . . 181
   Overall Mass Transfer Coefficient (K-Value) . . . . . . . . . . . . . . . 184
   Special Case for Overall K-Value. . . . . . . . . . . . . . . . . . . . . . 185
Selection of Most Appropriate AERR Models. . . . . . . . . . . . . . . . 186
   Surface Impoundment (SI) Model Selection . . . . . . . . . . . . . . . 186
   Landfill Model Selection. . . . . . . . . . . . . . . . . . . . . . . . . . . 188
   Open Dump Models. . . . . . . . . . . . . . . . . . . . . . . . . . . . . 190
   Landfarming Model Selection . . . . . . . . . . . . . . . . . . . . . . . 190
   Storage Tank Model Selection . . . . . . . . . . . . . . . . . . . . . . . 192

Waste Treatment Processes . . . . . . . . . . . . . . . . . . . . . . . . . . . . . 192
Open Tank with Mixing . . . . . . . . . . . . . . . . . . . . . . . . . . . . . . . . 194
    Biological Treatment Systems . . . . . . . . . . . . . . . . . . . . . . . . . 194
Waste Piles (Particulate Emissions) . . . . . . . . . . . . . . . . . . . . . . . 197

3. AERR MODEL VALIDATION EFFORTS REPORTED IN THE
    LITERATURE . . . . . . . . . . . . . . . . . . . . . . . . . . . . . . . . . . . . . 198
    Introduction . . . . . . . . . . . . . . . . . . . . . . . . . . . . . . . . . . . . . . 198
    AERR Model Comparison with Field Data . . . . . . . . . . . . . . . . . . 198
        Thibodeaux, Parker and Heck (1981d) . . . . . . . . . . . . . . . . . 198
        Cox, Steinmetz and Lewis (1982)—Draft Report . . . . . . . . . . . 200
        Thibodeaux, et al (1982) . . . . . . . . . . . . . . . . . . . . . . . . . . 200
        Ames, et al (1982) . . . . . . . . . . . . . . . . . . . . . . . . . . . . . . . 201
    Additional Data Available . . . . . . . . . . . . . . . . . . . . . . . . . . . . . 201

4. REVIEW OF SURFACE IMPOUNDMENT AERR MODELS . . . . . . . . 203
    Introduction . . . . . . . . . . . . . . . . . . . . . . . . . . . . . . . . . . . . . . 203
    Nonaerated Surface Impounds . . . . . . . . . . . . . . . . . . . . . . . . . 204
        Mackay and Wolkoff (1973) . . . . . . . . . . . . . . . . . . . . . . . . . 204
        Mackay and Leinonen (1975) . . . . . . . . . . . . . . . . . . . . . . . . 210
        Thibodeaux, Parker and Heck (1981d) Including Hwang (1982)
            and Shen (1982) Modifications . . . . . . . . . . . . . . . . . . . . . 212
        Smith, Bomberger and Haynes (1980–1981) . . . . . . . . . . . . . 218
        McCord (1981) . . . . . . . . . . . . . . . . . . . . . . . . . . . . . . . . . 222
    Aerated Impounds . . . . . . . . . . . . . . . . . . . . . . . . . . . . . . . . . . 222
        Thibodeaux, Parker, Heck (1981d) . . . . . . . . . . . . . . . . . . . . 222
        McCord (1981) . . . . . . . . . . . . . . . . . . . . . . . . . . . . . . . . . 226

5. REVIEW OF LANDFILL AERR MODELS . . . . . . . . . . . . . . . . . . . 230
    Introduction . . . . . . . . . . . . . . . . . . . . . . . . . . . . . . . . . . . . . . 230
    Farmer, et al (1978) for Covered Landfills . . . . . . . . . . . . . . . . . . 236
    Shen's (1980) Modification of Farmer's Equation . . . . . . . . . . . . . 239
    Thibodeaux's (1981a, 1981b) Landfill Equations . . . . . . . . . . . . . . 241
        Without Internal Gas Generation . . . . . . . . . . . . . . . . . . . . . . 241
        With Internal Gas Generation . . . . . . . . . . . . . . . . . . . . . . . . 241
        Barometric Pumping Effects . . . . . . . . . . . . . . . . . . . . . . . . . 241
    Shen's (1980) Open Dump Equation . . . . . . . . . . . . . . . . . . . . . . 242

6. REVIEW OF LAND TREATMENT AERR MODELS . . . . . . . . . . . . . 243
    Introduction . . . . . . . . . . . . . . . . . . . . . . . . . . . . . . . . . . . . . . 243
    Thibodeaux-Hwang: Modeling Air Emissions from Landfarming of
        Petroleum Wastes (1982) . . . . . . . . . . . . . . . . . . . . . . . . . . . 243
    Hartley Model (1969) . . . . . . . . . . . . . . . . . . . . . . . . . . . . . . . . 247

7. STORAGE TANK AIR EMISSION ESTIMATION TECHNIQUES . . . . . 249
    Introduction . . . . . . . . . . . . . . . . . . . . . . . . . . . . . . . . . . . . . . 249
    Fixed Roof Tanks . . . . . . . . . . . . . . . . . . . . . . . . . . . . . . . . . . . 250
    External and Internal Floating Roof Tanks . . . . . . . . . . . . . . . . . . 253

External Floating Roof Tanks . . . . . . . . . . . . . . . . . . . . . . . 253
Internal Floating Roof Tanks. . . . . . . . . . . . . . . . . . . . . . . 259
Recommended Emission Estimation Techniques for Floating
Roof Tanks . . . . . . . . . . . . . . . . . . . . . . . . . . . . . . . . 259
**Other Storage Tank Models**. . . . . . . . . . . . . . . . . . . . . . . . . 263
Storage Tank Model Developed in the USSR. . . . . . . . . . . . . . 263
Outdated/Nonrecommended Models for Floating Roof Tanks . . . . 267
**Special Considerations** . . . . . . . . . . . . . . . . . . . . . . . . . . . . 267
Storage of Mixtures. . . . . . . . . . . . . . . . . . . . . . . . . . . . . 267
Open Storage Tanks. . . . . . . . . . . . . . . . . . . . . . . . . . . . . 270

**8. AIR EMISSION ESTIMATION TECHNIQUES FOR WASTEWATER
TREATMENT PROCESSES** . . . . . . . . . . . . . . . . . . . . . . . . . . 271
**Introduction.** . . . . . . . . . . . . . . . . . . . . . . . . . . . . . . . . . . 271
**Open Tank System—No Mixing** . . . . . . . . . . . . . . . . . . . . . . . 272
**Open Tank-Mixing: Biological Treatment Systems**. . . . . . . . . . . . . 272
Hwang (1980) Activated Sludge Surface Aeration . . . . . . . . . . . 273
Biodegradation Kinetics . . . . . . . . . . . . . . . . . . . . . . . . 278
Determination of Biokinetics Rate Constants . . . . . . . . . . . . 281
Air Stripping Kinetics. . . . . . . . . . . . . . . . . . . . . . . . . . 282
Adsorption on Sludge. . . . . . . . . . . . . . . . . . . . . . . . . . 282
Determination of Adsorption Rate Constants . . . . . . . . . . . . 283
Freeman (1979) Activated Sludge Surface Aeration Model . . . . . . 284
Mass Transfer . . . . . . . . . . . . . . . . . . . . . . . . . . . . . . 284
Biokinetics. . . . . . . . . . . . . . . . . . . . . . . . . . . . . . . . 285
Model Solution. . . . . . . . . . . . . . . . . . . . . . . . . . . . . . . . 286
Biokinetic Rate Constant Determination . . . . . . . . . . . . . . . 288
Freeman (1980) Diffused Air (Subsurface) Activated Sludge
Model . . . . . . . . . . . . . . . . . . . . . . . . . . . . . . . . . . . . 288
Mass Transfer . . . . . . . . . . . . . . . . . . . . . . . . . . . . . . 288
Other Activated Sludge Models . . . . . . . . . . . . . . . . . . . . . . 290
Publicly Owned Treatment Works (POTWs) . . . . . . . . . . . . . 290
Selection of Biological AERR Model. . . . . . . . . . . . . . . . . . . . 292
Model Flexibility. . . . . . . . . . . . . . . . . . . . . . . . . . . . . 292
Model Simplicity. . . . . . . . . . . . . . . . . . . . . . . . . . . . . 293
Model Accuracy . . . . . . . . . . . . . . . . . . . . . . . . . . . . . 295
Model Assumptions . . . . . . . . . . . . . . . . . . . . . . . . . . . 295
Complete Mix Reactor . . . . . . . . . . . . . . . . . . . . . . . 295
Mass Transfer . . . . . . . . . . . . . . . . . . . . . . . . . . . . 296
Biokinetics Model . . . . . . . . . . . . . . . . . . . . . . . . . . 296
Sludge Adsorption. . . . . . . . . . . . . . . . . . . . . . . . . . 296
Model Selection . . . . . . . . . . . . . . . . . . . . . . . . . . . . . 296

**9. AIR EMISSIONS FROM DRUM STORAGE AND HANDLING
FACILITIES** . . . . . . . . . . . . . . . . . . . . . . . . . . . . . . . . . . 298
**Introduction.** . . . . . . . . . . . . . . . . . . . . . . . . . . . . . . . . . . 298
**Description of Drum Storage and Handling Facilities** . . . . . . . . . . . 298
**Source of Volatile Air Emissions** . . . . . . . . . . . . . . . . . . . . . . . 299

Literature Survey . . . . . . . . . . . . . . . . . . . . . . . . . . . . . . . 299
Other Considerations . . . . . . . . . . . . . . . . . . . . . . . . . . . . . 302

10. PARTICULATE EMISSIONS ESTIMATION TECHNIQUES FOR
WASTE PILES . . . . . . . . . . . . . . . . . . . . . . . . . . . . . . . . . . 303
Introduction. . . . . . . . . . . . . . . . . . . . . . . . . . . . . . . . . . . . 303
AP-42 Emission Factor Equation for Storage Piles . . . . . . . . . . . . . . . 303
MRI Emission Factor Equations for Storage Piles . . . . . . . . . . . . . . 304

APPENDIX A: REFERENCES . . . . . . . . . . . . . . . . . . . . . . . . . . . . 307

APPENDIX B: DERIVATION OF CURRIE WET SOIL CORRELATION
FOR EFFECTIVE DIFFUSIVITY . . . . . . . . . . . . . . . . . . . . . . . . . . 314

PART C
PROPERTIES AND CATEGORIZATION OF WASTES

1. INTRODUCTION . . . . . . . . . . . . . . . . . . . . . . . . . . . . . . . . . 318
1.1 Background . . . . . . . . . . . . . . . . . . . . . . . . . . . . . . . . . . 318
1.2 Purpose and Scope. . . . . . . . . . . . . . . . . . . . . . . . . . . . . . 319

2. METHODOLOGY . . . . . . . . . . . . . . . . . . . . . . . . . . . . . . . . . 322
2.1 Physical-Chemical Properties . . . . . . . . . . . . . . . . . . . . . . . . 322
Vapor Pressure . . . . . . . . . . . . . . . . . . . . . . . . . . . . . . . . 322
Solubility. . . . . . . . . . . . . . . . . . . . . . . . . . . . . . . . . . . . 322
Henry's Law Constant . . . . . . . . . . . . . . . . . . . . . . . . . . . . 323
Molecular Weight. . . . . . . . . . . . . . . . . . . . . . . . . . . . . . . 323
Relative Soil Volatility . . . . . . . . . . . . . . . . . . . . . . . . . . . . 323
Diffusion Coefficient . . . . . . . . . . . . . . . . . . . . . . . . . . . . . 324
2.2 Data Gathering/Estimation Methods . . . . . . . . . . . . . . . . . . . 325
Vapor Pressure . . . . . . . . . . . . . . . . . . . . . . . . . . . . . . . . 325
Solubility. . . . . . . . . . . . . . . . . . . . . . . . . . . . . . . . . . . . 326
Henry's Law Constant . . . . . . . . . . . . . . . . . . . . . . . . . . . . 326

3. DATA GATHERING AND CATEGORIZATION RESULTS . . . . . . . . 332
3.1 Physical-Chemical Properties and Relative Soil Volatility . . . . . . . 332
3.2 Categorization of RCRA Wastes According to Vapor Pressure . . . . 332
3.3 Categorization of RCRA Wastes According to Aqueous
Volatility . . . . . . . . . . . . . . . . . . . . . . . . . . . . . . . . . . . . 333
3.4 Categorization of RCRA Wastes According to Relative Soil
Volatility . . . . . . . . . . . . . . . . . . . . . . . . . . . . . . . . . . . . 333
3.5 Diffusion Coefficients of Highly Volatile RCRA Wastes . . . . . . . . 334

4. SUMMARY AND RECOMMENDATIONS . . . . . . . . . . . . . . . . . . 337

5. REFERENCES . . . . . . . . . . . . . . . . . . . . . . . . . . . . . . . . . . . 339

APPENDIX A: PHYSICAL-CHEMICAL PROPERTIES AND RELATIVE
SOIL VOLATILITY OF RCRA WASTES . . . . . . . . . . . . . . . . . . . . . . . 342

APPENDIX B: RCRA WASTE CATEGORIZATION BASED ON VAPOR
PRESSURE (VOLATILITY OF PURE SUBSTANCES) . . . . . . . . . . . . . . 384

APPENDIX C: RCRA WASTE CATEGORIZATION BASED ON
AQUEOUS VOLATILITY (HENRY'S CONSTANT) . . . . . . . . . . . . . . . . 393

APPENDIX D: RCRA WASTE CATEGORIZATION BASED ON
RELATIVE SOIL VOLATILITY . . . . . . . . . . . . . . . . . . . . . . . . . . . . . 402

APPENDIX E: DIFFUSION COEFFICIENTS IN AIR AND WATER
FOR RCRA WASTES IDENTIFIED AS HIGHLY VOLATILE FROM
WATER. . . . . . . . . . . . . . . . . . . . . . . . . . . . . . . . . . . . . . . . . . . . . 410

APPENDIX F: DIFFUSION COEFFICIENTS IN AIR AND WATER
FOR RCRA WASTES IDENTIFIED AS HIGHLY VOLATILE FROM
SOIL. . . . . . . . . . . . . . . . . . . . . . . . . . . . . . . . . . . . . . . . . . . . . . . 412

# Part A

# Evaluation of Emission Control Technologies

The information in Part A is from *Evaluation of Emission Controls for Hazardous Waste Treatment, Storage, and Disposal Facilities,* prepared by John R. Ehrenfeld and Joo Hooi Ong of Arthur D. Little, Incorporated for the U.S. Environmental Protection Agency, November 1984.

1

# I. Summary and Conclusions

Volatile emissions from hazardous waste treatment, storage and disposal facilities (TSDFs) represent a major source of organic contaminants entering the atmosphere. The emissions contain a wide range of organic constituents, which contribute to both oxidant generation and also to exposure to potentially hazardous air pollutants. Previous studies prepared for EPA estimate that volatile emissions from surface impoundments, storage and treatment tanks, landfills and land- treatment facilities range from 1.6 to about 5 million metric tons per year. The lower estimate represents about one-third of emissions of some 54 organic chemicals from all sources including industry and transportation. These 54 chemicals are the most volatile and toxic compounds appearing in wastes regulated under RCRA.

Surface impoundments are the largest source of emissions, providing about half of the total. Treatment and storage tanks make up about one-third of the total, with treatment tanks contributing about 50 times as much as storage tanks. Landfills produce about 12% and land treatment adds the remainder, about 3%.

TSDFs, particularly offsite commercial operations, generally treat complex mixtures of wastes. The wastes deposited in landfills historically have included a wide range of solvents, PCBs, industrial wastes, sludges and still bottoms, etc. The resultant emissions integrate volatile species from all of the wastes present in the landfill. Impoundments at large industrial facilities often serve the entire complex, receiving wastes of many different types. Treatment tanks and land-treatment facilities, on the other hand, are more often dedicated to a narrower mix of wastes.

The principal focus of the current Federal and state TSDF regulatory program has been on the prevention of groundwater contamination during and after the active life of disposal facilities, the prevention of exposure from atmospheric emissions of the products of incineration, and the prevention of accidental exposures in general. There has been little emphasis on prevention of exposure from direct emissions of volatile compounds in the wastes. This study examines the availability of controls for four principal types of TSDFs and evaluates the cost and effectiveness of each.

Four different classes of controls are examined. These are:

o    pretreatment
o    design and operating practices
o    in-situ controls
o    post-treatment techniques

2

Pretreatment includes technical controls and administrative measures (bans) that, in essence, prevent volatile wastes from reaching the TSDF in the first place. These types of controls obviously reduce the quantity of organics that will ultimately be emitted from TSDFs, and also reduce the rate of loss since the processes that control emission are proportional to waste concentration. The volatile materials that have been removed must be subsequently treated or disposed, however.

Design and operating practices and in-situ techniques can reduce the rate of loss from the facility. These controls include approaches such as floating covers for impoundments and tanks, improved temporary and permanent cover design and moisture control for landfills, and subsurface injection at landfills. The use of wind screens to lower wind speed over impoundments in order to reduce mass transfer from the surface also falls into this class.

Finally, post-treatment can be used to catch emissions that escape even if other approaches are used. This type of system involves some sort of collection system coupled with a process to remove or destroy the volatile components. Air-inflated structures have been installed at several locations including an aerated surface impoundment (about 0.4 acre). This system includes a carbon adsorption unit designed to remove the volatiles and recycle them to the aerated impoundment where biological degradation ultimately will destroy the wastes. Thermal incinerators could be used instead of adsorption system to destroy the volatile compounds.

The effectiveness of the four types of controls ranges from moderately effective to very high, potentially 99% or above (See Table I-1). The values in Table I-1 have been generated from first principles or have been taken from experience at related types of facilities. The validity at real hazardous waste TSDFs is unknown. The values shown may be overly optimistic or, in some cases, conservative. Substantial experimental work is needed, as noted further below, to verify the estimates developed and presented herein.

Pretreatment and post-treatment can achieve removal (and in some cases destruction) efficiencies of 90%. Wastes removed but not destroyed must be handled in additional steps to prevent emissions from occurring. Recycling as feedstocks, reclaimed solvents, etc. represents essentially complete control. The figures for pretreatment and post-treatment are quite reliable as the treatment technologies have been demonstrated on process streams comparable to hazardous wastes. Inflated covers as large as six acres have been shown to be practical. Systems of this size would fit most existing facilities. Future facilities could be designed with operating units compatible with the largest cover then available.

Pretreatment works best on wastes of reasonably well-defined and constant properties. The most likely opportunities are at on-site industrial plants and for solvent recovery. On the other hand, pretreatment may not fit well with the complex wastes going off-site

TABLE I-1

RANGES OF EMISSION CONTROL EFFECTIVENESS

Pretreatment        > 90% (removal)

Post-Treatment      > 90% (removal)

Operating Practices and In-Situ Controls (rate reduction)

Surface Impoundments

| | |
|---|---|
| Rafts/Spheres | 80 - 90% |
| Wind Screens | 10 - 15% |

Tanks (relative to open tanks)

| | |
|---|---|
| Rafts/Spheres | 80% |
| Fixed Roof | 90% |
| Floating Roof (external) | 90% |
| Floating Roof (internal) | > 95% |

Landfills

| | |
|---|---|
| Improved Cover Practices | > 90% |

Land Treatment

| | |
|---|---|
| Subsurface Injection | 20 - 90% |

Source:  Arthur D. Little, Inc.

to commercial TSDFs.  The volume and composition of the wastes may be so variable that pretreatment would not be technically feasible, or, if feasible, would cost more than the figures used in this and other studies.

In-situ controls and improved design and operating practices are promising.  The values shown in the table are, as noted above, based on theoretical principles as there are little or no data extant on emission controls at these types of facilities.  Several of the in-situ technologies (shades, wind screens, for example) have been used in similar, but distinctly different settings.  The validity of these techniques at hazardous waste TSDFs should be established. Using improved soils and controlling moisture in temporary and permanent covers at landfills can, according to the estimates made below, reduce emissions essentially to zero.  In practice, however, there will be limitations to achieving such efficiencies.  There are no data available to determine how closely this limit can be approached.

Pretreatment appears to be a cost effective control relative to the other types of controls particularly if only small quantities of volatiles are present in the wastes.  Pretreatment costs according to another EPA study are about $2-4,000 per megagram (Mg)($1-2 per pound) of volatiles, (Spivey et al., 1984).  Post-treatment costs are considerably more.  The treatment portion alone ranges from about $11,000 per megagram removed ($5 per pound) upwards.  The lower figures are for catalytic incineration or regenerative adsorption.  If the recovered wastes have market or fuel value the costs will drop substantially.  Thermal incineration without heat recovery and non-regenerative carbon adsorption cost upwards of $33,000 per megagram ($15 per pound) of volatiles removed.  The collection system will add to this figure, but for large facilities the added cost will be incremental to the treatment cost on a pound removed basis.

The cost effectiveness of in-situ approaches can potentially be very high for those operating practices that require little additional labor or equipment.  Moisture control at landfills or subsurface injection at land treatment facilities are good examples. The added cost of these practices is very small relative to the cost of conventional practices, so that the control potentially achievable per dollar would be high.  On the other hand, the absolute effectiveness that is achievable in practice may not attain acceptable levels.  In these cases, pre- or post-treatment may be used as an adjunct.

At the conclusion of the analysis there remain many unanswered questions generally as a result of the lack of good data character-izing emissions or the performance of controls at hazardous waste TSDFs.  It is recommended that field and laboratory programs be carried out to obtain such data.  In particular, data describing the effectiveness of cover practices at landfills; injection, cultivation, and application density practices at land-treatment facilities; and rafts, floating spheres and covers at impoundments and tanks should be developed.

The estimates developed in this report assume no losses occur through degradative processes in the TSDF or equivalently that volatilization and reaction are uncoupled. This assumption is not valid in reality at these facilities. Reducing emission losses may change the overall degradative efficiency per pound of volatile waste entering the system. The direction of such change is not clear at this time. Studies examining this coupling should be carried out both in the laboratory and field and also through theoretical modeling.

## Reference

Spivey, J.J., et al., Preliminary Assessment of Hazardous Waste Pretreatment as an Air Pollution Control Technique. Research Triangle Institute for the USEPA (February 1984).

# II. Introduction

A.   Objective

The purpose of this study is to evaluate controls for volatile emissions arising from the treatment, storage, and disposal of hazardous wastes.  For each principal type of hazardous waste management facility, sources of atmospheric emissions are described and controls representing different approaches, are examined and compared.  The evaluation is based on actual data and on theoretical models where data are lacking or where the control technologies have been borrowed from other types of applications or are novel concepts. The information developed in the study is intended to support the analysis of the regulation and control of these volatile emissions.

B.   Framework for the Analysis

Emissions from hazardous waste treatment storage and disposal facilities (TSDFs) have been estimated to be in excess of 1.5 million metric tons per year (Breton et al., 1983).  The worse case estimate provided in the above reference is in excess of 5 million metric tons per year.  To put this value in perspective, volatile organic compound emissions from industry, transportation, and other sources have been estimated to be 10.7, 7.7, and 3.9 million metric tons per year, respectively (Breton, et al., 1983).  In parts of the country where compliance with oxidant standards has not been attained or is difficult, control of the emissions from TSDFs may represent a significant opportunity for improvement in ambient air quality.

Federal and state air quality regulations for both industry and transportation sources contain a family of provisions designed to reduce or eliminate volatile organic compound emissions to the atmosphere.  Very stringent regulations have been adopted in regions with severe oxidant problems.  The regulations that govern the treatment, disposal, and storage of hazardous wastes, however, do not explicitly provide for control of volatile emissions.  The main thrusts in the regulations promulgated to date have been to protect surface water and groundwater.

The overall system for the treatment, storage, and disposal of volatile hazardous wastes is depicted in Figure II-1.  This figure schematically illustrates potential sources of atmospheric emissions and opportunities to impose controls to reduce those emissions. Starting at the left-hand side, wastes are created by industrial, commercial, and other generators.  The generators, by employing waste reduction and recycle practices within their process boundaries (that is, within their plant or facility) can reduce the quantities of wastes produced in the first place.  Although such techniques can provide a substantial potential impact on volatile emissions reduction, this report looks only at the processes that take place

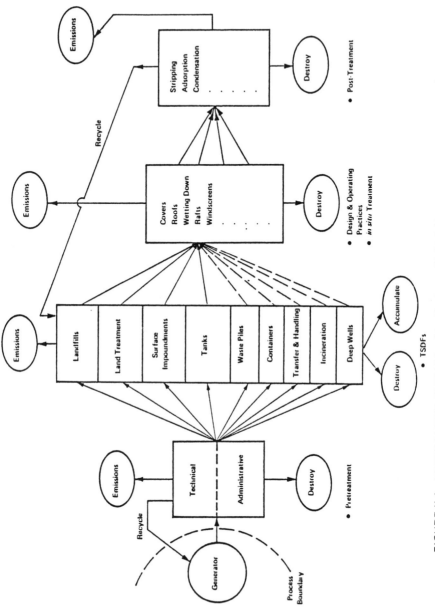

FIGURE II-1   GENERALIZED FLOW OF VOLATILE EMISSIONS (Source:   Arthur D. Little, Inc.)

once the wastes have left the process boundary and move either offsite or to the hazardous waste disposal portions of the generator's facility.

A variety of hazardous waste treatment, storage, and disposal techniques can be used. These are shown in the second column of the figure. The principal methods include landfills, land treatment, surface impoundments, waste piles, storage and treatment tanks, incineration, and deep well injection. The potential for emissions from incineration and deep well injection is considered to be very small compared to the other processes, and are not considered further in this study. Waste piles, containers, and transfer and handling operations are also omitted. GCA has estimated the relative emissions from the various types of TSDFs as shown in Table II-1 (Breton et al., 1983). The emissions calculation is based on some 54 chemicals listed specifically in the hazardous waste regulations. The total emissions of all volatile organics is estimated to be higher as is implied by the worst case estimate given above. Surface impoundments contribute over half the total emissions. Treatment tanks rank next, contributing about one-third of the estimated emissions. Land treatment adds the smallest quantity of the facilities included. The original reference to these data contains a number of caveats concerning the accuracy in the use of the numbers. The data are presented here to demonstrate the importance of this problem. For a more comprehensive examination of the sources of emissions, the original reference should be examined.

Control of emissions from the TSDFs can be accomplished through several means; pretreatment techniques, design and operating practices, in-situ techniques, and post-treatment techniques. Pretreatment controls are those administrative or technical procedures applied to wastes prior to being sent to a TSDF, which will reduce their emissions in those treatment, disposal, and storage facilities. Design and operating practices are techniques within control of the designer and operator that can reduce emissions without interfering with other, generally primary, objectives. In-situ treatments are defined as technologic means and are operating practices employed within, and as a part of the TSDF, that will reduce emissions from those facilities. Finally, post-treatment techniques are defined as procedures that can be used to remove volatiles from the streams that emerge after the imposition of in-situ treatment techniques to the atmosphere.

To evaluate the effectiveness of emission control methods, it is important to have a sense for the flow of volatile materials in the hazardous waste management process. Some controls may appear to be very effective, but only transfer the emissions from one part of the process to another. For example, an administrative ban of certain organic compounds from landfills will obviously reduce the potential of emissions of the prohibited compounds from the landfill to zero, but the wastes must then be treated or disposed of by some other procedure. If this other procedure is a non-emitting technique such

TABLE II-1

EMISSIONS FROM HAZARDOUS WASTE TREATMENT,
STORAGE, AND DISPOSAL FACILITIES

| Facility Type | Estimated Annual Emissions[*] ($10^3$ Metric Tons/Yr) |
|---|---|
| Treatment Tanks | 530 |
| Non-Aerated Surface Impoundment - Storage | 420 |
| Non-Aerated Surface Impoundment - Treatment | 310 |
| Landfill | 190 |
| Non-Aerated Surface Impoundment - Disposal | 66 |
| Aerated Surface Impoundment | 66 |
| Land Applications | 43 |
| Storage Tanks | 10 |
| Total | 1,635 |

[*] For 54 selected chemicals.

Source:  Breton et al., 1983.

as incineration, then such an administrative control can be very effective. If, on the other hand, the procedure is an alternative such as extended storage or disposal in a surface impoundment, then a ban would not be so effective when the potential emissions from the entire chain of events is accounted for.

As volatile materials move through the hazardous waste management process, they must be destroyed, accumulated, emitted or recycled to a prior step at each stage. This report examines in detail the effectiveness of design and operating practices and in-situ treatment techniques, and also the use of post-treatment methods to reduce subsequent emissions. A similar detailed analysis of pretreatment methods is not included. The effectiveness of pretreatment controls has been the subject of a study being carried out at Research Triangle Institute for the EPA (Spivey et al., 1984). Results of that study have been incorporated directly as a basis to show comparisons.

The studies referenced above approached the analysis from the perspective of the types of wastes, particularly the work of Spivey, et al., (1984). The approach herein is different, examining controls from the point of view of type of facility. The choice of control method is largely dictated by facility and environmental setting, not by waste properties. In certain cases, however, the effectiveness depends on waste properties. These cases are spelled out in the text.

Effectiveness is measured primarily by degree of reduction of the rate of emissions, not by reduction in total emissions over long periods of time. Many controls serve only to retard this rate of emission; for example, improved covers at a landfill. Others remove the volatile materials entirely, for example, pretreatment or post treatment, although the wastes removed may require subsequent treatment or disposal. Retarding the rate of loss by volatilization keeps the materials in the facility for longer periods. If other loss mechanisms occur, such as biological decay, then reducing the emission rate may change the absolute quantity emitted over time. It is not possible to predict the direction of the change, based on present data and theory. Mechanisms leading to either an increase or decrease can be invoked.

C.  Hazardous Waste Treatment, Storage and Disposal Facility-- Definitions and Descriptions

    1.  Landfills

Landfills are disposal facilities in which hazardous wastes are placed in containers or in bulk form, covered over with soils and left indefinitely. The present regulations require that free liquids must be mixed with a suitable bulking agent. Some state regulations now ban the disposal of certain types of organic liquids entirely. Other states are contemplating similar bans. Pre-RCRA and early RCRA regulations permitted the landfilling of free liquids. Emissions from landfill operations arise during the daily operation during which period wastes are placed within the landfill. Emissions can arise from liquids spilled on the ground or from solids, sludges, and bulk liquids which are exposed to the atmosphere. Once the material has been placed within the landfill and covered, emissions can occur from compounds vaporizing within the wastes, that diffuse toward the surface and escape to the atmosphere. A cover is generally applied at the close of each day, and is a permanent cap is installed on top when operations within a single cell have come to a close.

Landfills have three operational stages. The first is the active life during which waste is disposed within the landfill cell on a daily basis. The second is the closure period during which a completed cell is capped and taken out of operation and capped. Finally, the post-closure period is an extended period of monitoring and inspection that follows the closure of an individual cell and ultimately of the entire landfill. A single cell is generally designed to operate for a period of one to two years, whereas the active life of the entire landfill may be in the range of 10 to 20 years, and in the case of a few very large landfills, many decades.

## 2.    Land Treatment

Land treatment is a means of waste management in which wastes are deposited on and worked into soil so that natural processes can degrade and demobilize the hazardous constituents within the waste. Biological activity detoxifies some of the organic constituents. Inorganic constituents may be immobilized by adsorption or other processes that bind the elements to soil particles.

Emissions from land treatment facilities arise, first, by volatilization from the wastes that have been spread on the soil prior to being incorporated within the top layers. A second source is wastes which have been mixed into the soil, but which continue to release volatile materials which diffuse upward through the soil and reach the surface.

## 3.    Surface Impoundments

Surface impoundments are large basins created by excavation or by constructing dikes. Wastes generally containing large quantities of water are placed in the impoundments for either treatment or storage. Storage impoundments are generally nonaerated and are often used for the disposal of sludgy materials. Nonaerated impoundments are also used to evaporate water from aqueous wastes, leaving behind the residual solids and sludges. Atmospheric emissions are produced by the volatilization of compounds within the impoundment or at the surface followed by their diffusion into the atmosphere. In some cases, biological activity takes place under anaerobic conditions below the surface of the impoundment. When this occurs, the impoundments are often denoted as facultative. The biological activity can produce gases which bubble up through the impoundment. Under these circumstances the gas bubbles can also carry volatile materials to the surface.

Surface impoundments can also be aerated. Aeration is provided to enhance biological activity within the body of the impoundment, degrading the organic materials contained. The principal mechanism for emissions remains that of diffusion at the surface. The turbulence created by the aeration changes the rate of diffusion at the surface relative to the quiescent regions or to a nonaerated impoundment cover.

### 4.    Storage and Treatment Tanks

Treatment and storage tanks come in a wide variety of sizes and shapes. Wastes are stored and treated both in open tanks and in tanks with covers and roofs. In uncovered tanks the mechanism of loss to the atmosphere is similar to that in surface impoundments. Diffusion from the liquid surface to the atmosphere is the only means of escape. The details of diffusion at the surface, like surface impoundments, depend on whether the tank is quiescent or is being mixed by a impeller. The mechanism for fixed roof tanks is primarily the same mechanism. Volatile materials diffuse from the surface into the head space above the liquid, and then in general, are released to the atmosphere through some sort of vent. Vents are necessary in most storage tanks to control changes in pressure due to temperature variations and volumetric changes due to filling and emptying. On Pressure tanks are designed to withstand the changes of pressure without vents. No losses to the atmosphere would take place under normal operations from this type of tanks.

Floating roof tanks, commonly used to store bulk petroleum products, may be used for the storage of large volumes of organic solvents. In these tanks a roof floats on the surface of the liquid and is stabilized by means of a seal along the circumference. Losses to the atmosphere occur by leakage at the seal and by adherence of the stored fluids to the tank walls as the roof drops during emptying operations.

### D.    Controls/Definitions and Description

### 1.    Pretreatment Controls

Pretreatment controls include administration and technical measures. Administrative controls, such as bans or restrictions on the disposal of volatile materials in landfills, surface impoundments, or land treatment facilities have been instituted by several states and will be implemented in the future by the Federal government. Such controls have the effect of completely eliminating the potential of emissions from the facility in question. In evaluating the overall effectiveness of these controls, however, the ultimate means of treatment or disposal used must be considered.

Technical controls include methods which separate the volatile materials from the wastes and either recycle them back to the generator or other potential users, or destroy the potentially volatile materials through subsequent treatment. Separation techniques include:

o   Distillation
o   Stripping
o   Carbon Adsorption
o   Solvent Extraction

Each of these methods can be used to remove a fraction of the volatile materials from the waste stream. The choice depends on the particular composition of the waste. Once separated, the volatile fraction can be reused or destroyed by incineration with air oxidation and other more exotic methods. Some emissions may occur during the separation and treatment process, but they will be less than the emissions that would have occurred if the wastes had been deposited on the land or in treatment tanks in their original form.

Table II-2 shows a list of typical pretreatment processes and unit costs.

2.   Design and Operating Practices

This class of emission controls covers a great variety of approaches. Many are specific to one kind of treatment, storage, or disposal process. For example, emissions can be controlled by modifying the operating cycle at a land treatment site or by designing an impoundment to minimize surface area. Careful choice of cover materials and moisture control can, in theory, reduce emissions at a landfill by several orders of magnitude.

3.   In-Situ Controls

These are controls that can be used to provide efficiency beyond that which good practices can achieve. They include covers for most of the facilities, roofs, wind screens, and others. The opportunities to use such devices at hazardous waste management facilities are somewhat limited. In-situ controls may interfere with the basic functioning of the TSDF. The distinction between in-situ controls and operating practices is arbitrary. Both could be included in a single class for the purposes of this analysis.

4.   Post-Treatment Techniques

Several techniques are available to isolate or destroy the volatile components provided that they are in the form of a well-defined stream. Volatiles may be recovered, for example, from the vent of a storage tank by refrigeration/condensation or adsorption on activated charcoal. The recovered volatile material could be returned to storage or to the treatment facility.

TABLE II-2

PRETREATMENT PROCESSES AND COSTS

| Waste type | Applicable Pretreatment process | Example cost $/metric ton of organic material removed or recovered |
|---|---|---|
| Organic liquids | Distillation | 251 |
| Aqueous, up to 20% organic | Steam stripping | 151 |
| | Solvent extraction | 55 |
| Aqueous, less than 2% organic | Steam stripping | 151 |
| | Carbon adsorption | 1,600 |
| | Resin adsorption | 310 |
| | Air stripping with carbon adsorption | 1,400 |
| | Ozonation/radiolysis | 1,800 |
| | Wet oxidation | 1,390 |
| | Biological treatment | 68 |
| Sludge with organics | Air stripping with carbon adsorption | 1,400 |
| | Ozonation/radiolysis | 1,800 |
| | Wet oxidation | 1,390 |
| | Chemical oxidation | 91,000 |
| | Evaporation with carbon adsorption | 140 |
| Some sludge in organic or aqueous stream | Physical separation ($8/metric ton of waste treated for sludge removal) | |

Source:  Spivey et al, 1984.

Alternatively, post-treatment techniques that destroy the volatiles can be used. These consist of techniques such as flares, ozonation, thermal incineration, and the like.

### References

Breton, M. et al., 1983. Assessment of Air Emissions from Hazardous Waste Treatment, Storage, and Disposal Facilities (TSDF's) — Preliminary National Emissions Estimates -- Draft Final Report - GCA Corporation for the USEPA. GCA Report No. GCA-TR-83-70-G (August 1983).

Spivey, J.J. et al., 1984. Preliminary Assessment of Hazardous Waste Pretreatment as an Air Pollution Control Technique, Draft Final Report -- Research Triangle Institute for the USEPA (February 1984).

# III. Controls for Surface Impoundments

A.  Surface Impoundment Description

1.  Definition

Surface impoundment means a facility or part of a facility which is a natural topographic depression, man-made excavation, or diked area formed primarily of earthen materials (although it may be lined with man-made materials), which is designed to hold an accumulation of liquid wastes or wastes containing free liquids, and which is not an injection well. Examples of surface impoundments are holding, storage, settling, and aeration pits, ponds, and lagoons (Federal Register, May 19, 1980). Exceptions to the above definition include concrete-lined basins, which are, by definition, considered tanks.

2.  Types, Construction and Uses

As defined above, surface impoundments may be natural or man-made depressions. Acreages range from less than an acre to hundreds of acres and depths vary from 2 feet to as much as 30 feet below the land surface (EPA, 1982). Impoundments are generally built above the naturally occurring water table. Future impoundments handling hazardous wastes will have to be above the water table to comply with federal and state regulations. Some may be constructed on the land surface using dikes or revetments. Natural topographical features such as valleys or depressions may be diked on one or more sides to form the containment area. Dikes may also be required for impoundments in areas of high water tables or to take advantage of impermeable surface soils (EPA, 1982). The three major categories of impoundments are: (1) totally excavated; (2) filled; and (3) combination. Excavated impoundments are those dug from a surface so that the major portion of the capacity is below the grade of the surrounding land surface. Filled impoundments are built above grade such that the majority of the capacity is above the immediate surroundings. Combination impoundments result when material is both excavated and filled. Most impoundments are combination excavation-fill impoundments. Excavated material is used to build side walls, berms, basal areas, and for other miscellaneous construction needs (EPA, 1980).

Surface impoundments are used for a large variety of purposes. These uses can be generalized into temporary holding, treatment, or disposal of wastes. Surface impoundments are used for: treatment by biodegradation, stabilization, equilization, oxidation, evaporation, settling, and for sludges, tailings, and cooling water, among many varied uses. A settling pond is a very common use for a surface impoundment as a means to separate suspended solids from liquids by gravity. Chemical additives may be added to accelerate solids coagulation and precipitation. Settling can be a pre- or post-treatment operation. Surface impoundments may be periodically dredged to restore them to their original capacity. Settling is essentially a quiescent operation but emissions to the air can occur. Temporary storage of liquid wastes can be from a few days to quite an extended period. These wastes may be stored before appropriate treatment or stored as a post-treatment operation before disposal. As mentioned above, treatment operations may be performed in a surface impoundment. These treatment processes may involve aeration by mechanical aerators or may be non-aerated. Biodegradation may be aerobic, anaerobic, or facultative.

3.    Operation

There are three phases to the life of a surface impoundment after it is constructed. These three phases are: active, closure, and post-closure. The operation of a surface impoundment varies during each phase of its life-time and will be described separately.

3.1  Active

The active phase of a surface impoundment encompasses times prior to closure when the facility is accepting waste or is dormant for long periods. Because of the varied uses and large range of sizes of surface impoundments, the operation of a surface impoundment during its active phase will vary greatly from impoundment to impoundment. The surface impoundment may be receiving mainly sludge or mainly liquid hazardous waste, or mixtures of both. Each surface impoundment has a specified capacity for containing wastes depending on the design and on the volume already contained in the impoundment as shown in Figure III-1. The operating surface area, that is, the area exposed, varies with the depth of waste in the impoundment. The operating depth is defined as the distance from the surface to the lowest point of the bottom of the impoundment, and the freeboard is the depth from the top of the berm to the liquid surface. The surface impoundment may be designed with weirs, spillways, auto level controls or other means of preventing overtopping. The impoundment may be lined or unlined, and may contain a leakage detection system.

Impoundments may be operated individually or interconnected so that the flow moves from one impoundment to another in series or parallel. Depending on the use of an impoundment, it may be operated under steady-state or unsteady-state conditions, and it may be mechanically mixed to encourage aeration or not mixed at all. A steady-state operation is one in which the rate of inflow is equal to the rate of outflow and losses through evaporation or removal (in the case of solids).

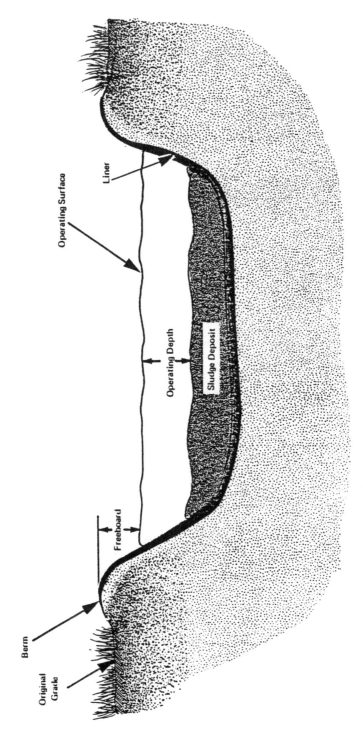

Source: Arthur D. Little, Inc., 1984

**FIGURE III-1    CROSS-SECTION-SURFACE IMPOUNDMENT**

Wastes are taken into the surface impoundment at a certain frequency. The average rate of flow and temperature of the influent varies greatly from impoundment to impoundment. The influent pipe may be submerged and wastes entering the impoundment from inside the bulk of the waste or the waste may be added onto the surface of the bulk waste. Wastes could enter the impoundment with or without pretreatment. The length of time wastes are kept in the impoundment before removal is called the retention time. If the surface impoundment is used for disposal, the total disposal capacity is important.

In certain uses, the solids in surface impoundments are removed at intervals so that the design capacity of the impoundment is restored. Such uses include settling ponds, precipitation lagoons, sludge holding, and tailing ponds. The rate of sludge or sediment accumulation is dependent on the inflow rate and the composition of the influent wastes. Dredging is the usual method for removing solids. The frequency of dredging will depend on sludge or sediment accumulation rate and the capacity of the surface impoundment.

Liquids are removed from surface impoundments periodically or continuously. The frequency of discharge of liquids and the method of discharge are dependent on the use of the impoundment. Some impoundments lose their liquids by evaporation or infiltration into the soil underneath. In the past, seepage of fluids were permitted for the purpose of percolation or infiltration. All new impoundments containing hazarous wastes must be lined to prevent any seepage. Discharge through evaporation is dependent on the physical and chemical properties of the wastes and the meterology and climatology of the site. Direct discharge may originate at the surface of an impoundment or within the bulk of the waste.

Surface impoundments may be totally drained and/or most of solids removed after some years of storage to clean out the impoundment, to restore the impoundment to full capacity, to make repairs if necessary, or to close the facility permanently as the next section describes.

### 3.2  Closure

Surface impoundments are generally constructed to be temporary containment structures. The service life of impoundments may vary and when the surface impoundment reaches the end of its active life, the impoundment undergoes a closure phase.

Closure of hazardous waste surface impoundments is discussed in detail in EPA (1982). There are two basic ways to close a surface impoundment: wastes and waste residuals are left in the impoundment after closure, and wastes and waste residuals are removed from the impoundment site. The key steps in surface impoundment closure are shown in Figure III-2. If hazardous wastes are to remain in the impoundment, procedures must be implemented to minimize the release of

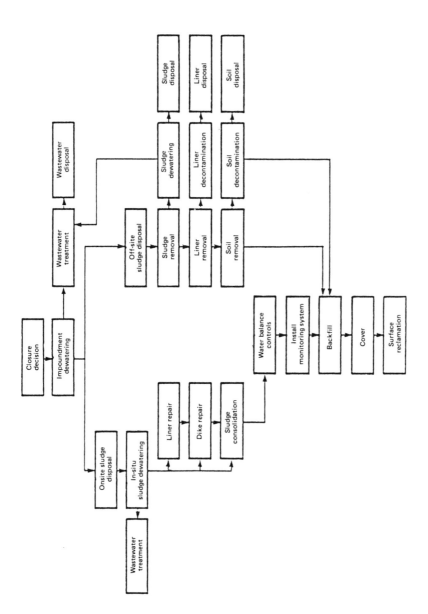

Figure III-2    Surface Impoundment Closure Key Steps

Source: U.S. Environmental Protection Agency, September 1982

contaminants into groundwater, soils, and the atmosphere. The remaining wastes must not contain free liquids. The wastes are consolidated and stabilized, and the impoundment closed with a cover like a landfill. Some impoundments may be best closed by removal of all the waste and residuals, including the liners and contaminated soils. These are transported and disposed in another acceptable site. The site may be reclaimed by filling, leveling, and revegetation. Site closure procedures are varied.

The first step in the closure of an impoundment is dewatering, involving the removal and disposal of standing liquid before the residual solids can be removed or the sediments dewatered. The various methods of dewatering are:

o    Decanting
o    Pumping and settling
o    Solar drying
o    Infiltration
o    Process reuse
o    Chemical neutralization
o    Absorbents

Several techniques may be used to remove the dewatered sediments as a slurry (wet) or as a dry solid (dry). Wet methods are to: resuspend as an air or water jet and then pump the slurry; or to excavate if sediments are hardened and non-flowing. Excavation techniques are high-pressure water or air jets, high-speed rotary cutter, or clamshell bucket. Dry methods require that the sediment be dried by evaporation and then removed by vacuum transport or excavation. The liner and contaminated soil may be removed from the impoundment as well, using normal excavation methods. The site is filled in and restored as far as possible by revegetation.

If the residual wastes are to be left in place, the wastes must have the consistency of nonflowable solids according to EPA guidelines. Further dewatering of the drained sediments may be needed either by passive or active dewatering. During the dewatering, the flowable sediment is consolidated to a nonflowable solid. Volume reduction may be further achieved by further consolidation physically, biologically, or chemically depending on the sediment characteristics. Biological oxidation can reduce the mass of solids by conversion to liquid and gaseous end products. Similarly, chemical consolidation can convert a portion of the waste solids to a dissolved or gaseous product which is removed with leachate or released to the atmosphere. The wastes can be further stabilized through solidification or encapsulation to produce a solid, chemically nonreactive material. After consolidation, the site can be closed as a landfill with a cover and vegetation.

3.3  Post-Closure

Once the surface impoundment is closed, it essentially is a closed landfill and has to be maintained as such (see section on landfills).

B.    Emission Sources and Models

1.    General Description of Factors Affecting Emissions

Emissions to the atmosphere may occur during all phases of the operation of a surface impoundment. Organic gases and fugitive dust may be emitted during the active life of a surface impoundment, during closure, and after closure.

Active

Vaporization of organics at the surface from floating immiscible layers or from the bulk aqueous wastes is the main process for emission losses during the active life of a surface impoundment. Some losses can also occur through vaporization of organics from the exposed walls of the impoundment if organics have migrated into the soil or liners. These losses are not considered in this analysis.

The rate of waste volatilization in impoundments is dependent on the physical and chemical properties of the waste and the surrounding environment. The natural factors affecting the rate of volatilization include wind, temperature, humidity, and solar radiation. The rate of emission into still air is slower than evaporation into the wind. Vapor pressures increase with increasing temperature and surface turbulence, either by wind or by mechanical agitation, increases the rate of volatilization. The characteristics of the waste affecting its rate of volatilization include its Henry's law constant and solubility.

Gases may be generated by reactions within the impoundment. Biological activity may decompose organics to produce methane, hydrogen sulfide, carbon dioxide, or other gases. These biological reactions may be aerobic or anaerobic depending on the conditions within the impoundment. These reactions may be encouraged by the operators of an impoundment through the addition of microorganisms to the impoundment and the aeration of the impoundment using mechanical aerators or agitators. Emissions would be increased in such cases due to the generation of more gases from decomposition and from increased turbulence on the impoundment surface. Chemical reactions may also increase emissions if gases are produced. Mixing used to increase the rate of reaction will increase emissions.

Another source of emissions during the active life of an impoundment are particulates generated by wind and human activity. Wind erosion of wastes depend on waste type, moisture content, wind velocity, and surface geometry. If a dry layer of waste forms on the surface of an impoundment, wind or human activity on top of the impoundment may induce erosion of this layer.

During discharging, pumping, dredging, or removing of impoundment sediments, additional emissions may be expected from several sources:

o    liquid surfaces can be renewed and vaporization is increased,

o    dried surfaces or crusts can be removed, venting entrained gases and creating a liquid surface,

o    pressures can be changed and residue components mixed, affecting gas venting and production,

o    turbulence of impoundment surface increasing vaporization,

o    generation of spray from operations,

o    drying of dredged material,

o    movement of equipment may cause fugitive emissions of contaminated soil.

### Closure

The closure process includes operations that will generate emissions. During dewatering, volatilization is expected to increase due to pumping and decanting. If solar drying is used, emissions will be greatly increased if volatile organics are present. During excavation, fugitive dust will be emitted. Exposed waste surfaces are susceptible to wind erosion and to the mechanical movement processes of the excavation equipment. Weather conditions affect the generation of fugitive dust emissions. During excavation, the waste will become drier due to volatilization of liquid from increased handling of the waste material.

If the impoundment is to be closed by leaving the wastes in the impoundment, further dewatering will increase gaseous emissions due to volatilization and wind erosion of dry waste will be increased. Consolidation is also expected to produce gaseous emissions.

### Post-Closure

If the impoundment is improperly closed, various organic compounds may still volatilize under the cover. Gases may be generated from biological and chemical reactions. Anaerobic biological activity may increase. These emissions may be released through the cover, or through cracks in the cover if not properly vented. More details of emissions after closure are given in the discussion on landfills.

2.   Emission Models

The basic relationship describing the flow of compounds in and out of a surface impoundment is the material balance around the impoundment. In the most general case, this relationship is:

$$\frac{dw}{dt} = \dot{W}_{in} - \dot{W}_{out} - Q_{vol} - Q_{seep} \pm Q_{react}$$

where      $w$ =  quantity of a waste compound in the impoundment at time $t$ (lb)

   $\dot{W}_{in}$ =  quantity of compound entering in influent (lb/hr)

   $\dot{W}_{out}$ =  quantity leaving in discharge (lb/hr)

   $Q_{vol}$ =  quantity volatilizing (lb/hr)

   $Q_{seep}$ =  quantity seeping through liner (lb/hr)

   $Q_{react}$ =  quantity lost or gained through reaction (lb/hr)

Each of the separate terms is a function of concentration. The solution is expressed as the function of concentration over time. Evaluation of the effectiveness of evaporative controls in the general case would involve solving the equation for two cases, one with and one without controls, for a unit mass of volatile compound entering the system. Comparison of the volatile emissions integrated over time would provide the desired measure of control efficiency.

In the general case, the equation is complex and difficult to solve. Non-aerated impoundments are poorly mixed so that concentration varies throughout the system. In this case, the equation shown is not quite proper; a partial differential equation should be used. Aerated ponds behave more like mixed systems. Reaction loss due to biological degradation is non-linear. Evaporative loss in the general case, depends not only on the efficiency of control, but on the relative rates of the other loss mechanisms.

For evaluative purposes, it is convenient to simplify the analysis by neglecting seepage and reactive (for example, biodegradation) losses and assuming steady-state. This set of assumptions reduces the governing equation to the simple form below. Mass is expressed in lb-mols. Under this set of assumptions, the effectiveness of a control is the reduction in instantaneous _rate_ of emissions relative to the no control case. But, given the no other losses and steady-state assumption, all of the incoming materials would eventually volatilize.

The basic relationship governing emissions from the surface of an impoundment is the instantaneous mass flow equation.

$$Q' = K_{oa} A (X_L - x^*)$$

where  $Q'$  =  mass flow ($\frac{lb - mol}{hr}$)

$A$  =  exposed area ($ft^2$)

$K_{oa}$  =  overall mass transfer coefficient ($lb\text{-}mol/ft^2\text{-}hr$)

$X_L$  =  mol fraction of diffusing component in liquid

$x^*$  =  mol fraction in equilibrium with gas phase concentration

The same functionality may be expressed in a different format using other variables in place of mol fraction.

Several additional relationships can be introduced to reflect dependence on specific properties of the diffusing component and of the overall system. One useful form of relationship between the equilibrium mol fraction and the concentration in the gas phase is:

$$y = Kx^*$$

where  $y$  =  mol fraction of component in gas phase

$K$  =  equilibrium constant, dimensionless

For dilute aqueous solutions, the equilibrium constant in this format is approximately related to the conventional Henry's Law constant as:

$$K = \frac{H \rho_w}{P_T MW_w}$$

where  $H$  =  Henry's Law constant ($\frac{atm - ft^3}{lb\text{-}mol}$)

$\rho_w$ = density of liquid water (lb/ft$^3$)

$MW_w$ = molecular weight of water

$P_T$ = total pressure (atm)

At one atmosphere,

$$K = 3.47H$$

The overall mass transfer coefficient, $K_{oa}$, can be expressed, using the conventional two-film theory of mass transfer, as:

$$\frac{1}{K_{oa}} = \frac{1}{k_L} + \frac{1}{k_G K}$$

where $k_L$ = liquid film mass transfer coefficient $(\frac{lb-mo.}{ft^2-hr})$

$k_G$ = gas-film mass transfer coefficient $(\frac{lb-mol}{ft^2-hr})$

Expressions for the individual liquid and gas film coefficients have been derived from theoretical and experimental bases (See GCA, 1983). The appropriate choice depends on the similarity of the situation being modeled to the conditions in the laboratory or theory from which the formula was derived. For surface impoundments, two regimes are important, aerated and non-aerated.

### Gas film, non-aerated

In the non-aerated case, the liquid surface is relatively quiescent, although waves may be created by wind action, increasing the effective area for transport. The following relationship selected for $k_G$ (see GCA, 1983) was developed by MacKay and Matsuga (1973).

$$k_G' = 0.0432 \ u^{.78} \ D_p^{-.11} \ (N_{sc})^{-.67}$$

$k_G'$ = gas film coefficient (ft/hr)

$u$ = wind speed (ft/hr)

$$D_p \quad = \quad \text{effective pool diameter (ft)}$$

$$N_{sc} \quad = \quad \text{Schmidt number} = \frac{\mu_g}{\rho_g \, D_{i,air}}$$

where:  $\mu_g$  = gas absolute viscosity $\left(\dfrac{lb}{ft\text{-}sec}\right)$

$D_{i,air}$  = diffusivity of ith component in air (ft$^2$/hr)

$\rho_g$  = gas density (lb/ft$^3$)

To convert $k_G$ to units consistent with the basic definitions above,

$$k_G = k_G{}' \, \frac{P_T}{RT}$$

where   R  = gas constant (ft$^3$-atm/°R-lb-mol)

T  = temperature (°R)

At one atmosphere and 77°F (25°C), the resultant equation is:

$$k_G = 1.1 \times 10^{-4} \; u^{.78} \; D_p^{-.11} \; (N_{sc})^{-.67}$$

### Liquid film, non aerated

The liquid phase coefficient, $k_L$, is taken from Owens et al (1964) as modified by Hwang (1982).

$$k_L = 3.12 \, (1.024)^{\theta-20} \; u_o^{.67} \; H_o^{-.85} \left[\frac{D_{i,\,H_2O}}{D_{0_2,\,H_2O}}\right]$$

$k_L$    =    liquid phase coefficient (lb-mol/ft$^2$-hr)

$\theta$    =    ambient temperature (°C)

$u_o$    =    wind speed at surface (ft/sec)

$H_o$    =    effective depth of impoundment (ft)


The wind speed at the surface may be approximated as a factor of the normal wind speed; the factor 0.035 is cited in the above references.

The original form of Owens relationship was applicable to flowing streams and showed a dependency on stream velocity and depth. The form used in this study, as modified by Hwang, accounts for upper layer movement induced by wind over the surface. The correlation developed by Cohen et al., (1978) is perhaps, more applicable but was not used. The choices of Hwang's form of the Owens relationship is consistent with several other studies related to volatile emissions from hazardous waste TSDFs (Breton et al., 1983; Spivey et al., 1984).

### Gas-film, aerated

For aerated portions of impoundments, the coefficients are quite different, reflecting the highly turbulent nature of mass transfer in this kind of system. The gas-film coefficient follows the work of Reinhardt (1977) as reported by Hwang (1982).


$$k_G = .00039 \frac{(\rho_g D_{i,air})}{D_T MW_i} (N_{re})^{1.42} (N_{fr})^{-.21} (N_p)^{.4} (N_{sc})^{0.5}$$

where    $k_G$    =    gas film coefficient$\left(\dfrac{lb\text{-}mol}{hr\text{-}ft^2}\right)$

$\rho_g$    =    gas density (lb/ft$^3$)

$D_{i,air}$    =    diffusion coefficient of ith component in air (ft$^2$/hr)

$D_T$    =    diameter of aerator turbine or impeller (ft)

$MW_i$    =    molecular weight of ith component

$N_{re}$    =    gas Reynolds number = $\dfrac{\rho_g D_T^2 \omega}{\mu_g}$

where    $\omega$   =   impeller rotational speed (radian/sec)

$\mu_g$   =   gas abolute viscosity $\left(\dfrac{lb}{ft\text{-}sec}\right)$

$N_{fr}$   =   Froude Number   =   $\dfrac{\omega^2 D_t}{g}$

where    g   =   gravitation constant (32.2  ft/sec$^2$)

$N_p$   =   power number   =   $\dfrac{P_r\, g}{\rho_L\, \omega^3\, D_T{}^5}$

where    $P_r$   =   power to impeller (ft-lb/sec)

$\rho_L$   =   liquid density (lb/ft$^3$)

$N_{sc}$   =   Schmidt Number   =   $\dfrac{\mu_g}{\rho_g\, D_{i,air}}$

The representation of the aerated portions of a surface impoundment used herein is highly idealized. In more sophisticated models, the aerated region is considered to have two portions, a central, highly turbulent region within the boundary of the aerator, and a second region outside the inner zone where flow moves radially outward (Lunney et al., 1984). The more simplified version used is sufficient to illustrate the general dependency of volatile emissions on a variety of potential controlling variables.

### Liquid film, aerated

The liquid phase coefficient for the aerated portions is taken from Thibodeaux (1978).

$$k_L = \frac{J\, P_r\, n\, (1.024)^{\theta-20}\, \alpha\, (10^6)}{165.04\, a_v\, V} \left[\frac{D_{i,\, H_2O}}{D_{O_2,\, H_2O}}\right]^{0.5}$$

where:

$k_L$   =   liquid phase coefficient $\left[\dfrac{lb\text{-}mol}{ft\; \text{-}hr}\right]$

$J$ = oxygen transfer rating of aerator $\frac{lb-O_2}{hr-hp} \approx 3$

$\eta$ = efficiency of power conversion ($\eta \approx .65 - .90$)

$\alpha$ = oxygen transfer correction factor ($\alpha \approx .8 - .85$)

$a_v$ = surface area to volume ratio ($ft^{-1}$)

$V$ = volume of impoundment within the effect of the aeration ($ft^3$)

The aerated portions of an impoundment generally cover only a fraction of the total surface area. Average mass-transfer coefficients are developed by using area-weighted coefficients for the aerated and non-aerated portions.

Referring back to the two film theory equation combining the contributions of both gas and liquid film coefficients, it can be seen that:

if $\quad k_L \quad >> \quad K\, k_G$

then $K_{oa} \approx K\, k_G$

and

if $\quad k_L \quad << \quad K\, k_G$

then $K_{oa} \approx k_L$

That is, if mass transfer through the gas film is very much slower than the liquid film (the first case above), the gas film controls the process. If the situation is reversed, (the second case above), the mass transfer is liquid-film controlled. For non-aerated surfaces, under normal conditions, the mass transfer is liquid-phase controlled, that is, diffusion through the liquid surface is slower than through the air at the surface. Under high wind speed conditions where the surface may become roiled, the situation can shift more toward a balance.

### Floating Immiscible Organic Layer

In this case, only the gas film, non-aerated coefficient is important.

$$K_{oa} = K\, k_G$$

where K = partial pressure of the pollutant in the air layer next to the organic layer.

$$K = P_{vp}/P_T$$

where    $P_{vp}$ = vapor pressure of pollutant, atm.

   $P_T$ = total pressure = 1 atm.

At atmosphere pressure measured in atmospheres,

$$K_{oa} = P_{vp}\, k_G$$

### 3.    Parameters That Control Emissions

The derivations above were developed in detail to provide a basis for describing mechanisms that lead to atmospheric emissions and to indicate the factors that can be controlled to reduce these emissions. The basic mass flow equation shows that the rate of emission depends on the overall coefficient, the exposed area, and the concentration or mole fraction in the liquid. The concentration in the gas phase in uncovered impoundments is effectively zero, as fresh air continually sweeps over the system.

Pretreatment controls function by reducing the mole fraction in the liquid. The effectiveness is linearly proportional to the degree of removal. If 50% of the potentially volatile materials are removed, then the rate of emission, and also ultimately the total quantity of wastes entering the atmosphere will be halved.

Another broad category of controls function by reducing the area exposed. The effectiveness of such controls is also by and large expressed as a linear relationship. If the area is reduced by 80%, then the emission rate would be reduced in proportion. The total quantity of materials volatilized, assuming no other loss mechanism, would not be changed over the long run. The time to volatilize a unit quantity of waste would increase inversely with the resultant emission rate. In the above example, the time during which emission would occur would be five (5) times that which would occur with no control.

A third approach to control of emissions is through the mass transfer coefficient. For the non-aerated portions the controlling mechanism is described by the Owens Equation above. The controllable parameters that determine the value of the mass transfer coefficient are the wind speed at the surface ($u_o$) and the effective depth ($H_o$). Barriers and fences, as described in the sections below, can be placed to reduce the wind speed. The relationship is not linear, and it is difficult to relate the reduction in wind speed to barrier design. The dependence on depth is inverse, that is, deeper impoundments have, in theory, lower mass transfer coefficients. The choice of depth is often dictated by overall capacity and land availability constraints, but given a choice, this relationship indicates that deeper impoundments would result in reduced rates of volatilization. This set of relationships is only approximate as real behavior departs from the idealization implicit in the Owens equation. But the general dependence on wind speed and depth is expected to be found in practice.

The above discussion of the two-film theory and of the controlling parameters is based on the assumption that the volatile materials are present as a dilute aqueous solution. It is also possible to have a layer of lighter-than-water, immiscible organic compounds floating on the surface. In this case, the controlling mechanism will be diffusion in the gas phase, not the liquid phase. The MacKay and Matsuga type of relationship should generally illustrate the dependence of mass transfer coefficient on the operating parameters for this case. Reducing the wind speed, as in the general aqueous case, should inhibit the rate of volatilization.

For gas-phase controlled situations, the rate of mass transfer depends on the Henry's Law constant as well as the individual mass transfer coefficient. This parameter is quite strongly temperature dependent, increasing with increasing temperature. Controls which reduce the impoundment surface temperature would, thus, inhibit the rate of volatilization, when this mechanism controls.

C.   Potential Controls

1.   Introduction

Controls to reduce atmospheric emissions from surface impoundments can be applied at different stages in the operation of an impoundment. If no volatile wastes were allowed into the surface impoundment in the first place, then volatile organics could not be emitted unless degradation forms volatile by-products from semi-volatiles. The surface impoundment can be designed and operated at each phase with the primary objective of reducing air emissions. Such practices may (and generally would) interfere with the basic functions of the impoundment. In situ technologies can be added on to a surface impoundment to control those parameters that influence emission rates; e.g., wind speed, mass transfer coefficient, effective surface area, etc. Finally, emissions can be collected and treated before being released into the atmosphere. For each control within these general approaches, the mode of action, the expected effectiveness and related costs are discussed.

### 2.   Pretreatment

Pretreatment basically removes the volatile components of a waste before it is put into a surface impoundment. Pretreating a waste may consist of air-stripping and carbon adsorption or condensation of the volatiles. The stripped waste may then be put into an impoundment with decreased potential of volatile organic volatilization. For a general discussion of pretreatment, see Spivey et al, 1984.

The effectiveness of this control is dependent on the efficiency of the pretreatment process. The concern here is that air emissions are not being transferred from the impoundment to the pretreatment system. If the volatiles that are generated in the pretreatment process are not properly recovered or disposed of, the final result may be that overall, air emissions would not have been reduced.

The costs associated with this control are (see Spivey et al, 1984):

o   pretreatment system costs.

o   cost of recovery of volatiles (e.g., carbon adsorption, condensation, etc.).

o   cost for treatment of recovered volatiles (if not reused).

o   disposal cost for all wastes associated with pretreatment, recovery (e.g., spent carbon), and treatment of recovered volatiles.

### 3.   Design and Operating Practices

At the present time the issue of air emissions is generally not a primary consideration in the design and operation of surface impoundments. Most of the concern is with the control of leaching from surface impoundments into groundwater and with other functions of the impoundment, such as evaporation or settling. It is conceivable that impoundments could be designed and operated differently if concerns about air emissions are incorporated into the initial decision-making process. Potential design considerations to control air emissions are: surface area minimization, freeboard depth, and inflow/outflow drainage pipe locations. Operations during the active life involving temperature of influent, dredging frequency, draining frequency, cleaning frequency, handling of sediments and sludge from dredging and types of wastes accepted at a facility could be designed to minimize emissions. Closure practices to prevent fugitive dust emissions during excavation and excessive volatilization during dewatering and consolidation may be instituted. After closure, the surface impoundment is essentially a landfill. Post-closure practices may be found in the section on landfills.

This discussion will be divided into: design, operating practices-active, and operating practices-closure.

### 3.1    Design

#### 3.1.1.    Surface Area Minimization

As seen from the model on air emissions from surface impoundments, the rate of emissions is directly proportional to the operating surface area. Theoretically, for a certain capacity surface impoundment, minimization of surface area with respect to depth would decrease air emissions. But, this approach would not be applicable to impoundments where other constraints determine surface area. For example, it would not be practical to use this approach for an evaporation lagoon since the purpose of the impoundment would be thwarted.

Costs probably will be higher than with conventional designs because of potential engineering difficulties. To reduce surface area, the side slopes of the impoundment could be increased as much as possible subject to erosion considerations, the ability to hold a liner, and the ease of construction (trafficability). Due to the greater care that is needed, costs are expected to be higher.

Other important considerations require that the efficiency of treatment in the impoundment should not be significantly decreased. Aeration and mechanical mixing would be reduced in efficiency because of increased depth.

#### 3.1.2    Freeboard Depth

Increase in freeboard depth will decrease wind and wave action on the surface of the impoundment. This will decrease:

o    volatilization,
o    turbulence on the impoundment surface,
o    spray formation, and
o    erosion of dust and the dried surface of the impoundment.

This control is expected to be effective in many settings. Field experiments with in-situ windbreakers have indicated that evaporation from reservoirs have been reduced significantly (see section on in-situ controls). The most important parameter in determining the effectiveness of the deeper freeboard is the ratio of freeboard depth to diameter of the impoundment (distance from edge to edge). The larger this ratio, the more effective this control will be.

To increase freeboard, the height of the berm could be increased or the impoundment could be filled to an average smaller depth. The cost factors are:

o    cost of material for berm
o    labor costs
o    costs for more frequent draining of impoundment

The draft RCRA Guidance Document (July, 1982) suggested at least 60 cm (2 feet) of freeboard to prevent overtopping. A way to increase freeboard is to minimize run-on into the impoundment by incorporating a run-on control system. For large lagoons, other in-situ controls to break the wind may be more effective since a very large freeboard may be necessary to achieve enough wind-breaking.

### 3.1.3    Inflow/Outflow Drainage Pipe Locations

Again, the idea of this control approach is to minimize disturbance of the surface of the impoundment. Inflow pipes discharging above the liquid surface create turbulence on the surface and may also cause spray formation. Any dry crust on the surface will be destroyed. This dry crust is helpful in reducing emissions by creating a barrier or a cover on the surface. Inflow pipes should be designed to discharge below the surface of an impoundment and as far as possible, into the bulk of the liquid in the impoundment. Similarly, outflow systems should pump out liquid from the bulk of the surface impoundment.

More power is required to pump a discharge into the bulk of a liquid. This translates into higher energy costs.

### 3.2    Operating Practices - Active

### 3.2.1    Temperature of Influent

The vapor pressure of a liquid increases with temperature. Also, when two liquids of different temperatures are mixed, convective currents are induced which cause mixing. This disturbance of the system will increase volatilization from the surface. To reduce air emissions, the influent should be discharged at as close a temperature to the bulk of the liquid as possible.

It is unknown how effective this control would be in reducing air emissions in actuality. Evaporation ponds would not be an appropriate impoundment for application since the efficiency of evaporation would be reduced. Another consideration is that treatment efficiencies will be reduced if temperatures are reduced. Most reaction rates are higher at higher temperatures. Oxidation ponds, aerobic and anaerobic lagoons are impoundments which may be affected by reduction of temperatures.

Costs are expected from cooling influent before discharging into an impoundment. A heat exchanger may be needed.

### 3.2.2    Dredging, Draining, and Cleaning Frequency

The more frequently a surface impoundment is disturbed, the more emissions will be released. One way to decrease emissions is to minimize dredging, draining, and cleaning frequency. The reduction in emissions will be proportional to reduction in frequency of dredging, draining, and cleaning. Costs would be reduced because both labor and equipment costs would be decreased. There is a trade-off in cost and efficiency of an impoundment. For cost reasons, most impoundments are already drained, dredged, and cleaned as infrequently as possible without overly impacting the performance of the impoundment.

### 3.2.3    Handling of Sediments and Sludge

After treatment and removal from an impoundment, sediments and sludges may be dewatered before disposal in another facility. Some methods for dewatering and disposal cause more air emissions than others. Solar drying will cause more air emissions and fugitive dust than mechanical drying and filter presses. The comparative reductions in air emissions using different dewatering and handling techniques are unknown.

There may be increased costs due to:

o    energy usage (solar drying has zero energy costs)
o    equipment costs
o    labor costs

### 3.3    Operating Practices - Closure

### 3.3.1    Dewatering

Using a different method for dewatering waste during the closure process can alter the amount of air emissions. Pressurized liquid pumping and disposal with the gases remaining entrained is a way to contain emissions. Rapid dewatering can limit biological and chemical production of gases. This can be accomplished by chemical fixation or sorption using soil, cement, or crushed coral (EPA, 1982).

If additional equipment is needed, there will be additional costs for capital, installation, maintenance, labor, and other associated costs. Materials, if used for rapid dewatering will increase the cost of closure.

### 3.3.2     Proper Consolidation

Proper consolidation before final cover reduces settling. Differential settling can cause the disruption of the integrity of the cover leading to uncontrolled emissions of gases from underneath the cover. Settling may be significant if remaining wastes are high in organic content or if organic sorbents are used to solidify the wastes. Inorganic chemical fixation processes are not prone to significant settling (EPA, July 1982). Wastes should be compacted as much as possible before installation of the final cover. This would reduce the possibility and extent of settling once the cover is installed.

### 3.3.3     Fugitive Dust Abatement

There are two common control techniques to reduce fugitive dust emissions: wetting and stabilizing. Wetting is the application of water as a short-term method to control dust in a confined site. Stabilization methods isolate dust sources from wind erosion and may be physical, chemical, or vegetative. Physically, a cover of stabilizer materials, e.g., rock, soil, crushed or granulated slag, bark, and wood chips, prevents the wind from disturbing the surface particles on the impoundment. Chemical stabilization involves the use of binding materials that cause smaller dust particles to adhere to larger surface particles. Many types are available, many of which are proprietary developments. They are applied in conjunction with water or separately. Most stabilizers work for a limited period of time, in general, no more than a few months. Vegetative stabilization provides permanent dust suppression, but the surface must be well prepared with fertilizers, organic matter, etc. Vegetative stabilization is not applicable to fugitive dust abatement during closure (EPA, 1982).

The effectiveness of wetting and stabilizing techniques are highly variable and depend greatly on site specific characteristics. Generally, these techniques would not be frequently applied during excavation since surface impounded waste materials usually possess sufficient moisture. Efficiencies also depend on the type of stabilizer used. Chemical stabilizers are extremely variable and particularly difficult to evaluate because of proprietary information. More information on fugitive dust statement is given in the section on waste piles.

Wetting is a cheap short-term method of dust control. Physical covers have high associated costs in their application, particularly if transportation is required. The costs of chemical stabilizers are variable.

4.    In-situ Controls

Technologies applied at an impoundment site during the active phase of its lifetime can be used to change one or more of the parameters that affect emission rates from the surface of the impoundment. Some of these parameters are exposed surface area, wind speed, mass transfer coefficient, etc. The in-situ technologies described below are: rafts, barriers, shades, floating spheres, and surfactant layers.

4.1  Rafts

Rafts reduce the surface area. They must be designed to remain afloat and must also be kept from being flooded by waves on the surface of the impoundment. Flexible membrane covers may be made into rafts by attachments on to a frame that is sufficiently high to keep the raft from being flooded. However, the higher the frame, the more wind the raft catches, and therefore, the stronger it has to be structurally.

A "moat raft" consisting of a small frame inside a large frame with the frames connected at the corners is a design that can reduce wind shear and also reduce flooding. The inner frame is membrane-covered and serves as the vapor barrier. The open area between the 2 frames traps the waves and prevents flooding (Cluff, 1967).

Rafts may lead to an increase in the temperature of the impoundment and thus a reduction in their effectiveness. A highly reflective top surface, for example, a surface of aluminum bonded to polyethylene or Styrofoam painted white on top would help in mitigating this problem. Crow (1973) showed that unpainted Styrofoam with 48 percent coverage reduced evaporation by 35 percent. When the top was painted white, a 45 percent coverage gave a reduction of 43-49 percent. Styrofoam not only floats and sheds rainwater but serves as insulation on the surface of the impoundment. Experiments have shown that one-inch thick Styrofoam panels significantly reduced daily variation in stored thermal energy (Crow, 1973).

Floating covers or rafts have been used on small water reservoirs to reduce evaporation. These rafts are small (around 8 feet by 8 feet) and can fit any size or shape reservoir. They are easy to install and do not have to cover the entire surface of the pond. A raft can be made of any kind of material. Styrofoam, polyethylene, aluminum bonded to polyethylene, butyl rubber, and floating concrete (made from cement, sand, and Styrofoam) have all been field tested. When used on a hazardous waste impoundment, the raft material would have to be chemically compatible with the waste components.

Another raft system uses floating covers of foamed glass blocks coated with fluoropolymer which have been used to reduce tank emissions and to maintain tank temperature. The blocks used in one case were rectangular, 2 inches thick, with 10 mils PTFE fluoropolymer coating, 9 x 18 inches and 12 x 18 inches in the main area of the tank. The purpose was to reduce sulfuric acid emissions. Triangular blocks were used to fill in the curved portions of the tank surface. The fluoropolymer coating was applied by a spray and bake process and gave a non-wetting surface with a low coefficient of friction. This surface was also chemically resistant to most organic and inorganic acids, solvents, salts, and mild caustics at temperatures up to 500°F. The glass blocks and coating are custom designed and commercially available. This technology can be easily adapted for hazardous waste applications. The size and shape of the blocks to be used depend on the configuration of the tank or impoundment surface. The coating to be applied on the blocks depends on the chemical characteristics of the waste contained in the tank. The cellular glass insulation blocks are made by Pittsburgh Corning and the coating was a proprietary process of W.L. Gore and Associates, Inc., Coatings Division (Sandberg et al, 1983). W.L. Gore does not presently provide the coating process (Personal Communications, June 1984).

The parameters that determine the effectiveness of rafts are:

o    percent of surface covered. Crow (1973) demonstrated that the reduction in evaporation from small reservoirs was approximately directly proportional to the percent of pond area covered.

o    wind speed over impoundment. Very high winds may cause rafts to be blown off the impoundment. Winds cause waves on the impoundment surface which might flood the rafts.

o    chemical compatibility of raft material with waste.

o    insulating capacity of rafts. This is related to thickness of raft and color of raft surface.

One inch thick white Styrofoam rafts covering 45 percent of a pond containing water reduced evaporation by 43-49% (Crow, 1973). It would be reasonable to assume for a surface impoundment, as in a water pond, that the percent reduction of emissions is proportional to the percent area covered by rafts. Also, assuming that diffusion is zero through the covered area, percent reduction emissions equals the percent of the area covered.

Rafts can be designed to fit any size or shape impoundment. They are also very easy to install.

## 4.2  Barriers

The main purpose of barriers is to serve as wind-breakers over the surface of an impoundment and thus reduce the mass transfer coefficient. They can be used in conjunction with other in-situ controls or by themselves as a means of reducing evaporation. Various kinds of barriers have been field tested in small reservoirs:

o    open-picket baffles which basically consist of a picket-type snow fence supported by floats. The floats can be adjusted for different barrier heights, varying the ratio of the barrier spacing to the barrier height (L/H).

o    closed wind barriers which consist of a picket fence with flexible membrane linings fastened to the pickets.

o    Styrofoam barriers consisting of Styrofoam strips joined to form a grid over the impoundment surface. No floats are necessary (Crow and Manges, 1967).

The parameters that determine the effectiveness of barriers as a means of minimizing wind over an impoundment surface and as ways of reducing emissions through vaporization are:

o    the higher the barrier, the more effective it is as a wind breaker. Within a certain range, the ratio L:H should be small for greater effectiveness. However, the higher the barrier is, the stronger it has to be structurally.

o    chemical compatibility of waste with barrier.

Costs would vary with material used to construct barriers, the dimensions of the barriers, and the size of the impoundment (surface area to be controlled).

Barriers are usable with other technologies, for example, rafts and floating spheres.

## 4.3  Shades

Shades serve as wind-breakers and also reduce solar irradiation over the surface of an impoundment. Shades have been tested in small water impoundments. In field tests, mesh shades made of polypropylene were stretched tightly slightly above impoundment surface (Crow and Manges, 1967). Other materials may also be suitable.

The effectiveness of a shade in reducing emissions through evaporation is dependent on the shade value (see Table III-1). The higher the shade value of the mesh shade, the greater the reduction in emissions. Crow and Manges (1967), found an evaporation reduction of 35 percent from a small reservoir using a polypropylene mesh of 6 percent shade value, 12 mesh per inch, and 0.95 density. When used together with a chemical film, the reduction was increased to 46 percent.

TABLE III-1

EVAPORATION SUPPRESSION BY SHADES

| Material Tested | Evaporation Reduction (percent) |
| --- | --- |
| Polypropylene Mesh<br>6 percent shade (natural) | 26 |
| Polypropylene Mesh<br>47 percent shade (black) | 44 |

Source:    Crow and Manges, 1967

The work by Crow and Manges was performed on water reservoirs. Evaporation from water surface is entirely gas-phase controlled and is quite sensitive to surface temperature. Volatilization from organic contaminated impoundments is generally liquid-phase controlled, as has been noted above. In this circumstance, the mass transfer is less strongly coupled to temperature. Thus, the effectiveness of shades may be lower for hazardous waste containing impoundments.

## 4.4 Synthetic Covers

A synthetic membrane cover over an impoundment reduces air emissions by reducing wind over the impoundment surface and by containing the emissions so that they can be treated further, if desired. One such cover system has been installed in September, 1983 over an aerated lagoon belonging to Upjohn, Inc. in New Haven, Connecticut. The lagoon (425 feet x 150 feet and 8 feet deep) has two 75 hp aerators and 25 7.5 hp floating aerators. Bacteria is added every day and also recycled from the clarifier. The air is maintained at 19 percent oxygen using 3 different fans. Exhaust is through either one of 2 carbon adsorbers installed at the vent outlet. The carbon adsorbers are steam regenerated every 2 days, and the condensate is recycled back to the lagoon to be further biodegraded.

The air structure at Upjohn is made of a vinyl-coated polyester membrane which is coated with Teflon on the inside. It is harnessed to the foundation around the impoundment by cables. The influent COD is 5,000 ppm and the effluent contains 700 ppm in solution. The structure was installed due to odors from the lagoon. Since installation, odor complaints have been decreased significantly (Upjohn Chemical Company, New Haven, Personal Communications, April, 1984).

A similar structure without a recovery system was installed in the fall, 1983, over glauber salt storage ponds used by American Natural Gas in Beulah, N. Dakota, to keep precipitation out. This company has two ponds that are alternately used. The air structure is movable from one pond to the other. The air structure covers around 2 acres and is made of vinyl reinforced material over a concrete foundation. Two air blowers are used to keep the structure inflated. The facility is totally enclosed except for one vent and a door for exit and entrance. Another ground-mounted, air-supported structure to be installed in Livingston, LA, by CECOS to keep precipitation out is the CECODOME, expected to be completed in September, 1984.

The Upjohn air structure cost around $400,000 with installation included. The various appurtenances (carbon adsorbers, fans, etc.) cost around $100,000. Up to the present time, total expenses are in excess of 0.5 million (around $600,000). Further expenditure is expected on roads, and other related expenses (Upjohn Chemical Company, Personal Communications, April, 1984). The American Natural Gas installation cost around $300,000 without installation and $400,000 including installation.

### 4.5  Floating Spheres

Hollow plastic spheres have been used in many applications in industry as a barrier to evaporative energy losses and also to reduce the generation of air pollutants in the work environment. These spheres have also been used for other purposes, for example, in the reduction of oxygen absorption, prevention of erosion and corrosion, etc. These spheres have been used for almost 20 years in industry throughout the world.

On a liquid surface that is completely covered with hollow spheres, each ball touches its six neighbors. The ratio of area covered by the balls to that not covered is independent of the ball diameter, 91 percent. Because the balls are hollow, they act as dead air space insulators. The spheres are designed with an anti-rotation collar which locks the spheres in place and prevents the turning of spheres which would expose the wet surfaces.

The hollow spheres are available in polypropylene and high molecular weight polyethylene. The sizes of the spheres are 20 mm (3/4 in), 38 mm (1½ in), 45 mm 1-3/4 in), and 150 mm (6 in) in diameter. Only the 45 mm is available in polyethylene. The balls are simply poured onto the surface of the liquid until it is completely covered. The approximate number of spheres required for an application is shown in Table III-2.

A cover of touching floating spheres reduces emissions proportionally to the area covered by the balls. The percentage area covered when a layer of spheres fills the surface is independent of the size of the spheres used. The maximum covered area is approximately 90 percent. Experimentally, evaporation from an open insulated tank containing water at 80°C was reduced by over 88 percent when a blanket of balls was used. Assuming that diffusion through the balls is negligible compared to the uncovered portions of the surface, the percent reduction of emissions when compared to a totally open surface is equal to the percent area covered. Adding another layer of spheres only marginally decreases the emissions from a surface relative to a single layer.

Normally, one layer of balls is sufficient. More layers may be used. However, the usefulness is limited when the bottom layer becomes submerged under the liquid surface. Using different size balls does not increase the effectiveness of the layer since a random array will be formed. The choice of the size of spheres depends on size of tank and its configuration. Because the maximum coverage, around 90%, is independent of the size of spheres, all blankets are equally effective.

TABLE III-2

SPHERE REQUIREMENT PER UNIT AREA AND VOLUME

| | 20 mm (3/4 in) | 38 mm (1½ in) | 45 mm (1-3/4 in) | 150 mm (6 in) |
|---|---|---|---|---|
| Quantity/$m^2$ | 2,500-3,000 | 750-800 | 500-575 | 45-50 |
| Quantity/$ft^2$ | 230-280 | 75-85 | 45-55 | 4-5 |
| Quantity/$m^3$ | 165,000-167,000 | 24,000-25,000 | 14,500-15,000 | 350-500 |
| Quantity/$ft^3$ | 4,670-4,730 | 690-710 | 410-425 | 10-11 |

Source: Capricorn Chemicals Corporation, Secaucus, NJ, 1984

The factors involved in the choice of spheres are:

o    chemical characteristics of waste.

o    compatibility with waste (Capricorn Chemicals literature include a long chemical resistance table). The spheres are resistant to most chemicals.

o    size of sphere.

o    number of spheres required.

The installation of these spheres is simple; they are dumped on a liquid surface. The spheres rapidly arrange themselves into a regularly formed cover and will adjust themselves around obstacles in the tank. Blockage or entry of spheres into pipes, etc. may be prevented by wire cages over inlets and outlets and by selection of the correct size spheres.

Removal of the spheres can be handled by using a net that is lifted up from the bottom of a tank. The contents of the tank are always accessible since the spheres are easily moved aside and reformed into a cover. Gas mixtures cannot accumulate under the blanket. Once applied, the system requires little maintenance, attention, or replacement.

4.6 Surfactant Layers

This category includes any system in which a thin layer is used as a cover on the surface of an impoundment to reduce evaporation from the impoundment. There have been many field tests on the use of monomolecular films of long chain alcohols (hexadecanol and octadecanol) on water reservoirs. The most important consideration in the effectiveness of a monomolecular layer is the maintenance of a continuous film on the water surface. In its use in reservoirs, a monomolecular film must retard the movement of water vapor but must allow sufficient oxygen, carbon dioxide, and other gases through. It must have an ability to seal itself after it has been disturbed. Because of losses by wind and biological attrition, a monomolecular film has to be maintained by additional dispensation.

Various ways of applying the film have been tested. The chemical has been applied in liquid, solid, powder, slurry, and emulsion form, using broadcasting, wire cages supported on rafts, wick drippers, and wind-controlled dispensers.

The major factor that destroys the continuity of the film is wind. Winds greater than 10-15 mph can push a film to the downwind edge so that there is almost no coverage.

There are several considerations in transferring this technology to reducing evaporation from surface impoundments:

o   Hazardous wastes are very seldom neat solutions. This complicates the search for the chemical that would work for the chemicals in a particular waste impoundment.

o   The chemical characteristics of hazardous waste impoundments vary from industry to industry and from impoundment to impoundment. A chemical that is effective as a surfactant in one situation may be useless in another. It also increases the difficulty of determining a rate of application that is sufficient to maintain a continuous layer on the impoundment surface.

o   The top layer of a hazardous waste impoundment may vary in composition spatially and temporally. This is especially true for active impoundments which are not in equilibrium.

The system may be effective for floating layers of organics and other situations where diffusion is normally gas-film controlled. If the layer reduces the mass transfer coefficient sufficiently to shift control to the liquid side, then the overall rate of emissions can be reduced significantly. This is the situation at the surface of water reservoirs. For dilute solutions of organics in water, the control general rests in the liquid-film. Adding additional liquid side resistance will not, in general, result in significant reductions in the overall rate of emission, although a thick floating layer of some immiscible liquid, such as mineral oil, may add sufficient resistance to be significant.

To retard the evaporation of a liquid, an adsorbed monolayer would have to be closely packed at the liquid-air interface. Bernett, et al (1970) tested some soluble surfactants and some slightly soluble or insoluble surfactants on toluene, 2,2,4-trimethylpentane, and nitromethane and concluded that the evaporation rate of these liquids was not significantly retarded even though the adsorbed molecules did spread spontaneously to form continuous monomolecular films. This seems to suggest that it would be difficult to find the suitable surfactant for any sample hazardous waste. Even if an adsorbed monolayer is closely packed at the surface, the rate of evaporation might not be retarded.

However, rates of evaporation may be reduced by relatively thick polynuclear films. Polydimethylsiloxanes reduced the evaporation of nitromethane by 28 percent when a film approximately 0.3 mm deep was used. Because of the low solubility of some organics in water, thin water layers may be effective as evaporation barriers. A variation of this may be aqueous foams like those used to extinguish fires (Bernett et al, 1970).

Important consideration in designing a surfactant layer system for the reduction of evaporation from surface impoundments are:

o   chemical characteristics of waste.

o   choice of a suitable surfactant based on the waste. This is most likely to be best resolved by tests with the actual waste on the top of the impoundment.

o   thickness of surfactant layer. This factor may not be significant. McCoy (1982) found that the rate of evaporation is independent of layer thickness. The thicker the layer, the more effective it is likely to be, although there is probably a critical thickness above which the marginal reduction of evaporation is very insignificant.

o   wind speed over the impoundment.

o   temperature profile in the impoundment. The use of a layer over an impoundment is likely to increase the temperature on the top of the impoundment and change the temperature gradient in the impoundment (Bartholic, 1967). This would somewhat decrease the actual evaporation reduction when compared to expected effectiveness.

The cost of using surfactant layers depends on the surfactant, the rate of application, the equipment and labor requirements.

As mentioned above, the effectiveness of a surfactant layer is dependent on the ability to maintain a continuous film. This means that a surfactant layer would not be very effective on an aerated impoundment since the surface is continually disturbed. This technology looks simple at the outset, but the choice of an effective surfactant would be a much less simple matter, considering the complex nature of hazardous waste impoundments. Surfactant layers are not included in further discussions of in-situ controls because of these uncertainties and possible limited applicability in surface impoundments.

5.   Post-Treatment

As discussed above in the section on synthetic membrane covers, collection and post-treatment of emissions have been used by at least one facility, Upjohn in New Haven, Connecticut. In this case, an air structure with a vent and post-treatment using regenerative carbon adsorption was used.

Basically, a post-treatment control requires a means of collecting the emissions by means of a cover and a vent, and treatment unit(s) at the vent. Other forms of treatment may be used besides carbon adsorption, including afterburning and condensation. Descriptions of two approaches, carbon adsorption and afterburners follow.

## 5.1  Gaseous Carbon Adsorption

Removal of volatile compounds from a gas stream by adsorption is a widely used process. Adsorption is a process in which the volatile components retained on the surface of granular and highly porous solids. Activated carbon is most commonly used as the adsorbent, although other materials are available for specialized applications. In the process, the gases to be treated are sent through a bed of adsorbent. The components to be removed adhere to the surface of the grains and, in addition, diffuse into and are trapped in the pores of the material.

The process can be reversed so that both the adsorbent material and the vapors that have been retained may be recovered. Carbon adsorption systems are available as complete package units from several manufacturers. These units can be installed with the minimum of on-site operations. Custom designed systems are also available for larger or special purpose applications.

The cost per unit quantity treated is generally considerably less in a regenerative configuration than in a non-regenerative system, even though the capital costs are higher for the former type. If the recovered material can be reused and has a high market value, recovery in carbon adsorption systems can produce positive cash flow. If the recovered material, on the other hand, has value only in terms of the heat content used as a fuel, then the overall costs will generally be higher.

In hazardous waste applications, particularly at commercial facilities which receive a wide variety of wastes, the recovered materials would not be expected to have appreciable market or fuel value. In this case generation and recovery will be considerably more expensive. If the recovered organic wastes must be subsequently disposed of by incineration or even by land disposal, then the overall costs may approach those of operating in a non-regenerative mode. Although the capital costs of non-regenerative systems are smaller than those which recover carbon and adsorbed materials, cost per pound of material removed are very much higher because large quantities of expensive carbon adsorbent would be required.

Adsorption is a very effective process. When gases containing organic vapors first contact fresh adsorbent the vapors are quickly and efficiently removed from the gas stream. Efficiencies of up to 95% are commonly achieved. As the gas being treated continues to flow over the bed of carbon, its adsorbtive capacity gradually decreases. At some point, the adsorbent will become saturated and no materials will be removed. In operating systems, the process is stopped before this point, called the breakthrough point, is reached and the carbon is either replaced or generated. Regenerative systems employ multiple beds so that gases can be continuously treated while one of the beds is being replaced or regenerated. Alternate approaches use moving bed adsorbers in which fresh carbon is continuously replaced. Spent sorbent is removed at one end of the system, passes through the regenerator, and is replaced at the gas inlet end. This type of system, in theory, has a more efficient utilization than does a fixed bed system using alternate units. Moving bed adsorbers are more expensive to construct and operate and more difficult to maintain.

In conventional treatment applications, carbon adsorption is generally more economical than the technique described below, incineration, at low concentrations below about 100 ppm. In applications for the recovery of volatile emissions from hazardous wastes, the economics may be reversed if the costs of secondary disposal of the recovered waste components is included. Incineration destroys the materials and no further disposal costs are required. If the recovered materials can be returned to a treatment system, then the carbon adsorption may retain its advantage for very dilute gases.

### 5.2 Afterburners

Afterburners are also called vapor incinerators because they are used to incinerate gases and vapors. Sources of information used in this description are Ehrenfeld and Bass, 1983; USEPA, 1978; USDHEW, 1970. Dilute concentrations of organic vapors are burned together with additional fuel to generate a high temperature of up to 870°C. The fuels used include natural gas, LPG, and distillate and residual fuel oils. Incoming gases and vapors are decomposed and oxidized as they pass through the afterburner. Combustion products include carbon dioxide, water and others, depending on the composition of the incoming gases. The residence time at temperatures less than 870°C is between 0.5 to 1.0 seconds. Generally, afterburners should only be used on those pollutants that will not produce undesirable oxidation products.

Lower oxidation temperatures may be obtained with the use of a catalyst. Catalytic afterburners operate at temperatures of between 540 to 870°C, although most combustion catalysts cannot be operated at temperatures greater than between 540 and 650°C. Catalysts used include platinum, platinum alloys, copper chromite, copper oxide, chromium, manganese, nickel, and cobalt. When a catalyst is used, care must be taken to prevent poisoning of the catalyst by limiting the concentration of incoming organics to prevent overheating of the catalyst. Most combustion catalysts cannot be operated at temperatures greater than 540 to 650°C. The maximum concentration of volatile organic carbon is limited to 25 percent of the lower flammability limit. Catalysts begin to lose their effectiveness as soon as they are used and need to be replaced when they are worn out. Catalyst life is usually between 1 and 5 years.

Afterburners can be operated with a heat recovery system. The hot combustion gases can be used to preheat process gases entering the afterburner. Less simple is secondary heat recovery with heat exchangers to transfer the heat energy elsewhere.

In designing an afterburner system, the following parameters must be specified:

o     gas and vapor volume, both average and extremes, also variations due to changes in seasonal temperature.

o     identify of contaminants in gas.

o     concentration of contaminants in gas stream.

o     expected destruction efficiency of the afterburner, from bench or pilot tests.

Afterburners can be designed to handle a range of gas inflow rates. The constituents of the gas stream must be restricted to those that will not produce undesirable combustion products, and must also not contaminate the catalyst if catalytic oxidation is used. If corrosive oxidation products are formed, the afterburner may have to be constructed from special materials. The contaminants in the gas stream must be destructible to required efficiencies at the operating temperatures and residence time of the afterburner.

The efficiency of an afterburner system depends on:

o     residence time of the gases in the combustion process; the efficiency increases with residence time for times less than 1 second.

o     temperature of combustion; efficiency increases with temperature.

o     degree of mixing of chamber; efficiency increases with flame contact and oxygen concentration.

o     nature of waste gas.

o     concentration of contaminants in waste gas.

o     catalyst type in the case of a catalytic process.

o     active surface area of catalyst which depends on how long the catalyst has been used.

Organics have been destroyed at efficiencies greater than 98 percent with well-designed and properly-operated afterburners.

The afterburner is a well established method for destroying volatile organics. It is a conventional and well-demonstrated technology. In its use in hazardous waste applications, the changes in gas flow rate and composition may decrease the operating efficiency of the afterburner when compared to the design efficiency.

D.    Effectiveness of Surface Impoundment Controls

1.    Methodology

1.1  Selection of Parameter Values

To evaluate the effectiveness of controls in reducing emissions from impoundments, waste, surface impoundment, and site (environmental) parameters necessary for quantifying emissions were defined. Some typical wastes found in surface impoundments were selected. The values of physical/chemical properties needed for calculating emissions were tabulated for the chemicals. Three sizes of surface impoundments were evaluated. The sizes were determined from WESTAT data and impoundments and the 10th, 50th, and 90th capacity percentiles were selected for evaluation. The corresponding surface areas and depths at full capacity were obtained from WESTAT data. It was assumed that the impoundments were operating at 75 percent capacity most of the time and the dimensions of exposed surface area, diameter and operating depth were calculated for symmetrical impoundments with circular cross-sectional areas. The wind speed and temperature were assumed to be 10 miles per hour and 25°C respectively.

1.2  Calculation of Mass Transfer Coefficients and Emissions

Three general types of impoundments were selected to represent the family of impoundments in use: non-aerated, aerated, and impoundments with a floating immiscible organic layer. The 10th percentile impoundment assumed not to be aerated because of its small size. Small aerated installations are treated in Chapter IV under tanks. The equations for calculating mass transfer coefficients are repeated in Table III-3.

Total emissions per hour were also calculated for each of the impoundments by:

$$Q = K_{oa} A (X_L - X^*) \text{ lb} - \text{mol/hr}$$

These emissions were calculated assuming that $X_L$ was much greater than $X^*$ and that $X_L = 1$. For any waste in an impoundment, $X_L$ is much less than 1 and emissions may be calculated by multiplying the emissions calculated here by the mole fraction of waste component in the impoundment. In essence, maximum emissions are provided in this methodology.

TABLE III-3
MASS TRANSFER COEFFICIENTS
(lb-mol/ft²-hr)

**Individual**

Non-Aerated:

$$k_G \; = \; 1.1 \times 10^{-4} u^{.78} D_p^{-.11} (N_{sc})^{-.67}$$

$$k_L \; = \; 3.12 \, (1.024)^{\theta-20} u_o^{.67} H_o^{-.85} \left( \frac{D_{i,H_2O}}{D_{O_2,H_2O}} \right)$$

Aerated:

$$k_G \; = \; .00039 \; \frac{\rho_g D_{i,air}}{\upsilon_T \, MWi} \; (N_{re})^{1.42} (N_{fr})^{-.21} (N_p)^{.4} (N_{sc})^{0.5}$$

$$k_L \; = \; \frac{J \, (P_r \eta) \, (1.024)^{\theta-20} \, \alpha \, (10^6)}{165.04 \, a_v \, V} \; \left( \frac{D_{i,H_2O}}{D_{O_2,H_2O}} \right)^{0.5}$$

Floating Immiscible Layer:

$$k_G \; = \; 1.1 \times 10^{-4} \, U^{.78} \, D_p^{-.11} (N_{sc})^{-.67}$$

**Overall**
Non-Aerated:

For a dilute aqueous solution at 1 atmosphere,

$$\frac{1}{K_{oa}} \; = \; \frac{1}{k_L} \; + \; \frac{1}{k_G \, (3.47 \, H)} \qquad \begin{array}{l} \text{where } H = \text{Henry's Law} \\ \text{Constant (atm-ft}^3/\text{lb-mol)} \end{array}$$

Aerated:

$$K_{oa} \; = \; (K_{oa})_c \; \frac{A_c}{A} \; + (K_{oa})_T \; \frac{A_T}{A}$$

Convective Zone

$$\frac{1}{(K_{oa})_c} \; = \; \frac{1}{(k_L)_c} \; + \; \frac{1}{K(k_G)_c} \qquad \text{where } K = 3.47H$$

TABLE III-3 Continued

**Turbulent Zone**

$$\frac{1}{(K_{oa})_T} = \frac{1}{(k_L)_T} + \frac{1}{K(k_G)_T} \qquad \text{where } K = 3.47H$$

**Floating Immiscible Layer:**

$$K_{oa} = P_{vp}k_G/P_T$$

Source:  Arthur D. Little, Inc.

## 1.3  Emission Reduction and Efficiencies of Controls

The reduction in emission rates is defined by:

$$R = Q_1 - Q_2 = K_{oal} A_1 (X_{L1} - X_1^*) - K_{oa2} A_2 (X_{L2} - X_2^*)$$

where the subscripts 1 and 2 refer to impoundments without controls and with controls respectively.

If $(X_{L1} - X_1^*) = (X_{L2} - X_2^*)$, and assumed to be 1, i.e., the highest emissions rates, then:

$$R = K_{oal} A_1 - K_{oa2} A_2.$$

The efficiency (E) of a control in reducing emissions is defined by 100 $(Q_1-Q_2)/Q_1$ or:

$$E = 100 \frac{[K_{oal} A (X_{L1} - X_1^*) - K_{oa2} A_2 (X_{L2} - X_2^*)]}{K_{oal} A_1 (X_{L1} - X_1^*)}$$

For our considerations, we assume no pretreatment and that $X_{L1} = X_{L2}$ and $X_1 = X_2$ and:

$$E = 100 (K_{oal} A_1 - K_{oa2} A_2)/(K_{oal} A_1)$$

## 1.4  Cost-Effectiveness of Controls

The in-situ controls for emissions reductions were compared using the cost of each percent of reduction efficiency per square feet of surface area ($/percent - ft$^2$). Pretreatment, in-situ controls, and post-treatment were compared by considering the curves of cost per pound removed (reduced) versus pounds per year removed (reduced). Similar curves were generated for cost per pound removed (reduced) versus throughput in pounds per year. For in-situ controls, the annual cost (A) is constant, irrespective of the pounds emissions reduced (R) per year or the throughput in the impoundment (T).

In-situ:

$$A = constant = yR (\$/yr)$$

where  $y$ = cost per pound reduction
$$y = \frac{A}{R} \quad (\$/\text{lb-reduced})$$

Post-treatment:

$$A = F + pR \quad (\$/\text{yr})$$

where  $F$ = cost of collection system per year
$p$ = cost per pound of post-treatment system
$$y = \frac{A}{R} = \frac{F}{R} + p \quad (\$/\text{lb-reduced})$$

In terms of throughput (T) and efficiency (E):

In-situ:

$$A = yET \quad (\$/\text{yr})$$
$$y = A/ET \quad (\$/\text{lb-reduced})$$

Post-treatment:

$$A = F + pET \quad (\$/\text{yr})$$
$$y = \frac{F}{ET} + p \quad (\$/\text{lb-reduced})$$

Pretreatment costs vary between $0.50 - $10.00 per pound of volatiles removed (Spivey et al, 1984). In-situ, post-treatment controls and pretreatment are compared below to different pretreatment costs.

Costs for pretreatment were obtained from Spivey et al (1984). Unit material costs for in-situ controls were obtained from and estimated for the three impoundment sizes by vendors of the materials. Installation, operation and maintenance, and other costs were only compared qualitatively since these were not easily available or estimated for the in-situ technologies. Collection costs were derived from a vendor (Air Structures International, Inc., Tappan, NY). Costs for post-treatment by carbon absorption and afterburners were obtained from EPA (1982).

2. Parameters

The values of waste characteristics for typical wastes are shown in Table III-4. As may be seen from Table III-4, the values of all the waste parameters varied only within about one order of magnitude except for the Henry's Law constant and vapor pressure which ranged in 5 orders of magnitude and 3 orders of magnitude for the chemicals listed respectively.

## TABLE III-4
## TYPICAL VALUES FOR WASTE PARAMETERS

| | Liquid Density (lb/ft³) | Liquid Viscosity (lb/ft-sec) | Henry's Law Constant (atm-ft³/lb-mol) | Molecular Weight | Diffusivity in Water (ft²/hr) | Gas Viscosity (lb/ft-sec) | Gas Diffusivity (ft²/hr) | Schmidt Number | Gas Density (lb/ft³) | $\Delta H_v$ ($\frac{Cal}{g\text{-}mol}$) | Vapor Pressure (atm) |
|---|---|---|---|---|---|---|---|---|---|---|---|
| Vinyl Chloride | 56.8 | | 1361.42 | 62.5 | | | | | 0.17 | 6263 | 3.5(25°C) |
| Methylene Chloride | 82.8 | $3 \times 10^{-4}$ | 129.79 | 84.93 | | | | | 0.24 | 7572 | 0.46(20°C) |
| Trichloroethylene | 91.6 | $3.9 \times 10^{-4}$ | 145.75 | 131.5 | | | | | 0.37 | 8315 | 0.079(20°C) |
| Benzene | 54.9 | $4.4 \times 10^{-4}$ | 88.09 | 78.11 | | $3.02 \times 10^{-5}$ | 0.29 | 1.71 | 0.22 | 10254 | 0.1(20°C) |
| 2-Hexanone | 50.6 | $3.9 \times 10^{-4}$ | | 100.2 | | | | | 0.28 | | |
| m-Xylene | 53.7 | $4.2 \times 10^{-4}$ | 83.29 | 106.16 | | | | | 0.29 | 9904 | 0.0079(20°C) |
| Methyl Ethyl Ketone | 50.3 | $2.8 \times 10^{-4}$ | 0.70 | 72.1 | | | | | 0.20 | 8150 | 0.102(20°C) |
| Tetrachloroethylene | 101.3 | $6.0 \times 10^{-4}$ | 132.94 | 165.83 | | | | | 0.46 | 9241 | 0.018(20°C) |
| Toluene | 54.1 | $4.0 \times 10^{-4}$ | 105.71 | 92.1 | | $3.5 \times 10^{-5}$ | 0.26 | 1.86 | 0.26 | 9369 | 0.029(20°C) |
| Ethyl benzene | 54.1 | $4.5 \times 10^{-4}$ | | 106.17 | | | | | 0.29 | 9309 | 0.092(20°C) |
| n-Butyl Alcohol | 50.6 | $2.0 \times 10^{-3}$ | | 74.12 | | | 0.27 | 1.88 | | | |
| Bromobenzene | 93.3 | $8.6 \times 10^{-4}$ | | 157.02 | | $1.9 \times 10^{-5}$ | 0.26 | 1.71 | 0.44 | 10158 | 0.043(20°C) |
| Carbon Tetrachloride | 99.5 | $6.5 \times 10^{-4}$ | 368.38 | 153.8 | | $6.05 \times 10^{-5}$ | 0.24 | 2.13 | 0.43 | 8272 | 0.15(25°C) |
| Chlorobenzene | 69.0 | $5.4 \times 10^{-4}$ | 59.26 | 112.56 | | $4.37 \times 10^{-5}$ | 0.24 | 2.13 | 0.31 | 10098 | 0.016(25°C) |
| Ethyl Alcohol | 49.3 | $8.1 \times 10^{-4}$ | | 46.07 | | | 0.38 | 1.30 | | 9674 | |
| Acetone | 49.3 | | 0.4 | 58.08 | $6.23 \times 10^{-5}$ | | | | | 7642 | |
| Average | 66.33 | $6.05 \times 10^{-4}$ | | 100.08 | $6.23 \times 10^{-5}$ | $3.77 \times 10^{-5}$ | 0.28 | 1.82 | 0.30 | 8872 | |
| Highest | 101.3 | $2.0 \times 10^{-3}$ | | 165.83 | | $6.05 \times 20^{-5}$ | 0.38 | 2.13 | 0.44 | 10254 | |
| Lowest | 49.3 | $2.8 \times 10^{-4}$ | | 46.07 | | $1.91 \times 10^{-5}$ | 0.24 | 1.30 | 0.17 | 6263 | |

Diffusivity of oxygen in water at 18°C, $(D_{O_2, H_2O}) = 6.59 \times 10^{-5}$ ft²/hr

Source: Arthur D. Little, Inc.

The impoundment sizes chosen are shown in Table III-5. The values for site and impoundment parameters are shown in Table III-6. Small impoundments are usually not aerated, and so the impoundment at the tenth percentile is considered to be a non-aerated impoundment only.

### 3.    Mass Transfer Coefficients and Emissions

The average, high, and low values for waste parameters from Table III-4 were substituted into the relationship for individual mass transfer coefficients. It was found that individual mass transfer coefficients did not vary greatly with changes in values of waste characteristics (within only about one order of magnitude). Since the primary interest is relative emissions under different control situations, the average values for the waste characteristics given in Table III-4 were used to define "typical" individual mass transfer coefficients.

The calculated individual mass transfer coefficients for non-aerated impoundments, aerated impoundments, and impoundments with a floating immiscible layer are shown in Table III-7. These values may be used for most wastes with errors within one order of magnitude.

Overall mass transfer coefficients and emissions, for $X_L = 1$, are shown in Table III-8 and III-9 respectively for a range of Henry's Law Constant and vapor pressure. In non-aerated impoundments with H much less than around 0.1 atm-ft$^3$/lb-mol, the waste volatilizes slowly at a rate dependent on H. The gas-phase resistance dominates over the liquid- phase (i.e. gas-phase controlled). For H much greater than 0.1 atm-ft$^3$/lb-mol, the volatilization is liquid-phase controlled and reaches a maximum of $k_\ell$ for wastes with very high H. From Table III-4, very few chemicals have low values for Henry's Law Constant. The values for acetone and methyl ethyl ketone are around 0.5 atm-ft$^3$/lb-mol. 3-Bromo-1-propanol and dieldrin are very non-volatile compounds with Henry's Law Constant around $8 \times 10^{-3}$ atm-ft$^3$/lb-mol (Lyman et al, 1982) and their volatilization is gas-phase controlled and dependent on Henry's Law Constant.

Emissions were calculated using a mole fraction of 1 of the volatile component in the impoundment. Using a median Henry's Law Constant of 100 atm-ft$^3$/lb-mol and median vapor pressure of 0.07 atm, the emissions without controls from aerated impoundments are the highest. The non-aerated emissions shown appear to be higher than those from floating-immiscible-layer impoundments. In reality however, the mole fraction of a waste component in an aqueous impoundment will be very low and emissions from non-aerated impoundments will be lower.

### 4.    Emissions Reduction and Efficiencies

The parameters that control $K_{oa}$ values in non-aerated impoundments are temperature, wind speed at surface impoundment, diameter and the effective depth of the impoundment. For aerated

TABLE III-5

SURFACE IMPOUNDMENT SIZES

| Percentile | Capacity $(ft^3)$ | (Full Capacity) | | (75% Capacity) | | |
| | | Surface Area $(ft^2)$ | Depth $(ft)$ | Volume $(ft^3)$ | Surface Area $(ft^2)$ | Depth $(ft)$ |
|---|---|---|---|---|---|---|
| 10% | 1,340 | 600 | 5 | 1,005 | 482 | 4 |
| 50% | 73,260 | $2x10^4$ | 9 | 54,945 | 16,635 | 7.6 |
| 90% | $4x10^6$ | $3x10^5$ | 15 | $3x10^6$ | 259,112 | 12 |

Source:  Arthur D. Little, Inc.

TABLE III-6

SITE AND IMPOUNDMENT PARAMETER VALUES

## Site

| | |
|---|---|
| Wind Speed, U ft/hr | 52,800 |
| Wind Speed at Surface, $U_o$ = 0.035 x U ft/sec | 0.513 |
| Ambient Temperature, $\theta$ °C | 25 |

| Impoundment (At 75%-capacity) | Percentile of Impoundments | | |
|---|---|---|---|
| | 10% | 50% | 90% |
| Effective pool diameter, $D_p$ ft | 24.77 | 145.54 | 574.38 |
| Effective depth, $H_o$ ft | 4 | 7.6 | 12 |
| Number of aerators | - | 6 | 6 |
| Diameter of aerator, $D_T$ ft | - | 1.97 | 6 |
| Impeller rotational speed, $\omega$ rad/sec. | - | 126 | 126 |
| Power to impeller, $P_r$ hp | - | 15.4 | 100 |
| Oxygen transfer rating of aerator, J $lb-O_2$/hr-hp | - | 3 | 3 |
| Efficiency of power conversion, $\eta$ | - | .78 | .78 |
| Oxygen transfer convection factor, $\alpha$ | - | .825 | .825 |
| Turbulent surface area to turbulent volume ratio $a_v$ $ft^{-1}$ | - | .09 | .0555 |
| Volume affected by aerators, V $ft^3$ per aerator | - | 3,119 | 39,710 |
| Effective surface area of turbulent zone, $A_T$ $ft^2$ (x number of aerators) | - | 1,704.0 | 13,236.0 |
| Effective surface area of convection zone,[*] | - | 14931.2 | 245876 |

[*]Assuming that surface impoundment is operating at 75% capacity.

Source:  Arthur D. Little, Inc.

TABLE III-7

## "TYPICAL" INDIVIDUAL k VALUES FOR SURFACE IMPOUNDMENTS WITHOUT CONTROLS

|  | 10% | 50% | 90% |
|---|---|---|---|
| **Non-Aerated** | | | |
| $k_G$ lb-mol/ft$^2$-hr | 0.254 | 0.205 | 0.177 |
| $k_L$ lb-mol/ft$^2$-hr | 0.654 | 0.379 | 0.257 |
| | | | |
| **Aerated** | | | |
| $k_G$ lb-mol/ft$^2$-hr | - | 0.42 | 1.78 |
| $k_L$ lb-mol/ft$^2$-hr | - | 703.53 | 581.87 |
| | | | |
| **Floating Immiscible Layer** | | | |
| $k_G$ lb-mol/ft$^2$-hr | 0.254 | 0.205 | 0.177 |

Source:   Arthur D. Little, Inc.

TABLE III-8

MASS TRANSFER COEFFICIENTS (lb-mol/ft$^2$-hr) FOR SURFACE IMPOUNDMENTS WITHOUT CONTROLS

| Vapor Pressure (Atmosphere) | Floating Immiscible Layer 10% | 50% | 90% |
|---|---|---|---|
| $7 \times 10^{-4}$ | $1.75 \times 10^{-4}$ | $1.44 \times 10^{-4}$ | $1.24 \times 10^{-4}$ |
| $7 \times 10^{-3}$ | $1.75 \times 10^{-3}$ | $1.44 \times 10^{-3}$ | $1.24 \times 10^{-3}$ |
| $7 \times 10^{-2}$ | $1.75 \times 10^{-2}$ | $1.44 \times 10^{-2}$ | $1.24 \times 10^{-2}$ |
| 0.7 | 0.175 | 0.144 | 0.124 |
| 7 | 1.747 | 1.438 | 1.236 |

| Henry's Law Constant (atm-ft$^3$/lb-mol) | Non-Aerated 10% | 50% | 90% | Aerated 50% | 90% |
|---|---|---|---|---|---|
| $10^{-3}$ | $8.65 \times 10^{-4}$ | $7.11 \times 10^{-4}$ | $6.11 \times 10^{-4}$ | $7.87 \times 10^{-4}$ | $8.96 \times 10^{-4}$ |
| $10^{-2}$ | $8.55 \times 10^{-3}$ | $7.00 \times 10^{-3}$ | $5.99 \times 10^{-3}$ | $7.76 \times 10^{-3}$ | $8.83 \times 10^{4}$ |
| 0.1 | $7.65 \times 10^{-2}$ | $6.00 \times 10^{-2}$ | $4.95 \times 10^{-2}$ | $6.89 \times 10^{-2}$ | $7.85 \times 10^{-2}$ |
| 1 | 0.3726 | 0.2474 | 0.1811 | 0.3702 | 0.4840 |
| 10 | 0.6080 | 0.3598 | 0.2467 | 1.777 | 3.086 |
| 100 | 0.6490 | 0.3770 | 0.2560 | 12.64 | 15.54 |
| 1000 | 0.6534 | 0.3788 | 0.2569 | 48.80 | 27.37 |
| 10000 | 0.6539 | 0.3789 | 0.2570 | 68.97 | 29.65 |

Source: Arthur D. Little, Inc.

## TABLE III-9

### EMISSIONS FROM SURFACE IMPOUNDMENTS WITHOUT CONTROLS (lb-mol/hr)*

| Henry's Law Constant (atm-ft³/lb-mol) | Non-Aerated 10% | Non-Aerated 50% | Non-Aerated 90% | Aerated 50% | Aerated 90% |
|---|---|---|---|---|---|
| $10^{-3}$ | 0.4169 | 11.8276 | 158.3174 | 13.0919 | 232.164 |
| $10^{-2}$ | 4.1211 | 116.4464 | 1552.0809 | 129.09 | 2287.96 |
| 0.1 | 36.8730 | 998.1120 | 12826.044 | 1146.17 | 20340.29 |
| 1 | 179.5932 | 4115.5485 | 46925.18 | 6158.35 | 125410.21 |
| 10 | 293.0560 | 5985.3450 | 63922.93 | 29560.75 | 799619.63 |
| 100 | 312.8180 | 6271.4704 | 66332.67 | 210268.93 | 4026600.48 |
| 1000 | 314.94 | 6301.41 | 66565.87 | 811797.76 | 7091895.44 |
| 10000 | 315.18 | 6303.08 | 66591.78 | 1147329.74 | 7682670.80 |

| Vapor Pressure (Atm) | Floating Immiscible Layer 10% | Floating Immiscible Layer 50% | Floating Immiscible Layer 90% |
|---|---|---|---|
| $7\times10^{-4}$ | 0.084 | 2.392 | 32.03 |
| $7\times10^{-3}$ | 0.844 | 23.921 | $3.203\times10^2$ |
| $7\times10^{-2}$ | 8.435 | 239.214 | $3.203\times10^3$ |
| 0.7 | 84.35 | 2392.14 | $3.203\times10^4$ |
| 7 | 842.05 | 23921.418 | $3.203\times10^5$ |

* Using $(X_L - X^*) = 1$

Source:  Arthur D. Little, Inc.

impoundments, all the controlling parameters are aerator related except for ambient temperature and Henry's Law Constant. The temperature however, cannot be changed by in-situ control without impairing the effectiveness of the aeration treatment. Henry's Law Constant in this case can only be decreased by pretreatment and/or by limitation and exclusion of volatile wastes. The alternative is collection and post-treatment of emissions. For the impoundment with a floating immiscible layer, the vapor pressure of the volatile component in the layer and wind speed directly affect $K_{oa}$. The effective pool diameter inversely affects $K_{oa}$. Vapor pressure can be reduced by pretreatment and/or by exclusion and limitation, or by decreasing the temperature of the impoundment surface. For very non-volatile waste (H << 0.1 atm-ft$^3$/lb-mol, decreasing the temperature would reduce $K_{oa}$ for non-aerated impoundments, but most chemicals have H greater than 0.1 atm-ft$^3$/lb-mol. In summary, the controllable parameters for each type of emission controls is shown in Table III-10.

The Henry's Law Constant decreases with decreasing temperature. At any temperature, the Henry's Law Constant of a compound is the ratio of the partial pressure (P) and the solubility of the compound at that temperature. The solubility could increase or decrease with temperature, but the sensitivity of solubility to temperature is not as large as that of vapor pressure to temperature which are related by the Clausius-Clapeyron equation:

$$\frac{d \ln P}{d T} = \frac{\Delta H_v}{\Delta Z RT^2}$$

where:  $\Delta Z$ = 1 for ideal gas (vapor)

$\Delta H_v$ = heat of vaporization, (cal/mol)

P = vapor pressure (atm)

R = gas constant, (cal/mol-$^\circ$K)

T = temperature, ($^\circ$K)

Neglecting the change of $H_v$ with temperature, the simplest solution to the above relationship is:

$$\ln P = A - B/T$$

where   A and B are constants;

$$B = \frac{\Delta H_v}{\Delta ZR} = \frac{\Delta H_v}{R} \quad \text{(for ideal gases)}$$

TABLE III-10

CONTROLLABLE PARAMETERS TO REDUCE EMISSION RATES

FOR CATEGORIES OF EMISSION CONTROLS

$$Q' \doteq K_{oa} A \ (X_L - X^*)$$

| Parameter | Proportionality | Effective on Impoundment Type |
|---|---|---|
| **Pretreatment** | | |
| Mol fraction of diffusing component, $X_L$ | direct | aerated, non-aerated, float. immiscible layer |
| **Design** | | |
| Effective depth, $H_o$ | $H_o^{-.85}$ | non-aerated |
| Effective pool diameter, $D_p$ | $D_o^{-.11}$ | float. immiscible layer, non-aerated |
| Wind Speed, U and $U_o$ | $U^{p.78}$ | float. immiscible layer, non-aerated |
| (by increasing freeboard) | $U_o^{.67}$ | non-aerated |
| Surface area, A | $_oA$ | non-aerated, float. immiscible layer |
| **Operating** | | |
| Temperature, $\theta$ | $(1.024)^{\theta-20}$ | non-aerated float. immiscible layer, aerated |
| Wind speed, U and $U_o$ (by increasing freeboard) | $U^{.78}$ $U_o^{.67}$ | float. immiscible layer, non-aerated non-aerated |
| Surface area, A | A | non-aerated, float. immiscible layer |
| **In-Situ** | | |
| Temperature, $\theta$ | $(1.024)^{\theta-20}$ | non-aerated float. immiscible layer, aerated |
| Wind speed, U and $U_o$ | $U^{.78}$ $U_o^{.67}$ | float. immiscible layer, non-aerated non-aerated |
| Effective pool diameter, $D_p$ | $D_p^{-.11}$ | float. immiscible layer, non-aerated |
| Surface area, A | $_pA$ | non-aerated, float. immiscible layer |

Source:  Arthur D. Little, Inc.

$$\ln P_1 - \ln P_2 = \frac{WH_v}{R} \frac{T_1-T_2}{T_1 T_2}$$

Using an average $\Delta H_v = 8872$ cal/g-mol and R = 1.9872 cal/mol-°K

$$\ln \frac{P_2}{P_1} = 4465 \frac{T_2-T_1}{T_1 T_2} \qquad (T \text{ in } °K)$$

$$\frac{P_2}{P_1} = \exp [4465 (T-_2T_1)/(T_1T_2)]$$

Neglecting the sensitivity of solubility with temperature, Henry's Law Constant at $T_1$ is related to Henry's Law Constant at $T_2$ by:

$$H_2 = \frac{P_2}{P_1} \cdot H_1$$

$$H_2 = H_1 \exp [4465 (T_2-T_1)/(T_1T_2)] \qquad (T \text{ in } °K)$$

Using the relationships of controllable parameters (temperature, wind speed and geometry) to emissions, the reductions in emissions and efficiencies of changing any of the parameters were calculated. Geometry changes were made by keeping capacity constant. The efficiencies of these changes on emission reductions from non-aerated and floating-immiscible-layer impoundments are shown in Figures III-3, III-4 and III-5 for the 50th percentile impoundment.

The figures above show efficiencies without relating them to real controls. Using the medians for Henry's Law Constant and vapor pressure (100 atm-ft³ lb-mol and 0.07 atm respectively), emissions reductions and efficiencies were derived for each of the in-situ controls as described below. The emissions reductions and efficiencies are shown in Table III-11.

### Rafts

The efficiency of rafts is approximately equal to the percent of area covered. White Styrofoam rafts 4 x 8 feet, 0.5 and 1.0 inches thick were chosen. It was assumed that 90 percent of the exposed surface area of an impoundment is covered by rafts at any time. Some

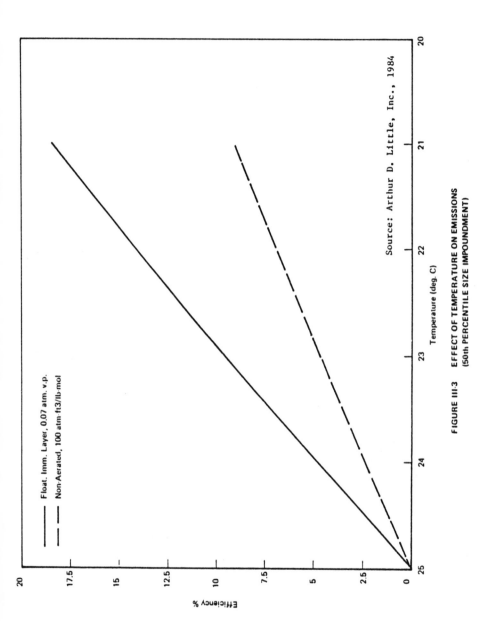

FIGURE III-3    EFFECT OF TEMPERATURE ON EMISSIONS
(50th PERCENTILE SIZE IMPOUNDMENT)

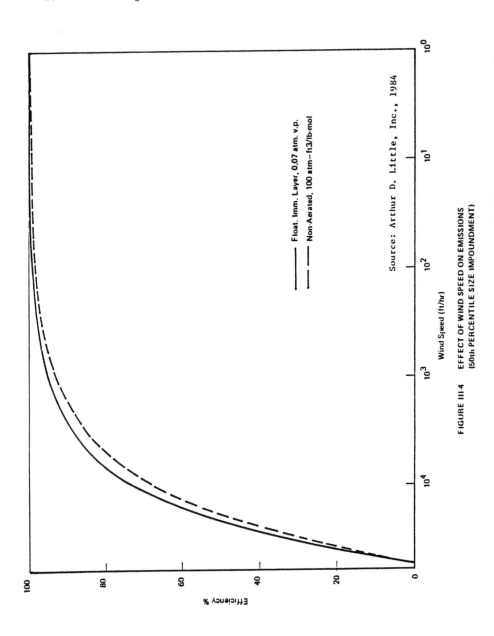

FIGURE III-4    EFFECT OF WIND SPEED ON EMISSIONS
(50th PERCENTILE SIZE IMPOUNDMENT)

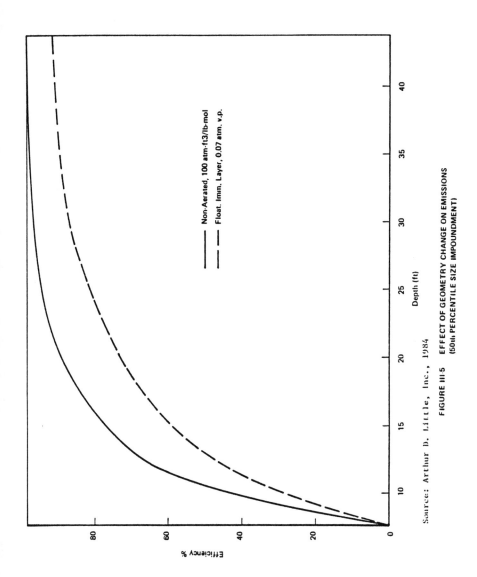

Non-Aerated, 100 atm-ft3/lb-mol
Float. Imn. Layer, 0.07 atm. v.p.

Source: Arthur D. Little, Inc., 1984

**FIGURE III-5    EFFECT OF GEOMETRY CHANGE ON EMISSIONS
(50th PERCENTILE SIZE IMPOUNDMENT)**

Depth (ft)

Efficiency %

## TABLE III-11

### EMISSIONS REDUCTIONS AND EFFICIENCIES OF IN-SITU CONTROLS

| In-Situ Controls | Reductions* (lb/yr) | | | Efficiencies (%) | | |
|---|---|---|---|---|---|---|
| | 10% | 50% | 90% | 10% | 50% | 90% |
| **Non-Aerated:** | | | | | | |
| Rafts | $2.468 \times 10^8$ | $4.948 \times 10^9$ | $5.234 \times 10^{10}$ | 90 | 90 | 90 |
| Barriers | $3.017 \times 10^7$ | $6.048 \times 10^8$ | $6.397 \times 10^9$ | 11 | 11 | 11 |
| Shades | $1.207 \times 10^8$ | $2.419 \times 10^9$ | $2.559 \times 10^{10}$ | 44 | 44 | 44 |
| Floating Spheres | $2.468 \times 10^8$ | $4.948 \times 10^9$ | $5.234 \times 10^{10}$ | 90 | 90 | 90 |
| **Floating Immiscible Layer:** | | | | | | |
| Rafts | $6.655 \times 10^6$ | $1.887 \times 10^8$ | $2.529 \times 10^9$ | 90 | 90 | 90 |
| Barriers | $8.134 \times 10^5$ | $2.307 \times 10^7$ | $3.089 \times 10^8$ | 11 | 11 | 11 |
| Shades | $3.254 \times 10^6$ | $9.228 \times 10^7$ | $1.236 \times 10^9$ | 44 | 44 | 44 |
| Floating Spheres | $6.655 \times 10^6$ | $1.887 \times 10^8$ | $2.527 \times 10^9$ | 90 | 90 | 90 |

* Using Henry's Law Constant = 100 atm-ft$^3$/lb-mol, vapor pressure = 0.07 atm, molecular weight = 100.08, $x_L - x^* = 1$

Source: Arthur D. Little, Inc.

of the area will be exposed due to difficulty of cutting the Styrofoam to fit the exact shape of the impoundment. The percent reduction will be assumed as 90 percent. Additional reduction may result from any effects on lowering wind velocity over the remaining open surface that are induced by the raft system.

### Barriers

The reduction of wind speed over an impoundment is dependent on the height: length of grid ratio (H/L). Using flat Styrofoam (expanded polystyrene) barriers 1 x 4 inch in a 8 x 8 feet grid (i.e., H/L = 1 inch/8 feet = $1.042 \times 10^{-2}$), Crow and Manges (1967) obtained 11 percent evaporation reductions in test ponds containing water. It is likely that the efficiency of a barrier also varies with the wind speed above the influence of the barriers. An 11 percent reduction was assumed for both non-aerated and floating-immiscible-layer impoundments.

The Crow and Manges studies were carried out on water reservoirs where diffusion is controlled in the gas phase. The relative effectiveness of barriers on liquid phase controlled impoundments would depend on the difference in dependence on wind velocity in the two regimes. The exponents on the wind velocity term in the two relationships used in this study are not very different (0.78 in the gas film, Mackay and Matsuga; 0.67 in the liquid film, Owens/Hwang). To the extent that these relationships are valid, barriers for impoundments should behave similarly to water reservoirs.

### Shades

The efficiency of a shade in reducing emissions from an impoundment is proportional to the percent shade of the material used, i.e., the amount of sunlight shaded out. Woven "black" polypropylene used in horticulture is chosen as the material for the shade. In pan tests, 42 percent shade black woven polystyrene mesh reduced evaporation of water by 44 percent (Crow and Manges, 1967). The reduction is due both to wind speed and temperature reduction. An efficiency of 44 percent was assumed for non-aerated and floating-immiscible-layer impoundments.

### Floating Spheres

The efficiency of floating spheres is approximately equal to the percent area covered. On a fully covered surface, the covered area is 91 percent. According to vendor literature (Capricorn Chemicals), efficiencies over 88 percent have been obtained experimentally for open tanks. An efficiency of 90 percent was assumed.

### Post Treatment

The rate of emissions is affected by the collection system since the collection system protects the impoundment from the environment. Emissions are drawn out of the system by fans through a vent to be post-treated.

Two post-treatment systems were used for comparison: carbon adsorption with no regeneration and afterburning. The efficiency of post-treatment control is equal to the efficiency of the post-treatment unit(s). Carbon absorption was assumed to work at an efficiency of 95 percent and afterburning at 98 percent.

### 5.   Costs

### Pretreatment

Pretreatment costs taken from Spivey et al (1984) were in the order of $1.00 per pound of volatiles removed from a wastestream ($2200/Mg). Costs for pretreatment depend on wastestream properties, treatment type, system design and size. The cost value chosen represented the range of most pre-treatment technologies within a factor of 2.

### In-situ Controls

Unit material costs for various in-situ controls are shown in Table III-12. Annual costs for in-situ controls and post treatment is shown in Table III-13. These costs were obtained from and estimated for the 3 impoundment sizes by vendors of the materials. Installation costs are not included in these estimates. Relative installation, and operation and maintenance costs are given qualitatively in Table III-14. The rationale used in obtaining material, installation, and operation/maintenance costs for each technology is described below.

#### Rafts

Two thicknesses, 0.5 and 1.0 inch of expanded polystyrene (Styrofoam) were chosen for the costing of rafts. These rafts are expected to have a lifetime of 1 year under the conditions of usage. The installation costs are expected to be relatively low since installation is simple, involving little labor in floating the Styrofoam panels on top of the impoundment. Operation and maintenance costs are also expected to be low. Rain will slide off the top of these Styrofoam rafts. Heavy snow may cause the panels to sink, but the panels should float automatically when the snow is melted. Another element of costs is the disposal of the Styrofoam panels upon disintegration of the foam or upon closure of an impoundment. These disposal costs may be high because of the volume of Styrofoam generated which may have to be incinerated or disposed in a landfill. The disposed volumes involved per year are 40 ft³ (10th percentile), 1,400 ft³ (50th percentile), and 21,600 ft³ (90th percentile) if 1-inch thick Styrofoam panels are used.

TABLE III-12

UNIT MATERIAL COSTS FOR SURFACE IMPOUNDMENT EMISSION REDUCTION IN-SITU TECHNOLOGIES

(Dollars per Square Foot, Summer 1984 Dollars)

| | 10th Percentile ($/ft²) | 50th Percentile ($/ft²) | 90th Percentile ($/ft²) |
|---|---|---|---|
| **EPS[1] Sheets** | | | |
| Size | | | |
| 4' x 8' x 1/2" | 0.08 | 0.06 | 0.05 |
| 4' x 8' x 1" | 0.13 | 0.12 | 0.10 |
| **EPS[1] Strips (8'x 8' grids)** | | | |
| Size | | | |
| 4" x 8' x 1" | 0.04 | 0.04 | 0.03 |
| 4" x 8' x 2" | 0.08 | 0.08 | 0.06 |
| 8" x 8' x 1" | 0.08 | 0.08 | 0.05 |
| 8" x 8' x 2" | 0.17 | 0.15 | 0.11 |
| 12" x 8' x 1" | 0.13 | 0.12 | 0.08 |
| 12" x 8' x 2" | 0.25 | 0.23 | 0.17 |
| **Shade Cloth[2]** | | | |
| Percent Shade | | | |
| 21% | 0.07 | 0.07 | 0.07 |
| 57% | 0.11 | 0.10 | 0.09 |
| 92% | 0.20 | 0.18 | 0.17 |

(1) Expanded polystyrene
(2) Woven "black" polypropylene

TABLE III-12 Continued

UNIT MATERIAL COSTS FOR SURFACE IMPOUNDMENT EMISSION REDUCTION IN-SITU TECHNOLOGIES

(Dollars per Square Foot, Summer 1984 Dollars)

| | 10th Percentile ($/ft²) | 50th Percentile ($/ft²) | 90th Percentile ($/ft²) |
|---|---|---|---|
| **Spheres** | | | |
| **Polypropylene** | | | |
| Diameter 1 3/4 inches | 6.55 | 5.05 | 4.65 |
| Diameter 6 inches | 8.69 | 7.50 | 7.50 |
| **High Density Polyethylene** | | | |
| Diameter 1 3/4 inches | 7.55 | 6.75 | 6.75 |

Sources: Reliable Plastics, Newark, NJ, 1984
X. S. Smith Company, Eatontown, NJ, 1984
Capricorn Chemicals Corporation, Secaucus, NJ, 1984

TABLE III-13

ANNUAL COSTS FOR SURFACE IMPOUNDMENT CONTROLS, $/YR (Summer 1984 Dollars)

| In-Situ[*] | 10th Percentile | 50th Percentile | 90th Percentile |
|---|---|---|---|
| Rafts | 38.56 - 62.66 | 998.11 - 1996.22 | 12955.60 - 25911.20 |
| Barriers | 19.28 | 665.41 | 7773.36 |
| Shades | 53.02 | 1663.52 | 23320.08 |
| Floating Spheres | 514.00 - 682.00 | 13672.00 - 20305.00 | 196087.00 - 316270.00 |
| Post-Treatment[**] | | | |
| Collection | | 24000 | 250000 |

[*]  Installation, O and M, and other costs not included.
[**]  Treatment costs are variable costs depending on the amount of
      emissions treated.  These costs are discussed in the text.  The
      costs shown here include basic structures and installation.

Source:  Arthur D. Little, Inc.

## TABLE III-14

## RELATIVE INSTALLATION AND OPERATION/MAINTENANCE COSTS

## OF IN-SITU CONTROLS

| | Installation | | | Operation/Maintenance | | |
|---|---|---|---|---|---|---|
| | 10% | 50% | 90% | 10% | 50% | 90% |
| Rafts<br>0.5 - 1 inch Styrofoam | Low | Low | Low | Low | Low | Low |
| Barriers<br>8' x 8' grids of Styrofoam<br>strips | High | High | High | Low | Low | Low |
| Shades<br>Black woven polypropylene | Medium | High | High | Low | Low | Low |
| Spheres<br>Polypropylene/HDPE | Low | Low | Low | Low | Low | Low |

Source:  Arthur D. Little, Inc.

### Barriers

A 8 x 8 feet grid made of Styrofoam strips 4" x 8' x 1" was chosen for the costing of barriers. The barriers may be 4 inches or 1 inch high. These barriers are expected to last a year. Barriers are expected to entail relatively high installation costs because the Styrofoam strips have to be constructed into a grid formation and installed on the impoundment surface. For small impoundments, the grid is small and consists of relatively few strips. The design and installation of the grid may be relatively easy. For large impoundments, the design and installation of the grid may have to be sub-contracted and will be expensive. Operation and maintenance costs are expected to be low under normal conditions. However, strong winds may effectively destroy a grid. Heavy snow may also damage the grid. The volume of Styrofoam generated that have to be disposed is lower than in the case of rafts but is another element of the costs in the use of wind barriers.

### Shades

Black woven polypropylene shade cloth used in horticulture is chosen as the shade material. There is not much difference in price within a small variation of shade value so the cost of a 57 percent shade value is used for that of a 47 percent shade value of the same material. The installation costs of shades are expected to be relatively high. For small impoundments, the shade may be stretched a few inches over the impoundment and secured at the edge. Shade cloth is sold in pieces and pieces need to be sealed together to obtain a large enough piece to cover impoundments. Larger impoundments may require the use of floats spread out over the surface of the impoundment to prevent the shade from dropping below the impoundment surface. In horticultural use, the shade cloth is used for the 3 summer months and lasts between 7-9 years. For year around use over a surface impoundment, the lifetime is estimated to be between 1-2 years. Operation and maintenance costs are expected to be low under normal conditions. However, strong winds and heavy snow may damage the shade or cause the shade to drop below the surface. The volume of shade cloth that needs to be disposed of at the end of its useful life is smaller than in the cases of rafts or barriers.

### Spheres

Polypropylene spheres of diameters 1-3/4 inches and 6 inches are used to provide a range of costs for spheres. The polypropylene spheres are cheaper than the high density polyethylene (HDPE) spheres for the same size spheres and are available up to 6 inches in diameter. HDPE spheres are available in only one size (1-3/4 inches diameter). Installation costs of spheres are expected to be relatively very low. Depending on the chemical(s) present in the impoundment, spheres are expected to last a relatively long period of use. An estimated lifetime of 10 years was used to annualize costs at 10 percent per year. Operation and maintenance costs are expected to be very low, or zero. The volume of spheres that need to be disposed of per year and at the end of its useful life is much smaller than Sytrofoam barriers or rafts.

### Post-Treatment

Costs for post-treatment consists of the costs for the collection system and the treatment. The collection system cost is a fixed annual cost for the design size as shown in Table III-15. This cost was estimated by Air Structures International, Inc. (Tappan, NY) for two air structures covering 17,000 square feet and 260,000 square feet corresponding to the 50th percentile impoundment and the 90th percentile impoundment respectively. The 10th percentile impoundment was too small for the installation of an air structure. The air structure for the 50th percentile impoundment was 170 feet long, 100 feet wide and 36 feet high made of translucent fabric with a cable system and anchors for support. It has an exit door and an entrance door, a $10^6$ BTU heating system, a 24,000 W lighting system and a vehicle air lock system. The 90th percentile impoundment air structure was 260 feet wide, 1,000 feet long, and 75 feet high, made of translucent fabric, with cables and anchors. It has 2 entrance doors and 16 exit doors, and 2 inflation blowers and 2 auxilliary blowers. The heating system was a $12.5 \times 10^6$ BTU system. Costs for lighting and the vehicle air lock system were not estimated because these varied greatly depending on the requirements. Installation costs were estimated to be between $1-2 per square foot, which includes foundations. For this analysis, installation was estimated to be $1.50 per square foot. The costs shown do not include maintenance (which is low) and energy costs. Costs for the collection system are shown in Table III-15 with costs annualized at 10 percent in 10 years.

Treatment costs are variable costs dependent on the total emissions treated. To estimate the treatment cost per pound of emissions removed, a 10,000 cubic feet per minute wastestream containing 50 ppm trichloroethylene was chosen as a representative stream through the post-treatment system. The emissions through the system would be around 20,000 pounds of TCE per year. Annual installed capital costs for both carbon adsorption and afterburning were at least one order of magnitude lower than annual operating costs (EPA, 1982). Total installed costs for carbon adsorption was $87,000 compared to an annual operating cost of $870,000. Total installed costs for afterburning was $230,000 and annual operating cost was $300,000 per year. After annualization, operating costs would be the dominant cost factor. Assuming 95 percent and 98 percent efficiencies for carbon adsorption and afterburning respectively, the costs per pound of TCE removed were around $45 and $16 respectively ($100,000/Mg and $35,000/Mg). Costs for regenerative carbon adsorption systems are an order of magnitude less than non- regenerative system. Catalytic afterburners are also less expensive than non-catalytic systems. If these two types of post-treatment systems are included, the treatment costs for post-treatment would be between $5-10 per pound TCE removed ($11,000-22,000/Mg).

TABLE III-15

CAPITAL COSTS FOR COLLECTION SYSTEM

FOR POST TREATMENT (Summer 1984 Dollars)

|  | 50th Percentile | 90th Percentile |
|---|---|---|
| Structure | 70000 | 985000 |
| Heating | 24500 | 175000 |
| Lighting | 15000 | Not estimated |
| Air lock vehicle | 15000 - 24000 | Not estimated |
| Installation | 25500 | 390000 |
| Total | ‾ 150000 | ‾1600000 |
| Annualized (10%, 10 years) | ‾ 24000 | ‾ 250000 |

Source:  Air Structures International, Inc. Tappan, NY, 1984

6.   Cost-Effectiveness of Controls

Table III-16 expresses the cost-effectiveness of in-situ controls in terms of the cost per percent of reductions per square foot of impoundment. Figure III-6 illustrates these values on a graph that compares the cost-effectiveness with impoundment size. Each set of points represent the effectiveness for the 10th, 50th, and 90th percentile impoundments. The figures show some slight economies of scale. Of the in-situ controls, rafts appear to be the most cost-effective. Shades are more cost-effective than barriers in terms of the pounds of emissions reduced. Floating spheres are the least cost-effective of the in-situ controls. When installation and operation/maintenance costs are included, the relative cost-effectiveness ranking of in-situ controls will not change, because rafts have low relative installation and operation and maintenance costs. Shades and barriers have higher values for these costs but the material costs for spheres are so much higher that the relative cost-effectiveness of spheres is still expected to be the lowest.

Figures III-7 and III-8 show the hyperbolic functions relating cost per pound of emissions reduced or removed to the throughput of volatiles per year and the total pounds of volatiles removed per year respectively. The horizontal curves represent the pretreatment cost per pound of volatiles removed. There are some economics of scale for pretreatment, but these have been neglected. Each of the hyperbolic curves are at a constant annual cost of in-situ controls. The curves for post- treatment are not shown because collection and treatment costs for post-treatment are very high and are over in the top right hand corner of the figure. For the most part, the post-treatment curves do not appear within the range of the other controls.

Figure III-7 incorporates the efficiency of each control technology by considering the throughput of volatiles. On the bottom left of the figure where the horizontal lines representing pretreatment are below and to the left of the curves, pretreatment is more cost-effective. Where the horizontal curves are above the hyperbolic curves, the in-situ controls are more cost-effective. As may be seen, pretreatment is more cost-effective for small systems, i.e. low emissions reduced and low throughputs. In a medium size impoundment, at pretreatment costs of $2.00 per pound removed ($4,400/Mg) or less, throughputs of less than 1,000 lb per year would favor pretreatment. At higher throughputs of volatiles, in-situ controls are more cost-effective. In terms of throughput, rafts are still the most cost-effective of the in-situ controls and spheres are the least. Shades when compared in terms of throughput are more cost-effective than barriers. This same comparison of effectiveness applies to small and large impoundments as well. Unless pretreatment costs for a wastestream are extremely low (much less than $0.50 per lb), in-situ controls are generally more cost-effective of all the controls.

TABLE III-16

COMPARATIVE COST-EFFECTIVENESS

FOR IN-SITU EMISSIONS CONTROLS

| | $\dfrac{\$}{\% \text{ Reduction}}$ per $ft^2$ ($\$/ft^2$) $(x10^{-4})$ | | |
|---|---|---|---|
| Rafts | 8.9 - 14.4 | 6.7 - 13.3 | 5.6 - 11.1 |
| Barriers | 36.4 | 36.4 | 27.3 |
| Shades | 25.0 | 22.7 | 20.5 |
| Floating Spheres | 118 - 157 | 91.3 - 136 | 84.1 - 136 |

*

Without installation, O and M, and other costs; assuming rafts, barriers, and shades last 1 year and spheres last 10 years. Costs are annualized for spheres at 10 percent for 10 years. The inclusion of installation, O and M, and other costs is not expected to change the relative cost-effectiveness ranking of each technology. (Please see text.)

Source:  Arthur D. Little, Inc.

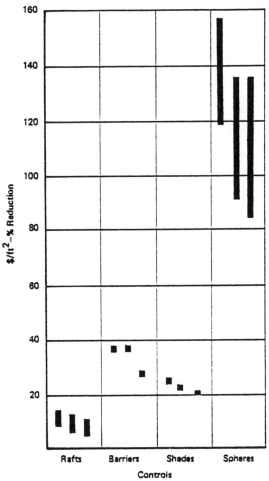

*Only material costs are used. Installation
O and M, and other costs are not included.

Source: Arthur D. Little, Inc., 1984

FIGURE III-6    COST-EFFECTIVENESS OF IN-SITU SURFACE IMPOUNDMENT CONTROLS*

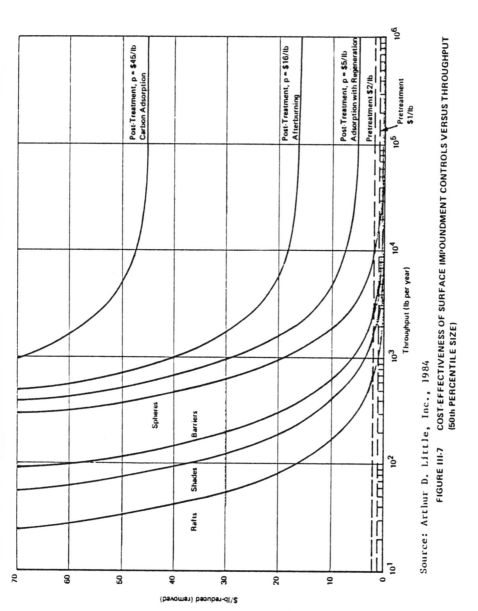

**FIGURE III-7    COST-EFFECTIVENESS OF SURFACE IMPOUNDMENT CONTROLS VERSUS THROUGHPUT (50th PERCENTILE SIZE)**

Source: Arthur D. Little, Inc., 1984

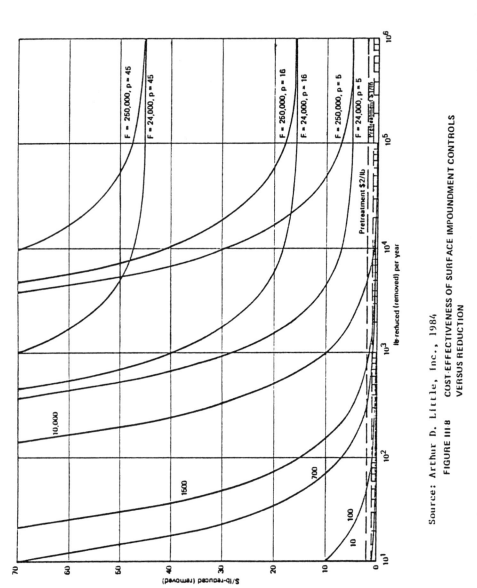

Source: Arthur D. Little, Inc., 1984

FIGURE III-8    COST-EFFECTIVENESS OF SURFACE IMPOUNDMENT CONTROLS
VERSUS REDUCTION

Figure III-8 shows the cost-effectiveness picture for various annual costs and a wide range of pounds of volatiles reduced or removed per year. Each of the hyperbolic curves represents an annual cost. As an example, for an in-situ system that costs $1,500 annually and if pretreatment is $1.00 per lb ($2,200/Mg), a removal rate or reduction of emissions of greater than 1,500 lb per year would favor the in-situ control. If pretreatment cost was increased to $2.00 per pound ($4,400/Mg), a rate of greater than about 800 lb per year reduced (removed) would favor the in-situ control costing $1,500 per year. The $17,000 annual cost curve which corresponds to using spheres in a 50th percentile situation illustrates the point. Floating spheres in this impoundment would be more cost-effective than pretreatment if the emissions reductions achieved was more than 17,000 lb per year and pretreatment was $1.00 or more per pound ($2,200/Mg) of volatiles removed.

E.   Summary

In-situ controls are generally experimental and in many cases are taken from some other different kind of applications. For example, rafts were experimentally studied in reservoirs as evaporation controls. Spheres are more directly transferable than the other in-situ controls since they are used in emissions control from treatment tanks. On the whole, data on in-situ controls are very poor. There also appears to be limits on the use of some in-situ controls on large impoundments.

In any in-situ control, there is some point at which it becomes more cost-effective than pretreatment. This tradeoff point is very sensitive to pretreatment costs. For a small change in pre-treatment cost, the trade-off point will shift relatively far to the left (with greater pretreatment cost) or to the right (with smaller pretreatment cost). Estimates of emissions from uncontrolled surface impoundments are of the order of 800,000 metric tons per year (GCA, 1983). This reference also estimates that there are about 2,000 operating impoundments. On the average, then, emissions are about 400 metric tons per year per impoundment. At this rate, in-situ treatment should be considerably more cost effective than pretreatment.

Post-treatment, including collection and treatment is the most expensive control technique. The curves for post-treatment would fall above and to the right of the in-situ controls in Figures III-7 and III-8. If the only concern is the rate of reduction of emissions, post-treatment would never be cost-effective. However, if there is concern about the absolute level of removal, the tradeoff to be considered is between pretreatment and post-treatment. Post-treatment is more expensive than pretreatment for each pound of volatiles removed primarily because very dilute mixtures in a gaseous medium are treated. For example, for carbon adsorption assuming no regeneration, the cost per pound (not including collection) post-treated is $45. A regenerative system would cost around $5 compared with a regenerative carbon adsorption pretreatment of around $1 per pound. If, however, the waste cannot be pretreated, post-treatment may be the only way to remove volatiles even though the cost is between $10 - 50 per pound, excluding collection costs. If there is a market to encourage recovery following post-treatment, this alternative may become less expensive. Also, the post-treatment alternative may be more attractive if, as in the Upjohn case, the volatiles can be replaced in a treatment impoundment.

If the design objectives offer any flexibility, there may be opportunities to reduce emissions by altering the design of an impoundment. Operating procedures also present possibilities in emissions reduction.

## F.   References

Air Structures International, Inc. Tappan, NY. July 1984. Personal Communications.

Bartholic, J.F., J.R. Runkles, and E.B. Stenmark. 1967. Effects of a Monolayer on Reservoir Temperature and Evaporation. Water Resources Research, 3:173-179.

Bernett, M.K., L.A. Halper, N.C. Jarvis, and T.M. Thomas. 1970. Effect of Adsorbed Monomolecular Films on the Evaporation of Volatile Organic Liquids. Industrial and Engineering Chemistry Fundamentals, 9:150-156.

Breton, M. et al, 1983. Assessment of Air Emissions from Hazardous Waste Treatment, Storage, and Disposal Facilities (TSDF's) - Preliminary National Emissions Estimates — Draft Final Report - GCA Corporation for the USEPA. GCA Report No. GCA-TR-83-70-G (August 1983).

Capricorn Chemicals Corporation. Secaucus, NJ. June 1984. Personal Communications.

Cluff, C.B., 1967. Rafts: New Way to Control Evaporation. Crops and Soils Magazine, 20:7-9.

Crow, F.R. and H.L. Manges, 1967. Comparison of Chemical and Non-Chemical Techniques for Suppressing Evaporation from Small Reservoirs. American Society of Agricultural Engineers, Transactions, 10:172-174.

Crow, F.R., 1973. Increasing Water Supplies by Suppression of Reservoir Evaporation. OWRR A-104-OKLA. Oklahoma Water Resources Institute, Stillwater, OK.

Ehrenfeld, J. and J. Bass. 1983. Handbook for Evaluating Remedial Action Technology Plans. EPA-600/2-83-076. U.S. Environmental Protection Agency, Cincinnati, OH.

Federal Register. May 19, 1980. Vol. 45, No. 98, p. 33075.

GCA. 1983. Evaluation and Selection of Models for Estimating Air Emissions from Hazardous Waste Treatment, Storage and Disposal Facilities. GCA-TR-82-83-G. Revised Draft Final Report. Prepared for the U.S. Environmental Protection Agency, Office of Solid Waste, Washington, D.C.

Hwang, S.T. 1982. Toxic Emissions from Land Disposal Facilities. Environmental Progress, Vol. 1, No. 1.

Lunney, P.D., C. Springer, and L.J. Thibodeaux, A. General Correlation for Predicting Liquid Phase Mass Transfer Coefficients for Surface Impoundments, Presented at 1984 National Summer Meeting. American Institute of Chemical Engineers, Philadelphia, Pa., August 1984.

Lyman, W.J., W.F. Reehl, and D.H. Rosenblatt, 1982. Handbook of Chemical Property Estimation Methods. McGraw-Hill, Inc., NY, NY.

Mackay, D. and R.G. Matsugu. 1973. Evaporation Rates of Liquid Hydrocarbon Spills on Land and Water. Canadian Journal of Chemical Engineering, 51:434–439.

McCoy, B.J. 1982. Evaporation of Water Through Surfactant Layers. AIChE Journal, 28:844–847.

Owens, M., R.W. Edwards, and J.W. Gibbs. 1964. Some Reaeration Studies in Streams. International Journal of Air and Water Pollution, Vol. 8. Pergamon Press.

Reinhardt, J.R., 1977. Gas-Side Mass Transfer Coefficient and Interfacial Phenomena of Flat-Bladed Surface Agitators. Ph.D. Dissertation, University of Arkansas, Fayetteville, AR.

Reliable Plastics. Newark, NJ. June 1984. Personal Communications.

Sandberg, H.W., C.P. Wickersham, and A. Gaines, 1983. "PTFE-Coated Foamed Glass Blocks Form a Floating Tank Cover That Prevents Acid Emissions." Reprint from Chemical Processing, February, 1983.

Spivey, J.J., C.C. Allen, D.A. Green, J.P. Wood, and R. L. Stallings. 1984. Preliminary Assessment of Hazardous Waste Pretreatment as an Air Pollution Control Technique. Draft Final Report. Research Triangle Institute, Research Triangle Park, NC. For the U.S. Environmental Protection Agency, IERL, Cincinnati, OH.

Thibodeaux, L.J., 1978. Air Stripping of Organics from Wastewater: A Compendium. Proceedings of the Second National Conference on Complete Water Use. Chicago, Illinois. May 4–8, 1978.

Upjohn Chemical Corporation, New Haven, Connecticut. April 1984. Personal Communications.

U.S. Department of Health, Education, and Welfare. 1970. Control Techniques for Hydrocarbon and Organic Solvent Emissions from Stationary Sources. National Air Pollution Control Administration Publication No. AP-68. U.S. Department of Health, Education, and Welfare, Public Health Service, Environmental Health Services, National Air Pollution Control Administration, Washington, DC.

U.S. Environmental Protection Agency. 1978. Control Technologies for Volatile Organic Emissions from Stationary Sources. EPA-450/2-78-022.

U.S. Environmental Protection Agency. 1980. Lining of Waste Impoundment and Disposal Facilities. SW-870. Office of Water and Waste Management, Washington, DC.

U.S. Environmental Protection Agency. June 1982. Handbook for Remedial Action at Waste Disposal Sites. EPA-625/6-82-006. Office of Emergency and Remedial Response, Washington, DC.

U.S. Environmental Protection Agency. July 1982. Draft RCRA Guidance Document Surface Impoundments Liner Systems, Final Cover, and Freeboard Control.

U.S. Environmental Protection Agency. September 1982. Closure of Hazardous Waste Surface Impoundments. SW-873. Office of Solid Waste and Emergency Response, Washington, DC.

W.L. Gore and Associates, Inc. Newark, DE. June 1984. Personal Communications.

X.S. Smith Co. Eatontown, NJ. June 1984. Personal Communications.

# IV. Controls for Tanks

A.  Tank Description

   1.  Definition

     Tank means a stationary device, designed to contain an accumulation of hazardous waste which is constructed primarily of non-earthen materials (e.g., wood, concrete, steel, plastic) which provide structural support (Federal Register, May 19, 1980).

   2.  Types, Construction and Uses

     The types and construction of hazardous waste tanks are similar to those used for the storage of petroleum liquids. The types of tanks are: open tanks, fixed roof tanks, floating roof tanks, variable vapor storage tanks, and pressure tanks. The minimum accepted standard for storage of petroleum liquids is the fixed roof tank. In the hazardous waste area, open tanks are still in use. Since detailed descriptions of tanks may be found in several readily available sources (U.S. EPA, 1977; U.S. EPA, 1978; U.S. EPA, 1980; American Petroleum Institute, 1962; 1964; 1980), we will only describe them briefly here.

     Open Tanks are essentially tanks without a roof. Concrete-lined basins are by definition considered tanks (Federal Register, May 19, 1980). They are examples of open tanks.

     Fixed Roof Tanks have a fixed roof equipped with some type of vent.

     Floating Roof Tanks consist of tanks with a roof that is free to float on the surfaces of the stored waste. External floating roofs are exposed on the surface. Internal floating roofs are covered by a fixed roof which protects the roof from the weather.

     Variable Vapor Storage Tanks work by storing expanding vapors emporarily in a gas holder. Venting occurs only when the holder capacity is exceed. During periods when vapors are contracting, the stored vapors are transferred back to the storage tanks.

     Pressure Tanks can withstand higher pressure variations before incurring emission losses.

     Tank material, configuration and auxiliary equipment must correspond to and be compatible with the stored waste. Materials of construction include carbon steel, stainless steel, corrosion resistant alloys, aluminum, concrete, or fiberglass reinforced plastics. Tanks vary widely in configuration, fabrication techniques, materials and operating conditions. Small tanks (between 1,300 to 21,000 gallons) may be shop-fabricated. Larger tanks are field-erected (Corripio et al, 1982).

Tanks are used for two primary purposes: storage and treatment. The uses of tanks are very similar to those of surface impoundments except that tanks are not used for disposal of wastes. Besides storage, tanks are used for treatment by biodegradation, neutralization, oxidation, among many other uses.

3.    Operation

The operations of a tank consist of filling and emptying. In treatment tanks, reagents may be added, and the tank may be mechanically mixed to encourage aeration. Each tank has a specified capacity for wastes depending on the design and on the volume already contained in the tank.

The frequency of filling and emptying is highly variable from tank to tank and depends on the capacity of the tank and the volume to be processed. Wastes could enter the tank with or without pre-treatment. The influent could be pumped into the bulk of the waste or waste could be added from the top onto the surface of the bulk waste. The total volume of waste processed through a tank per year is called the throughput per year. The annual throughput divided by the capacity of the tank represents the number of times per year wastes are totally turned over in the tank. This factor is called the turnover rate.

At the end of their lifetime, or when operations have ceased, tanks are disassembled and disposed of.

B.    Emission Sources and Models

1.    General Description of Factors Affecting Emissions

Emissions to the atmosphere occur when the tank is being filled or its contents withdrawn and also while the tank is standing.

The rate of waste volatilization is dependent on the physical and chemical properties of the waste and the surrounding environment. These factors (e.g., wind speed, temperature, etc.) are similar to those affecting waste volatilization in impoundments and will not be repeated here. An open tank is similar to a surface impoundment but with a very low wind speed at the liquid surface due to the higher operating freeboard and the smaller diameter of tanks. Please refer to the discussion on impoundments for a more detailed description.

Fixed roof tank emission losses are due to:

o    vapor expansion through the vent due to thermal expansion, barometric pressure changes, and/or increased vaporization (breathing losses), and o    vapor displacement due to filling and emptying (working losses).

Floating roof tank emission losses are due to:

o    losses due to imperfect fit of the seal and losses through the gap between the flexible seal and the inner wall (standing losses), and

o    losses due to the vaporization of the wet tank as the roof decends when the tank is being emptied (withdrawal losses).

### 2.    Emission Models

The relationships describing emissions from tanks are described in GCA (1983). For comparison, the base uncontrolled case will consist of two different uses of <u>open tanks</u>: storage and aerated treatment. Each of the different roofs (fixed roof and floating roof) will be considered as a control alternative. The relationships for the latter type are empirical. The models for emission losses from open tanks are the same as those describing surface impoundments. Table IV-1 summarizes these relationships.

The definition of each of the parameters in these relationships is shown in Table IV-2.

### 3.    Parameters That Control Emissions

The rate of emissions from an open tank depends on the overall mass transfer coefficient, the exposed area, and the concentration or mole fraction in the waste. These are the same factors that determine the rate from surface impoundments and are not repeated here.

## C.    Potential Controls

### 1.    Summary of Applicable Controls

The categories of potential controls are similar to those used in surface impoundments: pretreatment, design and operating practices, in-situ controls and post-treatment. Each control will be described under these approaches in terms of their mode of action, expected effectiveness and related costs. Only those approaches not already covered in surface impoundments will be discussed in detail.

### 2.    Pretreatment

Pretreatment basically removes the volatile components of a waste before it is put into a tank. A detailed discussion is given in the Surface Impoundments chapter.

TABLE IV-1

RELATIONSHIPS DESCRIBING EMISSION LOSSES IN TANKS

OPEN STORAGE:        Please see Surface Impoundments, Non-Aerated

AERATED:             Please see Surface Impoundments, Aerated

FIXED ROOF TANK:

$$\text{Losses} = Q_i = L_B + L_w$$

Breathing:  $L_B = 2.26 \times 10^{-2} \ M \ [\frac{P}{14.7-P}]^{0.68} \ D^{1.73} \ H^{0.51} \ \Delta T^{0.50} \ F_p \ CK_c \ \text{lb/yr}$

Working:  $L_w = 7.56 \times 10^{-4} \ M \ P \ K_N \ K_c \ Q \ \text{lb/yr}$

FLOATING ROOF TANK:

$$\text{Losses} = Q_i = L_s + L_w$$

Standing:  $L_s = K_S \ V^N \ P^* \ D \ M \ K_c \ E_F \ \text{lb/yr}$

Withdrawal:  $L_w = \frac{(0.943) \ Q \ C_L \ W_L}{D} \ \text{lb/yr}$

Source:  American Petroleum Institute

TABLE IV-2

PARAMETERS FOR TANK EMISSION RELATIONSHIPS

| PARAMETER | DEFINITION | UNITS |
|---|---|---|
| **Waste:** | | |
| M | Molecular weight | lb/lb-mol |
| P | True vapor pressure at bulk liquid | psia |
| $K_c$ | Product factor | - |
| $P^*$ (also environment) | Vapor pressure function | - |
| $C_L$ (also tank) | Shell clingage factor | bbl/1000 ft$^2$ |
| $W_L$ | Average organic liquid density | lb/gallon |
| **Tank:** | | |
| D | Tank diameter | feet |
| H | Average vapor space height | feet |
| $F_p$ | Point factor | - |
| C | Adjustment for small diameter tanks | - |
| $K_N$ | Turnover factor | - |
| $K_S$ | Seal factor | - |
| N | Seal related wind speed component | - |
| $E_F$ | Secondary seal factor | - |
| Q | Average throughput | bbl/yr |
| $H_o$ | Effective depth | feet |
| A | Exposed surface area | feet$^2$ |
| **Site (Environmental):** | | |
| V | Average wind speed | mph |
| $\Delta T$ | Average ambient diurnal temperature change | °F |
| $\theta$ | Average ambient temperature | °C |

$$P^* = \frac{(P/P_A)}{[1 + (1-P/P_A)^{0.5}]^2}$$

** where $P_A$ = average atmospheric pressure = 14.7 psia

3.    Design and Operating Practices

For a tank with fixed capacity and throughput, design considerations to control air emissions include surface area minimization and inflow/outflow pipe locations. The choice of a fixed roof tank, external or internal roof tanks, and other types of tanks are also viable design alternatives. An operational change that could decrease emissions is the temperature of the influent. These practices are discussed in Surface Impoundments.

4.    In-Situ Controls

Technologies applied at a tank can be used to change one or more of the parameters that affect emission rates from the surface of an open tank. The in-situ controls described below are: fixed roof, floating roof, rafts, and floating spheres.

4.1  Fixed Roofs

It is commonly accepted that fixed roofs over open tanks reduce emissions. Fixed roofs with a pressure/vacuum vent only release vapors when the internal pressure is exceeded. Fixed roofs are generally dome-shaped, and they are either welded or bolted onto the top of the tank wall.

Considerations in designing a fixed roof for an open tank include:

o    The pressure vacuum settings of the breather valve in a tank is determined by the structural strength of the tank for safety, and the maintenance of a vapor concentration below the lower explosive limit.

o    As shown in the equations above, the choice of paint on a tank is an effective means of reducing emissions, although the factor is more important for the fixed-roof tank than the floating roof tank. A highly reflective paint reduces the temperature of the tank and the liquid stored in the tank.

o    Roof material must be compatible with waste components.

Another kind of roof which may or may not be fixed is the aluminum dome. Information for this technology was obtained from Temcor in Torrance, California (1983). The material used in the construction of aluminum domes are aluminum alloys. The dome is formed using aluminum struts to form a triangular space truss. This is then covered with triangular panels. The dome does not require any columns for vertical support. Individual domes can be constructed alongside tanks and lifted into place. Hundreds of aluminum domes have been installed all over the country. Aluminum domes have been used to cover petroleum storage tanks, bulk storage areas, wastewater treatment tanks, water and other liquid storage tanks.  In wastewater treatment, for example, hydrogen sulfide or other gases from treatment can be collected and treated.  The dome also provides insulation to increase process efficiency.

Each aluminum dome is designed for the purpose at hand.  An important consideration in its use as a storage/treatment tank cover is the structural strength of the tank wall because it is the main support for the dome.  The waste in the tank has to be compatible with aluminum.

Vapor losses due to wind venturi action are eliminated because there are no seals used as in the case of a floating roof.  There are no floating roof weather problems.  The aluminum dome is maintenance free (according to Temcor) and aluminum is very resistant to corrosion.  Since there are no columns necessary in the design of an aluminum dome, there are few appurtenances.  According to Temcor, the cost of an aluminum dome is competitive with a steel cover.  Expensive downtime is eliminated because individual domes can be constructed separately and then lifted into place.

### 4.2  Floating Roofs

An excellent description of floating roofs and the types of seals available with floating roofs is given in U.S. EPA (1980).  There are basically two kinds of floating roofs, external and internal floating roofs.  Internal floating roofs are again divided into two types: contact where the roof floats on the liquid surface and non-contact when the roof is supported on pontoons several inches above the liquid surface.

Floating roofs are by far the most commonly used method of hydrocarbon loss control in the petroleum industry.  They are generally considered inherently effective in reducing emission losses. Most vendors claim reductions of around 95 percent over fixed roofs (U.S. EPA, 1976).  In the internal floating roof, the fixed roof is vented to allow sufficient air into the tank to maintain a vapor concentration below the lower explosive limit.  All internal floating roofs are designed to be retrofitted into existing fixed roof tanks. Seals for floating roofs are normally sold separately from the roof and are not necessarily dependent on the roof design (U.S. EPA, 1976).

According to Jonker et al (1977), the type of seals used is important in determining emission losses from tanks.  Maintenance of seals was also found to be important in reducing emissions.  Secondary seals were determined to be effective in reducing emissions in this study.  Gaps between tank walls and seals of greater than 1/8 inches were also found to be unavoidable for the vast majority of tanks.  As expected, these authors concluded that the amount of liquid exposed to a gap, the access of wind through the gap and to the vapor above the liquid, and the length of path for the vapors to reach the atmosphere were important parameters in controlling air emissions.

Runchal (1978) used a cyclone fence on the tank top to modify the aerodynamics above the roof of an external floating roof. He found significant reductions in the pressure differences on the two sides of the roof. Substantial impact on windflow was therefore achieved by the fence and wind-induced emissions were also probably reduced significantly as a result although emissions were not measured. Significant reductions were also obtained when the floating roof operated at a greater depth from the top of the tank (greater freeboard). These tests show the importance of the wind speed and flow in determining emissions from external floating roofs and are considerations in the design of a floating roof tank.

Other considerations include:

o    Stability of the floating roof under stresses of water and snow. A pan floater is less stable than a pontoon roof (Air Pollution Control Association, 1971). A covered floating roof does not have this problem.

o    Insulation of the roof can reduce temperatures in the tank. In a floating roof tank, a double-deck roof is not only more stable but has insulating qualities as well (Air Pollution Control Association, 1971).

o    Roof material and seals must be compatible with waste components.

o    Modifications to the tank may be necessary. Tank wall deformations and obstructions may have to be corrected so that seals will conform to the wall.

4.3    Rafts

Please see discussion in Surface Impoundments.

4.4    Floating Spheres

Please see discussion in Surface Impoundments.

5.    Post-Treatment

As discussed in post-treatment in surface impoundments, collection and post-treatment requires collection of emissions by means of a cover and a vent, and treatment units(s) at the vent. In the case of tanks, the collection system may consist of a fixed roof or an aluminum dome with a vent, and post-treatment with a variety of treatment technologies. Descriptions of two approaches, carbon adsorption and afterburners are given in the chapter on Surface Impoundments.

A post-treatment system has been installed at a facility owned by Waste Conversion in Hatfield, Pennsylvania. This facility treats wastes including acids, caustics, sewer sludges, coolants, food processing wastes in about 25 tanks. All of these are hooked up to scrubbers. The first scrubber was installed 4 years ago, and there have been numerous additions since. The post-treatment system includes carbon adsorption (sent to another company for regeneration), wet scrubbers with sodium hypochlorite and caustic soda, and precipitators. In this facility, there is some in-situ control as well, in the addition of activated carbon to treatment solutions themselves to adsorb volatiles before emission. The whole system is large, e.g., the carbon adsorption consists of 5 carbon drums 12 feet in diameter and 9 feet high each. The primary purpose of all the controls is the reduction of odors. Otherwise, the operators of this facility believe that they do meet point source air standards with controls. So far, the system costs around $500,000 (Waste Conversion, Personal Communication, July, 1984).

## D.    Effectiveness of Tank Controls

### 1.    Methodology

#### 1.1    Selection of Parameter Values

To evaluate the effectiveness of controls in reducing emissions from tanks, waste, tank and site (environmental) parameters necessary for quantifying emissions were defined. These parameters were shown in Table IV-2. Parameters required to calculate emissions from open tanks were the same as those used in surface impoundments except for wind speed at the surface of the liquid. This was varied because efficiencies from the baseline of using controls are very sensitive to the surface wind speed.

From various sources of data, GCA (1983) compiled typical ranges of values for input parameters used in relationships for calculating emission losses from tanks.

One tank size was chosen from these data as a representative tank. Other average or representative values were chosen for all parameters within reasonable expectations of tank design and operation. Some economies of scale are expected for larger sizes. The general cost-effectiveness relationships developed below for the single representative case are expected to hold for larger sizes.

#### 1.2    Calculation of Mass Transfer Coefficients and Emissions

The storage tank was selected to represent the tanks in use. Treatment tanks were not considered as different because technologies to control emissions would be similarly effective in treatment and storage tanks, unless the treatment tank was aerated in which case floating roofs, rafts and floating spheres would not be viable as control technologies. However, for each technology that is applicable to both types of tanks, the relative efficiencies and cost-effectiveness would be equivalent. It was also assumed that emissions were liquid-phase controlled.

The relationships for calculating mass transfer coefficients and emissions losses were shown in Table IV-1. Total emissions from open tanks were calculated as in surface impoundments. The floating roof tank chosen for comparison is the external floating roof tank. Emissions using rafts and spheres were calculated as 10 percent of baseline plus losses due to clingage during withdrawal using $L_w$ for floating roof tanks.

### 1.3  Emission Reduction and Efficiencies of Controls

The open tank was used as the basis of comparison for all the control technologies, i.e., it was considered the tank without controls.

The reduction in emission rates is defined by:

$$R = Q_1 - Q_2$$

where the subscripts 1 and 2 refer to tanks without controls and with controls respectively.

The efficiency (E) is defined by:

$$E = 100 \, (Q_1 - Q_2)/Q_1$$

The reduction in emission rates depends on the baseline selected in terms of surface wind speed in the open tank. The efficiencies of controls are extremely sensitive to surface wind speed. Curves of efficiencies of the controls with varying surface wind speeds will be shown.

### 1.4  Cost-Effectiveness of Controls

The in-situ controls for emissions reductions were compared using the cost of each percent of reduction efficiency per square feet of surface area ($/percent-ft$^2$). Curves were developed to show cost-effectiveness with varying surface wind speed based on the relationships in Table IV-1. Pretreatment, in-situ controls, and post-treatment were compared by considering the curves of cost per pound removed (reduced) versus pounds per year removed (reduced). Similar curves were generated for cost per pound removed (reduced) versus throughput in pounds per year. As in the discussion on surface impoundments the curves are:

In-Situ $\qquad y = \dfrac{A}{R}$ $\qquad$ ($/lb-reduced)

$\qquad\qquad\qquad y = A/ET$ $\qquad$ ($/lb-reduced)

Post-Treatment $\qquad y = \dfrac{F}{R} + p$ $\quad$ ($/lb-reduced)

$\qquad\qquad\qquad y = \dfrac{F}{ET} + p$ $\quad$ ($/lb-removed )

where:   A = annual cost of in-situ control ($/year)
$\qquad\quad$ R = reductions (removal) (lb/year)
$\qquad\quad$ E = efficiency of control
$\qquad\quad$ T = throughput (lb/year)
$\qquad\quad$ F = cost of collection per year ($/year)
$\qquad\quad$ p = cost per pound of post-treatment ($/lb)

We compared in-situ, post-treatment and pretreatment using different pretreatment costs as taken from Spivey et al (1984). In-situ costs were obtained from vendors for the tank size in this study. Post-treatment costs were obtained from EPA (1982).

2.   Parameters

The values for the parameters used in quantifying emissions are shown in Table IV-3. A tank 45.75 ft in diameter and 40 feet tall with a capacity of 11,700 bbl was taken as a representative tank. The turnover rate was 70 turnovers per year. Primary and secondary seals were assumed to be used in the floating roof and that they were in good condition. The waste parameter (P) was chosen as a representative value for wastes stored in tanks.

3.   Emissions

Emissions were calculated using a mole fraction of 1 for the volatile components in the impoundment. Emissions from open tanks were calculated for wind speeds at the liquid surface varying from $10^{-5}$ ft/sec to $10^{-2}$ ft/sec.

4.   Emissions Reduction and Efficiencies

In-Situ

Emissions reductions and efficiencies of each of the in-situ controls are shown in Table IV-4. Curves of efficiencies of controls with varying surface wind speeds are shown in Figure IV-1. Figure IV-1 shows that at very low surface wind speeds, rafts and floating spheres are more efficient than fixed roofs. External floating roofs, the most efficient controls, are much more efficient than fixed roofs at low surface wind speeds but at higher surface wind speeds, the efficiencies of both approach each other. At higher surface wind speeds, rafts and spheres become less efficient than fixed roofs.

TABLE IV-3

PARAMETER VALUES FOR TANKS[**]

| PARAMETER | VALUE | UNITS |
|---|---|---|
| **Waste:** | | |
| $M$ | 100.08 | lb/lb-mol |
| $P$ | 2.5 | psia |
| $K_c$ | 1.0 | - |
| $P^*$ | 0.0466 | - |
| $C_L$ | 0.3 | bbl/1000 ft$^2$ |
| $W_L$ | 9 | lb/gallon |
| **Tank:** | | |
| $D$ | 45.75 | feet |
| $H$ | 10 | feet |
| $F_p$ | 1.3 | - |
| $C$ | 1.0 | - |
| $K_N$ | 0.6 | - |
| $K_S$ | 0.7 | - |
| $N$ | 0.4 | - |
| $E_F$ | 0.75 | - |
| $Q$ | 8.2 x 10$^5$ | bbl/year |
| $H_o$ | 30 | feet |
| $A$ | 1643.9 | feet |
| **Site (Environmental):** | | |
| $V$ | 10 | mph |
| $\Delta T$ | 15 | °F |
| $\theta$ | 25 | °C |

[**] Assume tank is 40 feet tall capacity = 11,700 bbl, and turnovers
per year = 70. Primary and secondary seals, both the good condition.
Source:  Arthur D. Little, Inc.

## TABLE IV-4

### EMISSIONS REDUCTIONS AND EFFICIENCIES OF CONTROLS IN TANKS

| Surface Wind Speed $(U_o/ft/sec)$ | Reductions (lb/yr)* | | | Efficiency | | |
|---|---|---|---|---|---|---|
| | FR** | EFR*** | Rafts/ Spheres | FR | EFR | Rafts/ Spheres |
| $10^{-5}$ | 16,107 | 72,603 | 61,032 | 13.6 | 61.3 | 51.5 |
| $2 \times 10^{-5}$ | 86,431 | 142,656 | 124,080 | 45.8 | 75.6 | 65.8 |
| $10^{-4}$ | 451,944 | 508,440 | 453,286 | 81.5 | 91.7 | 81.8 |
| $2 \times 10^{-4}$ | 779,609 | 836,105 | 748,184 | 88.4 | 94.8 | 84.8 |
| $10^{-3}$ | 2,490,508 | 2,547,004 | 2,287,993 | 96.1 | 98.2 | 88.2 |
| $10^{-2}$ | 12,025,586 | 12,082,082 | 10,869,563 | 99.2 | 99.6 | 89.6 |

\*     Using unit mole fraction of volatile component

\*\*    FR - Fixed Roof

\*\*\*   EFR - External Floating Roof

Source:  Arthur D. Little, Inc.

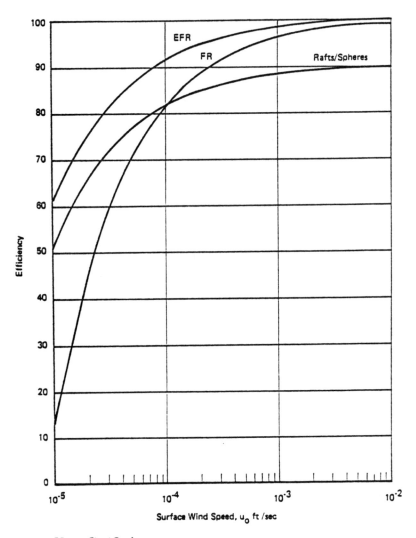

FR  =  Fixed Roof
EFR =  External Floating Roof

Source: Arthur D. Little, Inc., 1984

FIGURE IV-1    EFFICIENCIES OF IN-SITU CONTROLS FOR TANKS

## Post-Treatment

Two post-treatment systems were chosen for comparison: carbon adsorption with no regeneration, and afterburning. The rate of emissions is affected by the collection system and the pressure through the post-treatment unit(s). Carbon adsorption was assumed to work at an efficiency of 95% and afterburning at 98%.

## 5.    Costs

### Pretreatment

Pretreatment costs taken from Spivey et al. (1984) were in the order of $1.00 per pound of volatiles removed ($2,200/Mg) and represented the range of most pretreatment technologies within a factor of 2.

### In-Situ Controls

Costs for fixed and external floating roofs were obtained from a vendor, the Chicago Bridge and Iron Company (Personal Communications, July 1984). Costs for aluminum domes were obtained from Temcor, Torrance, CA (Personal Communications, October 1983). The material costs for rafts and spheres were those used in surface impoundments.

In order to estimate installation costs for roofs, the location was assumed to be the Midwest. Costs for roofs were derived by taking the difference in a new tank with a roof and a tank without a roof. According to the vendor, the costs estimated independently would be within the same magnitude of error (± 10 percent) as the figures derived by taking the difference.

Installation costs were included in the cost estimates for roofs. The cost for aluminum domes includes all costs. Rafts and floating spheres have very low installation and maintenance costs. These costs are not included in the estimates given in Table IV-5. Disposal costs for rafts and spheres were not included.

Roofs have a long lifetime (over 25 years). It was assumed that rafts last 1 year and that floating spheres had a lifetime of 10 years. Costs for roofs and spheres were annualized at 10 percent for 10 years. The cost per year is around 16 percent of the total cost after annualization. Total costs and annualized costs are shown in Table IV-5.

### Post-Treatment

Costs for post-treatment consists of the costs for collection and treatment. The collection system is either the fixed roof or the aluminum dome. These costs were estimated for in-situ controls and shown in Table IV-5. Treatment costs are variable costs dependent on the total emissions treated. These costs for carbon adsorption

TABLE IV-5

TOTAL COSTS AND ANNUALIZED COSTS FOR CONTROLS (Summer 1984 Dollars)

| Tanks[1] | Total ($) |
|---|---|
| Open top with wind girder | 95,000 |
| With column supported cone roof | 115,000 |
| With double deck external floating roof incl. primary and secondary seals | 215,000 |

| In-Situ | Total ($) | Annualized ($/yr)[2] |
|---|---|---|
| Column supported cone roof | 20,000 | 3,200 |
| Aluminum dome[3] | 16,000 – 25,000 | 2,600 – 3,900 |
| Double deck external floating roof incl. primary and secondary seals | 120,000 | 19,000 |
| Rafts | 130 – 210 | 130 – 210 |
| Floating spheres | 10,800 – 14,300 | 1,700 – 2,300 |

Post-Treatment

| | Total ($) | Annualized ($/yr) |
|---|---|---|
| Collection (excl. treatment) | 16,000 – 25,000 | 2,600 – 3,900 |

1. For tanks 45.75 feet in diameter, 40 feet high. Including erection and installation, assuming Midwest location. Not including land or foundations, piping, valves, and other appurtenances. Source: Chicago Bridge and Iron Co., Boston, MA 1984.
2. Annualized for roofs, domes, and spheres at 10% in 10 years.
3. Source: Temcor, Torrance, CA. 1983. Estimated at $10-15 per $ft^2$ with installation.

without regeneration and afterburning were estimated in the discussion on surface impoundments and were around $45 and $16 per pound, respectively ($100,000/Mg and $35,000/Mg). If carbon regeneration and catalytic afterburner systems were included, treatment costs would be between $5-10 per pound ($11,000-$22,000/Mg).

6.   Cost-Effectiveness of Controls

Table IV-6 and Figure IV-2 express the cost-effectiveness of in-situ controls in terms of the cost per percent of reductions per square foot of tank for surface wind speeds ranging from $10^{-5}$ to $10^{-2}$ ft/sec. Since aluminum domes and column supported cone roofs are similar, only one curve is shown for these two controls. Only the average for rafts and spheres are shown.

Rafts appear to be the most cost-effective of all the in-situ controls. External floating roofs are the least cost-effective, being much more expensive than the other controls. When the tank is operated in a very quiescent situation (e.g., very high freeboard or in a site with extremely low wind speed), external floating roofs and fixed roofs (or aluminum domes) become more comparable. Under these conditions, rafts and floating spheres are more cost effective than any of the roofs. As wind speed increases (e.g., if the tank is consistently operating at very low freeboards or if wind speed is high in the site of operation), fixed roofs and aluminum domes become more comparable with rafts and spheres.

Figures IV-3 and IV-4 show the hyperbolic functions relating cost per pound of emissions reduced or removed to the throughput of volatiles per year and the total pounds of volatiles removed per year respectively. The horizontal lines represent the pretreatment cost per pound of volatiles removed. The hyperbolic curves are at a constant annual cost of in-situ controls and for three levels of post-treatment costs.

Figure IV-3 incorporates the efficiency of each control technology by considering the throughput of volatiles. An efficiency of 90 percent was used for external floating roofs, and a corresponding efficiency of around 76 percent and 80 percent were used for fixed roof tanks and rafts and spheres respectively. The most cost effective of the controls is rafts except where throughput is very small (below 200 lb/yr at pretreatment cost of $1/lb ($2,200/Mg). At the bottom left of the figure where the horizontal pretreatment lines are below the hyperbolic curves, pretreatment is more cost-effective. Above a throughput of around $5 \times 10^3$ lb/yr, spheres and fixed roofs are more cost-effective than pretreatment at around $1/lb ($2,200/Mg). Spheres are more cost-effective than fixed roofs. The use of post-treatment is expensive compared to all the other controls except when throughput is very low (around less than $4 \times 10^5$ lb/yr for carbon adsorption at $45/lb ($100,000/Mg) and around $10^3$ lb/yr for afterburning at $16/lb ($35,000/Mg). Generally, post-treatment becomes less cost-effective as throughput increases.

**TABLE IV-6**

COMPARATIVE COST-EFFECTIVENESS FOR TANK IN-SITU CONTROLS

Cost-Effectiveness[*]

| Surface Wind Speed (ft/sec) | Column Supported Cone Roof | Aluminum Dome | Double Deck Ext. Floating Roof | Rafts | Spheres |
|---|---|---|---|---|---|
| $10^{-5}$ | 14.2 | 11.7 – 17.5 | 18.9 | 0.2 – 0.3 | 2.0 – 2.7 |
| $2 \times 10^{-5}$ | 4.2 | 3.5 – 5.2 | 15.3 | 0.1 – 0.2 | 1.6 – 2.1 |
| $10^{-4}$ | 2.4 | 2.0 – 2.9 | 12.6 | 0.1 – 0.2 | 1.3 – 1.7 |
| $2 \times 10^{-4}$ | 2.2 | 1.8 – 2.7 | 12.2 | 0.1 – 0.2 | 1.2 – 1.6 |
| $10^{-3}$ | 2.0 | 1.7 – 2.5 | 11.8 | 0.1 | 1.2 – 1.6 |
| $10^{-2}$ | 2.0 | 1.6 – 2.4 | 11.6 | 0.1 | 1.2 – 1.5 |

[*]Cost-Effectiveness is expressed in $x10^{-2}$ per percent emissions reduction per square foot. Multiply values in the table by 100 to get values in dollars.

Source: Arthur D. Little, Inc.

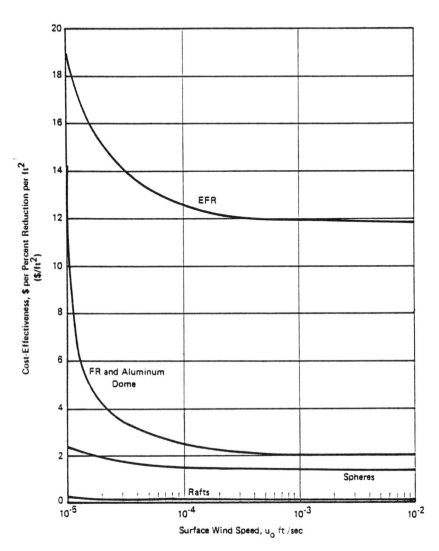

FR  = Fixed Roof
EFR = External Floating Roof

Source: Arthur D. Little, Inc., 1984

FIGURE IV-2  COMPARATIVE COST-EFFECTIVENESS OF IN-SITU CONTROLS FOR TANKS

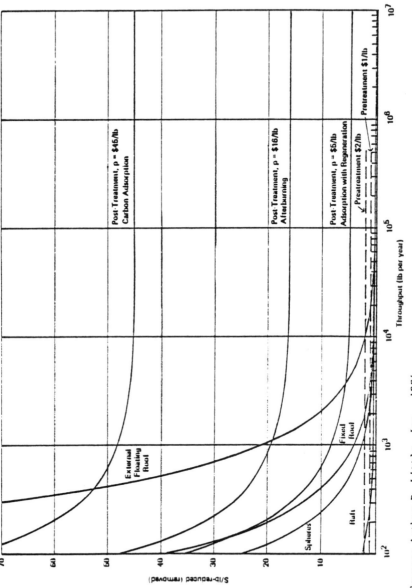

Source: Arthur D. Little, Inc., 1984

FIGURE IV-3   COST EFFECTIVENESS OF CONTROLS ON TANKS VERSUS THROUGHPUT

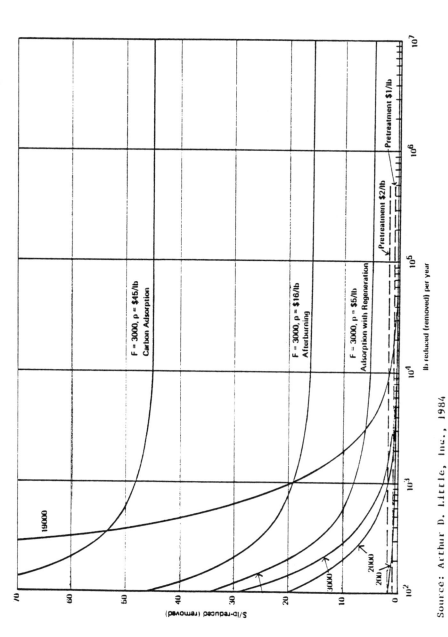

Source: Arthur D. Little, Inc., 1984

FIGURE IV-4    COST-EFFECTIVENESS OF CONTROLS ON TANKS VERSUS REDUCTION

If the price of post-treatment can be reduced, it may become more cost effective than external floating roofs at throughputs of less than around $10^3$ lb/yr (e.g., if post-treatment costs $5/lb ($11,000/Mg) with carbon regeneration). Under low post-treatment costs, post-treatment may even become more cost-effective than fixed roofs at low throughputs.

In the analysis, the efficiencies and the cost-effectiveness of roofs are probably underestimated with respect to spheres and rafts. Fixed roofs are likely to be very comparable to using spheres. However, because of the low cost of rafts, they are still the most cost-effective.

Figure IV-4 shows the cost-effectiveness versus the pounds per year reduced or removed per year. Each of the hyperbolic curves represents an annual cost. In-situ controls are favored to pretreatment as the pounds per year is increased. If an in-situ control has an annual cost of $3,000/yr (corresponding to fixed roof or aluminum dome), a reduction rate of 3,000 lbs/yr is the trade-off point at which the in-situ control becomes more cost-effective than pretreatment at the cost of $1/lb ($2,200/Mg). If pretreatment cost was increased to $2.00 per pound ($4,400/Mg), a rate of reduction (removal) of greater than 1500 lbs/yr would favor the in-situ control costing $3,000 a year.

E.    Summary

Roofs are well-used, proven emission control technologies from petroleum storage facilities. Floating spheres, and to a much lesser degree, rafts have been used in treatment tanks in the chemical industry for emissions control. Pretreatment and post-treatment are taken from proven, similar applications and are easily applicable in the case of tanks. Waste Conversion in Hatfield, PA, is a good example where post-treatment has been used effectively to control emissions from treatment tanks. As an example of a commercially available product, Calgon Corporation offers Ventsorb, a catalogue item. Ventsorb is a 55-gallon drum containing activated carbon specifically developed to be used at vents in tanks, landfills, and process applications. Each unit costs between $535-$635, and is normally disposed of after usage. The units can also be taken apart and the carbon replaced. Replacement carbon costs around $300 per unit. (Calgon Corporation, Personal Communication, September 1984).

Certain technologies may be applicable in storage tanks but not in treatment tanks. Floating roofs are generally not applicable in treatment tanks. Spheres and rafts are not applicable in aerated tanks. Fixed roofs, pretreatment and post-treatment are applicable to both kinds of tanks.

If an in-situ control that reduces emissions by decreasing volatilization (e.g., spheres, rafts, fixed roofs) is used in a treatment tank, the absolute removal can be increased by the additional time volatiles are retained in the tank, and therefore, the additional amount of treatment obtained. This was not included in our analysis. The hyperbolic curve in this case would be shifted to the left and the in-situ control will be more cost-effective at a lower trade-off point.

In a similar way, in-situ controls which reduce emissions by reducing volatilization when used in storage tanks containing wastes in transition before treatment increase the absolute removal by increasing the volume treated. In treatment tanks which are at the final stage of the process, post-treatment which removes emissions from these tanks may be the only means of absolute removal in some cases, e.g., in aerated tanks or in tanks containing waste before disposal.

The trade-off point at which a control is equally cost-effective with pretreatment is very sensitive to pretreatment costs as in the case of surface impoundments. For a small change in pretreatment cost, the trade-off point will shift relatively far to the left (with greater pretreatment cost) or to the right (with smaller pretreatment cost).

In practice, the choice of one control over another will probably not be based on cost-effectiveness as the only evaluation criterion. The use of the tank, its size, the downtime required for retrofitting, the structural strength of its walls, and the amount of structural changes that have to be made will probably be important determining factors as well. As an example, floating roofs are normally used on large tanks because they are very expensive and significant economies of scale exist. In situations where the tank walls are deformed due to age or other reasons, they cannot be easily retrofitted into a tank without a large amount of extra expenses. In a case of small tanks and particularly tanks with wall deformities, spheres and rafts would be more viable, maintenance-free solution with no expenses needed in retrofitting. In a tank with insufficient wall strength, aluminum domes with no columnar supports will not be a good alternative.

The design and operation of a tank may present opportunities of emissions reductions. In a new tank where retrofitting and downtime do not present problems, installation of roofs and other major changes in design (e.g., pressure tanks or variable pressure tanks) may be viable solutions.

## F.   References

Air Pollution Control Association, 1971. Control of Atmospheric Emissions from Petroleum Storage Tanks. Journal of the Air Pollution Control Association, 21:260-268.

American Petroleum Institute, 1962. Evaporation Loss from Fixed Roof Tanks. Bulletin 2518. American Petroleum Institute, Evaporation Loss Committee, Washington, DC.

American Petroleum Institute, 1964. Use of Variable-Vapor-Space Systems to Reduce Evaporation Loss. Bulletin 2520. American Petroleum Institute, Evaporation Loss Committee, Washington, DC.

American Petroleum Institute, 1980. Evaporation Loss from External Floating Roof Tanks. Bulletin 2517 (Revised). American Petroleum Institute, Evaporation Loss Committee, Washington, DC.

Calgon Corporation, Pittsburgh, PA. September 1984. Personal Communications.

Chicago Bridge and Iron Company, Boston, MA. July 1984. Personal Communications.

Corripio, A.B., K.S. Chrien, and L.B. Evans, 1982. Estimate Costs of Heat Exchangers and Storage Tanks via Correlations. Chemical Engineering, 89(2): 125-127.

Federal Register, May 19, 1980. Vol. 45, No. 98, p. 33076.

GCA, 1983. Evaluation and Selection of Models for Estimating Air Emissions from Hazardous Waste Treatment, Storage and Disposal Facilities. GCA-TR-82-83-G. Revised Draft Final Report. Prepared for the U.S. Environmental Protection Agency, Office of Solid Waste, Washington, DC.

Jonker, P.E., C.B. Scott, and W.J. Porter, 1977. Pollution Regulations from Floating-Roof Tank-Seal Study. Oil and Gas Journal, 75 (24):72-75.

Runchal, A.K., 1978. Hydrocarbon Vapor Emissions from Floating Roof Tanks and the Role of Aerodynamic Modifications. Journal of the Air Pollution Control Association, 28(5): 498-501.

Spivey, J.J., C.C. Allen, D.A. Green, J.P. Wood, and R.L. Stallings, 1984. Preliminary Assessment of Hazardous Waste Pretreatment as an Air Pollution Control Technique. Draft Final Report. Research Triangle Institute, Research Triangle Park, NC. For the U.S. Environmental Protection Agency, IERL, Cincinnati, OH.

Temcor, Torrance, CA., October 1983.  Personal Communications.

U.S. Environmental Protection Agency, 1976.  Evaluation of Methods for Measuring and Controlling Hydrocarbon Emissions from Petroleum Storage Tanks.  EPA-450/3-76-036.  U.S.  Environmental  Protection  Agency, Research Triangle Park, NC.

U.S.  Environmental  Protection  Agency,  1977.  Compilation  of  Air Pollution  Emission  Factors  -  Third  Edition.  U.S.  Environmental Protection Agency, Research Triangle Park, NC.

U.S. Environmental Protection Agency, 1978.  Control Techniques for Volatile Organic Emissions from Stationary Sources.  EPA-450/2-78-022. U.S. Environmental Protection Agency, Research Triangle Park, NC.

U.S. Environmental Protection Agency, 1980.  Benzene Emissions from Benzene Storage Tanks.  Background Information for Proposal Standards. EPA-450/3-80-034a.  U.S.  Environmental  Protection  Agency,  Research Triangle Park, NC.

U.S. Environmental Protection Agency, 1982.  Handbook for Remedial Action  at  Waste  Disposal  Sites.  EPA-625/6-82-006.  Office  of Emergency and Remedial Response, Washington, DC.

Waste Conversion, Hatfield, PA, July 1984.  Personal Communications.

# V. Landfills

A.  Landfill Description

A landfill is a waste disposal facility where wastes are placed in or on the land for custodial safekeeping. A variety of control techniques is required by regulation as well as good operating practices to isolate the wastes from the environment. A typical landfill facility contains three types of areas: an unused portion for future operations, a currently operating portion or set of cells, and one or more closed cells.

Operations consist of placing wastes arriving in various modes into the landfill in a manner to obtain maximum utilization of the volumetric capacity and minimize exposure to workers and ultimately to the environment. Wastes arrive in containers, or as bulk solids or liquids. Containers are placed directly in the landfill, generally in regular patterns. Bulk solids are deposited directly in the landfill and are distributed over the operating area by use of earth-moving equipment. Bulk liquids are mixed with solidifying agents prior to being placed in the landfill. After being mixed with these agents, liquids are handled essentially in the same manner as bulk solids.

At the end of each operating day or at less frequent intervals, a soil cover is placed over the wastes deposited during that day. The soil cover generally consists of from 6 to 12 inches of soil, fly ash, or other aggregates capable of being spread, compacted and supporting subsequent layers (lifts) of wastes and covers.

The area of an operating cell ranges from a few tenths of an acre at many currently operating sites to as much as several acres at a few large operations.

Eventually the wastes placed in a working cell together with the daily cover materials will reach the cell's capacity. When this occurs, operations at that cell cease and it is closed permanently. Closure consists mainly of placing a permanent cover over the cell followed by monitoring the integrity of the system for a period of many years. The cover may consist of several layers, the uppermost of which is vegetated to prevent erosion and in some cases to permit use of the land surface following closure.

Long-term degradative processes continue in the waste mass after the cell has been closed. Infiltration of surface water can generate leachate. Containers will generally deteriorate and may release their contents after long exposure to mechanical stresses and the normally corrosive environment in a landfill. The waste materials may be degraded by chemical processes such as hydrolysis or by anaerobic bacterial decay. The next section describes potential sources of emissions to the atmosphere associated with the several major stages of operation at a typical landfill.

114

B.    Emission Sources and Models

1.    General

During operations emissions may arise from activities at the uncovered area (or operating face), and through the area with the daily (or temporary) cover.    In addition, volatile compounds can permeate the final (permanent) cover over closed sections of the landfill.    The rate of emission depends, generally, on the wastes and the form in which the wastes are deposited, ambient conditions and operating practices at the facility.

At the operating face, waste can be deposited as bulk solids, containerized wastes, or bulk liquids.    Containers are assumed to be leak tight so that no emissions are generated.    If the containers lose integrity at some later date, then the volatile constituents can ultimately permeate through the entire waste mass and diffuse through the cover into the air.    Volatile constituents in bulk solids diffuse to the surface and into the atmosphere.    The rates are generally quite low as many volatile compounds become absorbed on the solid surface and exhibit low vapor pressure.

Bulk liquids may be deposited in landfills only after they have been solidified with a bulking agent.    In some operations bulking is carried out by adding liquids directly to the bulking agent near the operating face and then placing the mix in the landfill.    It is possible to have pools of liquid present for short periods, although good practice would avoid this situation.    In this case emissions would be similar to those at a surface impoundment with a floating organic layer or a land treatment site immediately following a surface application of wastes.    Since this situation (pooled liquids) should be avoided for many reasons, including safety, good operating practice, etc., it will not be considered further in this analysis.

The mix of liquid and bulking agent may be exposed for significant times at the operating face.    During this time, emissions occur by mechanisms related to those at land treatment facilities where wastes are placed on and tilled into soil surfaces.

Emission through daily or permanent covers results from diffusion of vapors arising from the bulk of the wastes upward and through the cover.    The process occurs in stages.    Vapors from the lowermost lifts (stages in the filling sequence) diffuse through the first cover into the second lift and so on until the vapors reach the uppermost lift. In active, operating cells the uppermost temporary cover will be the same as all the others; in the closed portion the uppermost cover will be the permanent seal and generally will be less permeable than the daily covers.    The rate of loss will be controlled by the uppermost cover, whether it is final or temporary, as vapors will gradually equilibrate among the various lifts.

In addition to the basic diffusion mechanisms, emission through the covers are also influenced by several other processes.  In hazardous waste landfills with codisposed, mixed biodegradable wastes, methane can be generated through anaerobic decay.  The methane will build up and gradually flow upwards (and laterally as well) carrying volatile compounds with it.  Barometric pumping, due to fluctuations in ambient pressure, can also increase flow relative to purely diffusive mechanisms.

Lateral diffusion will occur as well as vertical diffusion. Vapors diffusing sideways out of the fill may then turn and diffuse to the surface.  The requirement for liners at the sides and bottom of hazardous waste landfills will minimize emission via this mechanism.

Emissions may also arise from the leachate which is generated within the landfill and is removed through the leachate control system.  Leachate may contain free, immiscible organic contaminants and dissolved volatile species.  Leachate may be exposed to the atmosphere in sumps or in tanks where the leachate is stored pending treatment and disposal.  Controls for leachate-based emissions are not considered explicitly in this study.  Leachate is essentially a new liquid hazardous waste.  Emission controls depend on the types of TSDF used to manage it.

2.    Emission Models

Emission models are developed below to illustrate how emissions depend on various controllable parameters and to provide a means for quantitative analysis.  The models cover only the parts of the emission processes that are most significant.  Models are given for diffusion through covers and for volatilization from bulk solids and solidified bulk liquids.

### Covers

The basic steady-state relationship for diffusion through covers is:

$$Q = \frac{A\, D_{eff}(C_i - C_o)}{t_c}$$

where:

| | | |
|---|---|---|
| $Q$ | = | volatilization rate (lb/hr) |
| $A$ | = | area of cover (ft$^2$) |
| $D_{eff}$ | = | effective diffusivity (ft$^2$/hr) |
| $C_i, C_o$ | = | concentration of vapor on underside and top of cover, respectively (lb/ft$^3$) |
| $t_c$ | = | thickness of cover (ft) |

For the top cover, the concentration in the atmosphere can generally be neglected relative to the concentration below the cover. Effective diffusivity expresses the diffusive behavior of a single compound through the pores of the cover for soils or through the fabric of a synthetic cover. Of several equations available for soil-type covers, Farmer's (1978) equation appears most appropriate in terms of the objectives of this analysis. It is relatively simple but indicates how diffusion relates to soil properties. Farmer's equation is:

$$D_{eff} = D_{air} \frac{(\varepsilon_a)^{10/3}}{(\varepsilon_t)^2}$$

where:  $D_{air}$ = diffusivity of compound in air ($ft^2/hr$)
$\varepsilon_a$ = porosity of air-filled pores
$\varepsilon_t$ = total porosity

Further, porosity can be related to the density to which the soil has been compacted as:

$$\varepsilon_t = 1 - \frac{\beta}{\rho_s}$$

where:  $\beta$ = soil bulk density ($lb/ft^3$)
$\rho_s$ = particle density ($lb/ft^3$)

If the soil is completely dry, then total porosity and air-filled porosity are the same and the effective diffusity becomes the normal diffusivity in air. If the soil is wet, part of the pore volume becomes filled by water. In this case, the air-filled porosity becomes:

$$\varepsilon_a = \varepsilon_t - \frac{\omega}{\rho_w}$$

where:  $\omega$ = gravimetric soil water content, $lb/ft^3$
$\rho_w$ = water density ($lb/ft^3$)

For synthetic covers, the basic form of the mass flow relationship is similar.

$$Q = \frac{A\, D_m\, C}{t_m}$$

where:  A    =   area of cover ($ft^2$)
        $D_m$  =   permeation rate of a given diffusing compound through synthetic cover material ($ft^2$/hr)
        C    =   concentration of diffusing vapor at the top of the landfill ($lb/ft^3$)
        $t_m$  =   thickness of synthetic cover (ft)

The concentration term in the basic equation represents an equilibrium either between vapor and a volatile organic adsorbed on a solid surface or between vapor and a free liquid in the interstices of the wastes or soil. In the first case, adsorption, the particular form of the relationship depends on many factors; chemical structure, quantity, temperature, etc. One common form of equation, the Freundlich equation relates vapor concentration to quantity adsorbed as (Weiser, 1949):

$$c = k\, x^n$$

where:  c    =   vapor concentration ($lb/ft^3$)
        k    =   empirical constant
        x    =   mass of volatile compound absorbed per unit mass of soil
        n    =   empirical constant

For free liquids, the vapor concentration follows the conventional relationship for solutions:

$$C = \frac{P_o\, MW}{RT}\, \gamma x$$

where:  $P_o$  =   vapor pressure of pure component (atm)
        MW   =   molecular weight of diffusing component (lb/lb-mol)
        R    =   gas law constant (0.73 atm-$ft^3$/lb-mol-°F)
        T    =   temperature (°R)
        $\gamma$  =   activity coefficient
        x    =   mole fraction

For dilute solutions, Henry's Law can be used as an alternate form. In all of these relationships, vapor concentration increases with increasing temperature, all other factors being equal.

### Bulk liquids

The mix of free liquids and solidifying or bulking agents can be represented by the models developed for land treatment. In land treatment, also called land-farming, oily wastes are deposited on and tilled into soil surfaces. The rate of emission displays a complex time-varying relationship reflecting depletion of the volatile materials near the surface (Thibodeaux as reported in Hwang, 1982). The characteristic time for land treatment cycles is quite long relative to the period during which mixed liquids are exposed at the working face of a landfill. At landfills, bulk liquids are mixed with solidifying materials such as kiln dust, prior to deposition in the landfill. Pooled free liquids may be present, but only for short periods. The bulked liquids, in the solidified matrix, may be exposed to the atmosphere for periods ranging from hours to several days.

In this situation, it is reasonable to assume that losses occur at the exposed surface through diffusion from free liquid in the pores of the bulking agent. In this case, the mass flow equations simplify to:

$$Q = Ak_{oa} C^{*}$$

where:   $A$   =   area of exposed face ($ft^2$)
         $k_{oa}$   =   overall mass transfer coefficient (ft/hr)
         $C^{oa}$   =   equilibrium concentrations above free liquid ($lb/ft^3$)

For pure or mixed organic compounds, the overall coefficient is generally gas-film controlled and follows a law similar to those discussed in the surface impoundment chapter. The coefficient depends generally on wind speed, some characteristic dimension of the exposed surface, and diffusivity in air.

### Bulk solids

Bulk solids containing adsorbed volatiles would behave very similarly to bulk liquids as described above. The equilibrium concentration ($C^{*}$) would refer to the solid surface instead of the liquid surface.

3.    Controlling parameters

### Covers

The factors that determine mass flow out of a soil cover are area, effective diffusivity, vapor concentration, and cover thickness. Area is generally constrained by factors considered in the overall design of a landfill. For a given volume of wastes, area can, in theory, be minimized by starting with a deeper excavation. But in practice, depth is determined primarily by hydrogeological factors (such as the water table level), geological setting, and liner and leachate collection system requirements. Thickness of both daily and final cover are matters of design and operating practice and can be set over a wide range of values depending on the objectives, including economy, infiltration prevention, emission minimization, or otherwise.

Vapor concentrations can be reduced by pretreatment, similar to the case of surface impoundments. Again, removal of the volatile compounds will reduce not only the rate of emissions, but also the total quantity released over long periods of time. Since vapor concentration is a very strong function of temperature, any means of lowering the internal temperature of the landfill would reduce volatile emissions. Diffusivity of soil cover can be modified by selection of the basic cover material, compaction, and moisture content. For synthetic covers, diffusivity can be set by choice of material and membrane thickness.

### Working faces

Minimizing time of exposure and area exposed would reduce emissions. Enclosing areas in which solidification of bulk liquid is carried out would provide a means to control emissions. Ventilation and volatiles control, such as carbon adsorption, could be added. Requiring daily cover emplacement would reduce emissions at facilities that presently cover the wastes at less frequent intervals. Modifying mass transfer coefficients, as in the surface impoundment case, is not practical.

## C.    Controls

1.    Introduction

There is a limited set of controls available for landfills compared to surface impoundments. Pretreatment results and costs are the same for all treatment and disposal modes, including landfills. Design and operating practices may reduce emissions considerably without interfering with normal operations. Discussion of permanent covers including vents is included in the design and operating practices section, although this approach could be considered as an in-situ control. With this assignment there are no so-called in-situ controls for landfills. The last group, post-treatment approaches, includes two distinct types of controls.

## 2.    Pretreatment

Pretreatment controls are essentially the same as those described above the surface impoundments. In addition, regulations that would ban disposal of volatile organic wastes in landfills would have the same effect as pretreatment, i.e., prevent the disposal in the first place.

## 3.    Design and Operating Practices

### 3.1  Handling Bulk Liquids or Solids

Emissions from bulk liquids, if permitted in landfills, can be minimized by mixing the  liquids and solidification  agents  in closed vessels or in enclosures. Emission from closed vessels can be captured and treated with a high degree of effectiveness. Practices where liquids are poured onto the ground prior to the addition of the bulking agent should be avoided. If direct, open mixing is employed, liquids should be added to or into the solids and quickly mixed so that liquid pools at the surface are avoided.

Some bulking agents are more effective than others in adsorbing the liquids into their pores. These agents tend to minimize emissions by reducing the free liquid area relative to those bulking agents whose particles remain coated with the liquid even after the mixture has been solidified.

In general, bulk liquids or solids should be covered as quickly as is practical. Operations should be planned to minimize the exposed area and time of exposure. Temporary covers (tarpaulin or plastic covers) can be put in place to reduce emissions until it is time to place the daily cover layer.

### 3.2  Temporary Cover Practices

The next set of operating practices refers to means of reducing emissions through the temporary cover. One fairly obvious means is to utilize thicker covers than required by other constraints. Emission through a cover is inversely proportional to the thickness. Doubling the thickness would have the emission rate. It would also double the material cost of the cover and would reduce the effective volume of the landfill by taking up space that would otherwise be filled with waste materials.

A second and perhaps more effective approach to reduction of emissions is to reduce the air filled porosity, and thereby, the effective diffusivity of volatile materials. Substantial reductions over normal operating practice can be obtained by choice of cover material, compaction practices, and moisture control, as diffusion through the cover depends on the percentage of the void volume that is air filled. Voids are empty spaces remaining in compacted soils because the shape of the individual particles prevents every space from being filled.  Table V-1 shows a ranking of a variety types of soils according to several different functions (EPA, 1979).  The most effective soils are those with very high clay content.  Unfortunately, clays may not be found at or near the site, may be expensive to import, and have poor handling properties.  Clays can be mixed with naturally occurring soils so that the overall permeability is decreased substantially without loss in the workability of the resulting materials.  Figure V-1 and Table V-2 show the effective permeability relative to the permeability in air, based on Farmer's equation, for a series of soils as a function of water content.  The different curves correspond to soils with dry porosity ($\epsilon_T$) ranging from 0.4 to 0.8.  The curve for the least porous soils ($\epsilon_T = 0.4$) correspond to clays and mixes of clays and other soils.

The curves show several other features.  Diffusivity can be decreased by compacting the soil at a given water content to more dense or less porous levels.  Compaction is a function of the soil grain size distribution and types of materials present, the weight of the earth-moving equipment used to compact the soil, the number of passes, and the soil moisture content.  Figure V-2 shows the relationship, for several soil types, between density and moisture content corresponding to a standard laboratory test method (EPA, 1979).  The curves shown correspond to the so-called standard compaction test.  The maximum density shown in the curve is called the standard Poctor density.  It is also called the standard AASHTO density, because the American Association of State Highway and Transportation Officials has adopted the same test procedure used to generate these standard curves.  The moisture content corresponding to the maximum density point is denoted the standard optimum moisture content.  The line noting the zero air void curve (S = 100%) represents the dry density (and water content) of soil completely saturated with water at various degrees of compaction.

Soils can generally be compacted to a higher degree than shown in the standard Proctor test through the use of heavier compacting machinery.  Depending on the type of soil, increases in density of between 10 and 20% can be obtained by using the modified technique relative to the standard technique.  The laboratory test results should be reproducible in the field through the use of heavier compacting machinery.  Changes in density of the order of 10 to 15% can be achieved.  At a water content of about 15 lbs/cubic foot, this level of compaction can reduce the effective diffusivity by from 50 to 100%.

Even more dramatic reductions in effective diffusion can be obtained by adjusting water content to reduce the air filled porosity to essentially zero.  The appropriate water content is that along the zero air voids curve in Figure V-2.  Although it is in theory possible to reduce porosity to zero, this objective may simultaneously produce deleterious effects in handling and workability properties.  On the other hand, the sharp increase in diffusivity as water content

**TABLE V-1**

**RANKING OF USCS SOIL TYPES ACCORDING TO PERFORMANCE OF COVER FUNCTION**

| USCS Symbol | Typical Soils | Trafficability | Water Infiltration | | Gas Migration | | Erosion Control | | Crack Resistance | Support Vegetation |
|---|---|---|---|---|---|---|---|---|---|---|
| | | | Impede | Assist | Impede | Assist | Water | Wind | | |
| GW | Well graded gravels, gravel-sand mixtures, little or no fines | E | F | G | F | E | E | E | E | F |
| GP | Poorly graded gravels, gravel-sand mixtures, little or no fines | E | P | E | F | E | E | E | E | F |
| GM | Silty gravels, gravel-sand-silt mixtures | G | F | F | F | G | G | G | G | F |
| GC | Clayey gravels, gravel-sand-clay mixtures | G | G | F | G | F | G | G | G | G |
| SW | Well graded sands, gravelly sands, little or no fines | E | F | G | F | G | E | E | E | F |
| SP | Poorly graded sands, gravelly sands, little or no fines | E | P | E | F | G | E | E | E | F |
| SM | Silty sands, sand-silt mixtures | E-G | F | G | F | G | F | G | E | E |
| SC | Clayey sands, sand-silt mixtures | G-F | F | F | G | F | F | F | G | E |
| ML | Inorganic silts and very fine sands, rock flour, silty or clayey fine sands, or clayey silts with slight plasticity | G-F | G | F | G | F | P | F | G | G |
| CL | Inorganic clays of low to medium plasticity, gravelly clays, sandy clays, silty clays, lean clays | F | E | P | E | F | P | F | F | F |
| OL | Organic silts and organic silty clays of low plasticity | F | – | – | – | – | P | F | F | G |
| MH | Inorganic silts, micaceous or diatomaceous fine sandy or silty soils, elastic silts | F | G | F | – | – | F | F | F | G |
| CH | Inorganic clays of high plasticity, fat clays | F | E | P | E | F | F | F | F | F |
| OH | Organic clays of medium to high plasticity, organic silts | P | – | – | – | – | F | – | F | F |
| Pt | Peat and other highly organic soils | P | – | – | – | – | G | – | – | G |

Key:  E = excellent; G = good; F = fair; and P = poor.

Source:  EPA, 1979.

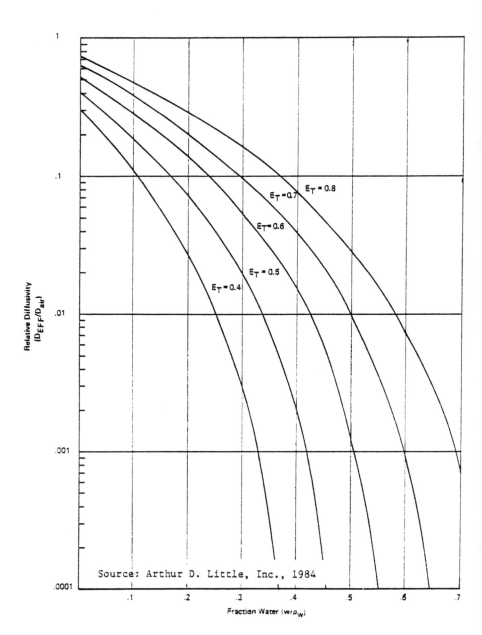

FIGURE V-1   RELATIVE DIFFUSIVITY IN SOIL COVERS

TABLE V-2

RELATIVE DIFFUSIVITY[1] AS A FUNCTION OF SOIL POROSITY

| Water Content(w) $(lb/ft^3)$ | Relative Water Fraction $(w/\rho_w)$ | Total Porosity $(\varepsilon_T)$ | | | | | |
|---|---|---|---|---|---|---|---|
| | | 0.3 | 0.4 | 0.5 | 0.6 | 0.7 | 0.8 |
| 0 | 0 | .20 | .29 | .40 | .51 | .62 | .74 |
| 6.24 | 0.1 | .052 | .11 | .19 | .28 | .37 | .48 |
| 11.5 | 0.2 | .005 | .029 | .072 | .14 | .20 | .28 |
| 18.7 | 0.3 | 0 | .0029 | .019 | .052 | .096 | .16 |
| 25.0 | 0.4 | − | 0 | .0019 | .014 | .037 | .073 |
| 31.6 | 0.5 | − | − | 0 | .0012 | .0096 | .028 |
| 37.4 | 0.6 | − | − | − | 0 | .00095 | .0072 |
| 43.7 | 0.7 | − | − | − | − | 0 | .00072 |

1. $$\frac{D_{eff}}{D_{air}} = \frac{(\varepsilon_a)^{10/3}}{(\varepsilon_T)^2} \quad \text{and} \quad \varepsilon_a = \varepsilon_T - \frac{w}{\rho_w}$$

Source:  Arthur D. Little, Inc.

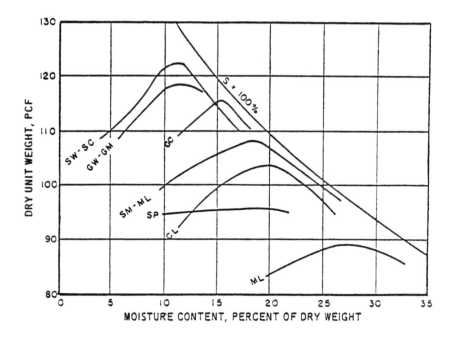

FIGURE V-2   EXAMPLE STANDARD COMPACTION CURVES
FOR VARIOUS SOIL TYPES

Source:   U.S. Environmental Protection Agency, 1979

decreases demonstrates the need to prevent drying out. Very large relative increases in diffusivity occur as the moisture content begins to fall much below the optimum point.

### 3.3  Final or Permanent Covers

When a working cell at a hazardous waste landfill has been filled up to its capacity, that portion of the landfill is closed. Closing involves the installation of a cover or cap over the top of the collected wastes to serve many purposes. These include the prevention of infiltration of surface water, the reduction of gaseous emissions, general site security, and others. The current regulations for landfills focus on the problem of groundwater protection and emphasize infiltration control and leachate collection as a means to achieve this objective. Fortunately, from an emissions point of view, this objective also serves as an implicit gaseous emissions control. Covers that are effective in reducing water infiltration are also effective in reducing emissions. Both gas and liquid permeability of porous media such as compacted soils depend in a complex manner on the porosity. Similarly, synthetic membrane covers which are essentially impermeable to liquid water are resistant to volatile vapor permeation. They are not, however, impermeable, and thus, there will always be some residual level of emissions even when a synthetic membrane cover has been installed.

The installation of covers does not eliminate the generation of gases. Gas generation continues through the volatilization of organic wastes and in some settings through the generation of methane as a product of anaerobic degradation. These gases will gradually build up in the landfill and may, if sufficient pressure is generated, damage the cover. At the same time because of the increased pressure, the gases will tend to diffuse laterally out of the landfill and ultimately to the surface. Some sort of vent system is often installed in conjunction with the cover in order to prevent either of these two situations from arising.

The discussion above on temporary covers describes the basic behavior of soils relative to permeability. In designing and installing a permanent cover, there is considerably more flexibility than is possible in temporary covers. The economics permit a much wider choice of materials and complexity of design. In particular, layered designs are becoming quite commonplace. The designer can incorporate favorable trafficability, water impermeability and gas barrier properties are using different materials rather than suffering the trade-offs intrinsic in the use of a single material. Figure V-3 (EPA, 1979) illustrates the concept of layering. In even more complex systems, synthetic membranes may be added in addition to the various soil layers.

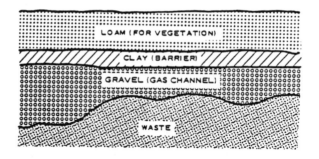

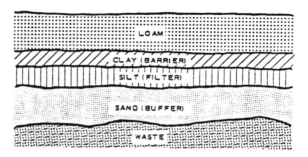

FIGURE V-3  TYPICAL LAYERED COVER SYSTEMS

Source: U.S. Environmental Protection Agency, 1979

The gas channel shown in the figure is installed for use with some sort of vent system to control lateral gas migration. Several technologies can be used:

o   Trench vents. Trench vents are narrow trenches backfilled with sand, gravel and/or stone. They are often lined on one side and can be open to the atmosphere or capped with clay and fitted with laterals and riser pipes vented to the atmosphere. They can also be connected to a negative pressure fan for forced withdrawal.

o   Vertical barriers. Vertical barriers include slurry walls, grout curtains, and synthetic liners.

o   Forced wells. Forced wells for lateral migration control are identical to those for vertical control, except that they are placed around the perimeter of the site and spaced such that all gas is drawn to the wells before crossing the site boundary.

o   Injection trenches and wells. These are similar to forced induction trenches and wells, except that a blower is used to force air into the system. This creates a pressure gradient in the landfill, causing gas to flow away from the system, and thereby preventing lateral migration across the boundary.

In installations where the principal purpose of the gas control system is to prevent damage to the cover, the vapors released directly into the atmosphere. For emission control purposes, some sort of treatment would be added.

4.    In-Situ Controls

None of the controls considered are classed as in-situ controls.

5.    Post-Treatment

Post-treatment involves the collection of gaseous emissions from the landfill and treatment of those emissions to remove or destroy the volatile organic components. Such systems consist of the collection means and the treatment means. Combinations of external covers and treatment techniques such as carbon adsorption or incineration have been discussed above in the chapter on Surface Impoundments. This approach can be used on landfill operations in essentially the same manner as that for surface impoundments. As previously discussed, air inflated structures can be installed over closed or operating landfill sections covering the typical range of sizes from a fraction of an acre to as much as 5 or 6 acres. These types of structures can include complicated vehicle air locks to permit access during the operating phase.

The air introduced to support the structure, which in conventional systems escapes by permeating through the material, would in this application be exhausted through a series of vents connected to the treatment system. The exhaust gases containing vapor released from the surface of the landfill would be released to the atmosphere after passing through the treatment system.

The installation of a cover serves other purposes as well as providing a means for emissions control. The principal reason for the installation of a cover at the CECOS facility in Livingston, Louisiana, is to keep precipitation out. An alternative control approach with comparable performance would be a treatment system connected directly to gas vents installed within the landfill and exiting through the cover. Forced vents would be used to provide a driving mechanism for introducing gases into the treatment unit and for achieving high collection within the landfill.

D.   Effectiveness

1.   Introduction

This section presents information pertaining to the effectiveness and costs of the various control approaches. The methodology is slightly different from that used for surface impoundments. For landfills there is no simple way to establish an uncontrolled reference case. Every landfill has some sort of cover but the emission rate through the covers varies potentially over several orders of magnitude. Therefore, comparisons based on percent reduction are arbitrary and depend markedly on the choice of reference. Cost-effectiveness is, however, examined on the basis of cost per pound of emissions removed.

2.   Effectiveness

2.1  Covers

Little or no data describing emissions through landfill cover systems under field conditions have been collected. As an alternate means to examine the effectiveness of covers, data based on the Farmer equation described above have been developed to demonstrate the relative effectiveness of different types of cover systems. Table V-3 and Figure V-4 present data for a number of combinations of soil covers of different total porosity at varying water contents. Both are expressed in terms of unit emissions through a 12-inch cover. Unit emissions are the pounds per hour of emissions per square foot per pound per cubic foot of vapor concentration on the bottom side. Diffusivity was assumed to be 0.28 ft$^2$/hr, a figure typical of a wide range of organic compounds in air. The porosity and percent water vary over the conditions that would be expected at a wide range of landfill operations and locations. The unit emissions are inversely

TABLE V-3

## UNIT EMISSIONS[1,2] THROUGH 12-INCH SOIL COVERS FOR SOILS OF DIFFERENT DRY POROSITY

| | Unit Emission (ft/hr) | | | |
|---|---|---|---|---|
| | Dry Porosity (Bulk density)[4] | | | |
| % Water[3] | 0.3(118) | 0.4(101) | 0.5(84) | 0.6(67) |
| 0 | .056 | .083 | .111 | .142 |
| 5 | .016 | .039 | .069 | .10 |
| 10 | .0020 | .015 | .039 | .073 |
| 15 | .0000032 | .0036 | .020 | .050 |
| 20 | - | .00033 | .0084 | .032 |
| 25 | - | - | .0027 | .020 |
| % Moisture at Zero Porosity | 15.8 | 24.7 | 37.1 | 55.8 |

1. Units are $\dfrac{lb}{ft^2\text{-}hr}$ / $\dfrac{lb}{ft^3}$ or ft/hr; $\dfrac{Q}{AC} = \dfrac{D_{air}}{t}(\epsilon_a)^{10/3}/(\epsilon_t)$

2. Diffusivity in air = 0.23 $ft^2$/hr; $t$ = 1 ft;

3. Percent of dry weight

4. Bulk density ($lb/ft^3$) in the parentheses, based on particle density = 2.7

Source:  Arthur D. Little, Inc.

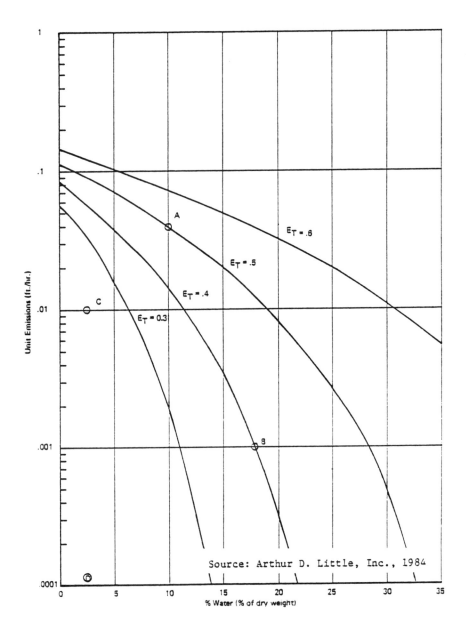

FIGURE V-4    UNIT EMISSIONS THROUGH 12" SOIL COVER OVER A LANDFILL

proportional to cover thickness. For a 6 inch cover, the data shown in the table and figure should be multiplied by 2. Conversely, for a 24-inch cover, these data should be divided by 2.

Figure V-4 shows the extreme sensitivity of emissions to conditions in the soil. For a given soil type compacted to a given porosity, the emission rate can vary over several orders of magnitude depending on the percent of water. In the steeper part of the curves, a change in only a few percentage points in water content can make a substantial change in emission rates.

The reduction in emissions along any one of the curves in Figure V-4 is due to the displacement of air filled voids in the soil by water. At some point, the pores will become completely filled with water and the emissions will become essentially zero. The moisture content corresponding to pore saturation is indicated at the bottom of Table V-3. Under saturated conditions, organic compounds can diffuse through the water in the pores. The rate of this process is very slow compared to the rate through air and has been assumed to be negligible.

The extreme sensitivity to conditions illustrated in the curves indicates the critical significance of choice of cover material, compaction practices, and moisture control in achieving high levels of emissions control when using soil covers. In most cases, the choice of daily cover materials is made on economic grounds with the result that materials excavated from the site or available from nearby sources are generally used. It may be fortuitous that these materials have properties desirable from an emissions control point of view. If not, the properties can be altered to provide lower porosity and higher moisture retaining properties by mixing in relatively small amounts of clay or clay materials.

Moisture control is perhaps the most important factor in emissions reduction. But little attention is generally given to moisture control. Cover materials are often piled up uncovered in an inactive portion of the site. The moisture content, under this practice, depends on the recent weather patterns. To minimize emissions, daily cover materials should be as wet as possible relative to the maintenance of mechanical properties appropriate to workability and structural integrity.

Soil, synthetic membrane, or combinations of the two can be used for permanent covers. As is the case for soil materials, there are little data available for vapor permeation through synthetic membranes of the type and configuration used as covers or liners. Table V-4 presents data recently reported on permeation rates for a variety of materials for several different organic chemicals. The data shown in this table have been converted into the same units used to display the soil cover characteristics above. Permeation rates through all polymeric materials is highly dependent on the chemical nature of the diffusing compound. Polar compounds such as acetone generally behave

TABLE V-4

UNIT EMISSIONS THROUGH MEMBRANE CAPS[1]

### HIGH DENSITY POLYETHYLENE - 30 MILS

| | | |
|---|---|---|
| Xylene | - | 0.011 |
| Acetone | - | 0.00028 |
| Chloroform | - | 0.0071 |

### TEFLON - 4 MILS

| | | |
|---|---|---|
| Xylene | - | 0.000008 |
| Acetone | - | 0.00063 |
| Chloroform | - | 0.0027 |

1.  Units are $\dfrac{lb}{ft^2\text{-hr}}$ / $\dfrac{lb}{ft^3}$

Source:   Arthur D. Little, Inc. based on data in Haxo, H. et al., 1984.

quite differently from hydrocarbons such as xylene or chlorinated hydrocarbons such as chloroform. In the 30 mil polyethylene membrane the permeation rate of xylene is about 50 times that of acetone, but in a 4 mil Teflon membrane, the ratio is reversed. Acetone passes through the Teflon membrane at about 100 times more rapidly than xylene, but Teflon remains more effective in absolute terms.

It is important to note that the magnitude of these permeabilities fall within the range obtainable with soil covers. Four points have been indicated on the Figure V-4 above to represent typical temporary and permanent soil cover conditions and 2 membranes. Points A and B respectively are located at points corresponding to conditions that might be expected in a temporary soil cover and at a permanent cover. Point C and D represent the permeability of a neoprene and a higher performance material such as Teflon, respectively. The synthetic materials which are often characterized as impermeable are in theory of the same order of effectiveness as properly designed and maintained soil covers. And, if water content is maintained at the saturation point, soil covers can theoretically reduce emissions essentially to zero (neglecting liquid phase diffusion), a level unattainable with the polymeric materials commonly used in today's practice.

Composite membranes including a layer of Mylar, a polyester polymeric material, can achieve vapor permeation rates that are effectively unmeasurable. These materials have been used in developmental applications in food containers and in special protective fabrics. (Personal communication, A. Schwope). Laminates in forms suitable for field application as permanent cover materials are presently unavailable.

As components of permanent covers, synthetic membranes have several advantages relative to soils. Once installed, they require little maintenance to maintain their efficacy. Performance over long periods of time is, however, uncertain. The materials may degrade and if this occurs, would have to be replaced. In practice, cracking and nonhomogeneities in the materials could significantly reduce the effectiveness relative to the design values. Small fissures in a cover would act as conduits for the vapors generated within a large area of the landfill in the vicinity of the crack. Careful maintenance of the cover including vegetation to prevent erosion, and installation of sprinklers to maintain moisture content at uniform levels will reduce the probability of the formation of cracks. The roots of plants must be prevented from penetrating the gas barrier.

Maintaining appropriate moisture levels is, however, inconsistent with groundwater protection objectives at a landfill. Downward infiltration of water and prevention of leachate formation is a primary objective for covers from this point of view. Combinations of a soil cover and membrane can be designed to achieve both sets of objectives.

Although with soil covers zero emission is possible and desirable, it is more likely that the cover system will exhibit a finite emission rate. Table V-5 shows the quantity of emissions per year through an acre of surface corresponding to the four cover systems depicted as points A, B, C, and D, in Figure V-1 above. Emissions are given for a range of vapor concentrations in the landfill. The table indicates that, even with high performance covers in place, considerable loss of high vapor pressure compounds can occur. Volatile solvents such as MEK or benzene have equilibrium vapor pressures of the order of 100,000 parts per million at ambient conditions. Such compounds if present in relatively undiluted form could be emitted in the range of several metric tons per year for soil of quite low permeability. Neoprene which is often considered a suitable cover material (EPA, 1979) could permit emissions of about an order of magnitude greater.

Emissions in existing closed facilities where cover systems are not as effective as those corresponding to the characteristics discussed above could be emitting substantial quantities of volatile organics at present. These emissions could be reduced by replacement of cover systems with more effective designs or by the installation of one of the post-treatment systems discussed in the next section.

2.2  Post-Treatment

Covers alone may not be capable of achieving desired performance levels. Covers do not reduce the total quantity of emissions ultimately entering the atmosphere. As long as there is some finite permeability, emissions will continue to enter the atmosphere until all the materials in the landfill have diffused into the atmosphere. For very effective covers the time period may stretch out over hundreds of years. Post-treatment offers a means to reduce the quantity of volatile materials reaching the atmosphere by removing them from the gases that are formed in the landfill or from the atmosphere above the landfill. Two types of post-treatment were discussed above, the systems employing vents to collect the vapors and systems employing external structures to capture the gases. Either configuration plus adsorption system, incineration, or other gaseous control means can be used. The effectiveness depends primarily on the collection system. The treatment portion can achieve high levels of efficiency exceeding 95%. The effectiveness of an external cover depends on the integrity of the cover and permeability of the materials used.

Forced induction trench vents with risers and laterals are the most effective lateral control method if the trenches extend to groundwater or to bedrock. Forced well systems may be somewhat less effective, but may also be less expensive and useful at greater depth. Natural induction wells and trenches have not been effective in many situations. The effectiveness of vertical barriers (including lined trenches) and injection systems is expected to be good, but experience is limited. The integrity of grout curtains and vertical liners as gas barriers may be difficult to achieve and maintain.

TABLE V-5

ANNUAL EMISSIONS

| Vapor Concentration (ppmv) | Daily Cover | Emissions (lb/year-acre) | | |
|---|---|---|---|---|
| | | Final Cover | Neoprene Cap | Laminate Cap |
| | A | B | C | D |
| 10 | 42 | 1.0 | 10.5 | 0.1 |
| 50 | 210 | 5.3 | 53 | .53 |
| 100 | 420 | 10.5 | 105 | 1.05 |
| 500 | 2100 | 53 | 530 | 5.3 |
| 1000 | 4200 | 105 | 1050 | 10.5 |
| 5000 | 21000 | 530 | 5300 | 53 |
| 10000 | 42000 | 1050 | 10500 | 105 |
| 50000 | 210000 | 5300 | 53000 | 530 |
| 100000 | 420000 | 10500 | 105000 | 1050 |

Notes:

1.  A, B, C, D refer to points on Figure V-4.

Source:  Arthur D. Little, Inc.

Controls for lateral gas migration should extend to bedrock or the water table to be effective. This prevents gas from migrating underneath the control mechanism. When depth of bedrock and the water table is greater than 20 to 30 feet, forced induction wells may be the only effective control strategy.

Forced induction trenches and wells can be designed in a manner similar to wells for control of vertical migration. Forced trenches (i.e., forced induction through riser pipes) are effectively the same as wells completed in gravel. The radius of influence of the wells will therefore be greater, allowing greater spacing.

### 3. Costs

#### 3.1 Pretreatment

The same unit costs used in the section and surface impoundments were used. Pretreatment costs taken from Spivey et al., (1984) are in the range of $1.00 per pound ($2,000/Mg) of waste removed from an incoming waste stream. Cost of other controls are compared to a range of pre-treatment costs to reflect variations due to waste stream properties, treatment type, size, and overall system design.

#### 3.2 Design and Operating Practice

Installed cover system cost estimates are shown in Table V-6. (EPA, 1979). Membrane cover layers are 4-8 times as expensive as soils. Costs depend on many site specific factors. This table should be considered only as a basis for comparative analysis, not as a basis for design and selection of a cover system at a particular site.

Costs for moisture control have not been estimated. Incremental costs for operating practices that will reduce atmospheric emission should be quite small. Costs may include annualized capital costs for heavier earth-moving equipment than would conventionally be used. Extra labor might be required to achieve higher compaction if extra passes were performed. Moisture control should require little additional labor. Using special imported temporary cover materials or additives would increase the costs of soil covers over that shown in the table.

#### 3.3 In-Situ Controls

No in-situ controls are included in the analysis of landfills.

#### 3.4 Post-Treatment

Costs for cover structures plus post-treatment were developed above in the discussion of surface impoundments. The costs are summarized in Table V-7. Treatment costs are based on non-regenerative carbon adsorption or incineration without heat recovery.

TABLE V-6

ESTIMATED UNIT COSTS FOR SOME COVER LAYERS

| Layer Type and Thickness | Installed Cost dollars/yd$^2$ |
|---|---|
| Loose soil (2 ft) | 0.35 |
| Compacted soil (2 ft) | 0.70 |
| Cement concrete (4 in.) | 9.00 |
| Asphalt concrete (4 in.) | 2.50-3.50 |
| Soil-cement (7 in.) | 1.50 |
| Soil-asphalt | 1.50 |
| Polyethylene membrane (10 mil)+ | 1.00-1.50 |
| Polyvinyl chloride membrane (20 mil) | 1.30-2.00 |
| Chlorinated polyethylene membrane (20-30 mil) | 2.40-3.20 |
| Hypalon membrane (20 mil) | 2.50 |
| Neoprene membrane | 5.00 |
| Ethylene propylene rubber membrane | 2.70-3.50 |
| Butyl rubber membrane | 2.70-3.80 |
| Paving asphalt (2 in.) | 1.20-1.70 |
| Sprayed asphalt membrane (1/4 in.) and soil cover | 1.25-1.75 |
| Reinforced asphalt membrane (100 mil) and soil cover | 1.50-2.00 |
| Bentonite layer (2 in.) | 1.40 |
| Bentonite admixture (9 lb/yd$^2$) in soil | 0.75 |

+ Not recommended because of thinness.

Source: U.S. Environmental Protection Agency, 1979

TABLE V-7

POST-TREATMENT COLLECTION SYSTEM COST ESTIMATES[1,2]

|  | Size of Facility | | | |
|---|---|---|---|---|
|  | 0.4 Acre | | 6 Acre | |
|  | Capital | Annualized[3] | Capital | Annualized |
| Air-Supported System | $150,000 | $24,000 | $1,600,000 | $250,000 |
| Pipe Vents | 80,000 | 12,000 | 403,000 | 60,000 |
| Compacted Soil Cover | 2,000 | 500 | 28,000 | 5,000 |
| 20-Mil Hypalon Cover | 6,500 | 1,500 | 96,000 | 15,000 |

Note:  1.  Treatment costs must be added to the figures shown.

2.  Cost in 1982 dollars.

3.  Annualized costs include a capital contribution (based on a 10-year life and 10% interest) and typical O&M costs.

Source:  Arthur D. Little, Inc.

Costs for a post-treatment system based on collection by means of vents coupled to carbon adsorption or incineration are also shown in Table V-7. The costs are based on a pipe vent system. Pipe vents include the blowers required to drive a control system. Passive trench vent systems might be used in situations where cover protection is the only function of the gas control. Pipe vents appear better suited for post-treatment applications. The costs presented do not include the cover.

### 4.   Cost-Effectiveness

As noted in the introductory paragraph to this section, definitions of an uncontrolled, reference landfill is rather arbitrary since covers are required by regulations, but vary widely in performance. Thus, it is not meaningful to compare cover performance to other operating controls on the basis of pounds of volatiles removed.

Notwithstanding that analytic problem, operating practices involving covers may be extremely cost-effective, that is, providing significant emissions reduction for little incremental cost. The earlier discussion describing emission rate as a function of cover properties indicates that several orders of magnitude or more improvement in performance may be attainable with careful practice. The costs to achieve this performance can be quite low. Moisture control of daily cover and grain size modifications by adding soil conditioners, clays, etc., are not expensive relative to the basic operating costs at a landfill. Under these circumstances, the cost-effectiveness would be high, compared to pretreatment, for example, expressed as pounds of volatile reduction per dollar. Even the added cost of heavier earth-moving equipment to achieve denser compaction should not change the cost-effectiveness very much.

The use of permeability reducing techniques for permanent covers should be similarly cost-effective compared to conventional design approaches. Low vapor permeability soil and synthetic membrane covers do not represent significant incremental costs. Design requirements for covers are established currently by regulations designed to reduce or prevent surface water infiltration. Incremental costs to maximize gas control performance are small.

Combination covers appear most effective. Although permeability in soil covers can, in theory, be reduced to zero, it would be extremely difficult to achieve perfect performance in practice. Imperfections in soil covers, particularly in large cells, are more or less inevitable. Maintaining high moisture content runs counter to control of water infiltration. Adding a synthetic membrane could offset these practical difficulties without significant incremental costs. The cover would prevent water infiltration from penetrating into the closed cell, and would act as a seal to inhibit permeation through cracks in the soil layer.

Even as performance approaches the theoretical limit, there are two factors which suggest the potential applications of alternative controls. First, as noted, practical performance may not reach theoretical levels. There are no data currently available to estimate the departure from theoretical performance. Second, covers only retard the loss of materials from the landfill. Little decay occurs in hazardous waste landfills so that, with any finite permeability, wastes will continue to enter the atmosphere. The applications of pre- or post-treatment can remove volatiles permanently.

Pretreatment costs are less than post-treatment for landfills as they are for other kinds of disposal facilities. Thus, unless there is some technical reason that pretreatment would not be practical, post-treatment would not be the cost-effective choice. Pretreatment costs are expected to be less than five dollars per pound of waste removed (1-2 dollars/per pound have been used in comparison, above). Post-treatment costs are expected to exceed about five dollars per pound removed plus the costs of the collector (cover or vent system). Thus, pretreatment would always be preferred. Covers serve more than one purpose; keeping out rain as well as keeping in volatiles. If the costs can be allocated to several purposes, post-treatment, using covers, would appear more cost-effective.

E.    Summary

Cover design and maintenance present the most cost-effective and highest degree of emission reduction potential of all the controls. In theory, emissions can be reduced to essentially zero by maintaining the pores of a soil cover full of water. Synthetic membranes in current use as cover materials all exhibit finite permeability to organic vapors.

Combination cover systems with both soil layers and synthetic membranes may achieve very high performance levels for both vapor and water infiltration control. Combinations can offset practical limitations in soil covers due to cracks and inhomogeneities.

Very little data are available that characterize current practice at operating landfills or at closed, previously active sites. There is, even in this situation, a reasonable expection that past and current practices are poor with respect to vapor control, and so there is substantial opportunity to reduce vapor emissions on a national scale.

Pretreatment eliminates the potential for emissions, but may have limited application to the highly variable wastes that are placed in landfills. Much of the wastes may come from widely dispersed small generators. These conditions limit the practicality of pretreatment.

Post-treatment systems have not been used at landfills, but the recent installation of an air inflated structure at a landfill in Louisiana indicates that this approach can be applied. The critical element in a post-treatment system is the collectors means. Treatment technology is proven but expensive. The covering structure includes air locks to permit vehicle access and prevents rainfall from falling on the site.

## F.  References

EPA, 1979, Design and Construction of Covers for Solid Waste Landfills (Prepared by R.J. Lutton, et al.). S. Report No. EPA-600/2-79-165.

Farmer, W.J. et al., 1978, Land Disposal of Hazardous Wastes: Controlling Vapor Movement in Soils, in Proceedings of 4th Annual EPA Research Symposium. EPA Report No. EPA-60019-78-016.

Haxo, H. et al., 1984, Permeability of Polymeric Membrane Lining Materials, Proceedings, International Conference on Geomembranes, Denver, CO.

Hwang, S.T., 1982, Toxic Emissions from Land Disposal Facilities. Envir. Progr., Vol. 1, No. 1 (Feb. 1982).

Schwope, A., July 1984, Arthur D. Little, Inc., Cambridge, MA. Personal Communications.

Spivey, J.J., C.C. Allen, D.A. Green, J.P. Wood, and R.L. Stallings, 1984. Preliminary Assessment of Hazardous Waste Pretreatment as an Air Pollution Control Technique. Draft Final Report. Research Triangle Institute, Research Triangle Park, NC. For the U.S. Environmental Protection Agency, IERL, Cincinnati, OH.

Weiser, H.B., 1949, Colloid Chemistry, John Wiley and Sons, New York.

# VI. Land Treatment Facilities

A. Description

Land treatment involves application of wastes to soil in which natural processes degrade or immobilize constituents in the waste. For organic contaminants, the principal operating mechanisms are volatilization and biological decay supported by microorganisms that are ubiquitously present in all soils. Inorganic constituents are immobilized within the soil through several processes which lead to absorption of the constituents on the surface of soil particles. Land-treatment facilities come in a variety of sizes and shapes. The medium area derived from EPA survey data is approximately 12 acres (Spivey et al., 1984). Sites range from approximately one acre at the low end to several hundred acres at the upper end. A few land treatment facilities are around 1,000 acres in area. EPA survey data indicate approximately 150 sites were operating in 1981, but other sources indicate closer to 200 sites exist (EPA, 1983).

The area at a land treatment facility can be used in several ways. Major site designs are:

o   Single Plot - wastes are applied to the entire area at each application.

o   Progressive Plot - wastes are applied to a few small subplots until their capacity is reached. These subplots are then closed, and the application progresses to a few more subplots until the entire area is exhausted.

o   Rotating Plot - waste is applied to small subplots, which are then cultivated and revegetated. After the wastes have degraded, the plots are reused.

The wastes may be applied to the soil surface in a variety of techniques. The choice of the technique can quite significantly influence the degree to which wastes volatilize. Application technique is the most important operational variable at a land treatment facility. The four major application techniques are:

o   Spray Application - liquid waste is sprayed on the surface of the site.

o   Surface Application - wastes are dumped on the site surface either in small quantities (that is, waste is spread as it is applied) or in bulk (waste is dumped in one place and spread by plow or other equipment).

o   Subsurface Injection - waste is injected at a shallow depth generally ranging from 4-8 inches (10 to 20 cm) into the soil.

144

o    Overland Flow - wastewater is caused to flow over relatively
     impermeable soil with a slope from about 2 to 8%. This
     technique is used for treating contaminated run-off or
     wastewater effluents from industrial processes.

Following the application of wastes to the soil, the wastes may
be incorporated into the top layer of soil by standard cultivation
techniques. The soil may be tilled several times following a single
application of waste before the next application. Typically, the
wastes are tilled into the top 4 to 8 inches (10 to 20 cm) of soil.

In some settings the site is revegetated. In this case, the
surface is not tilled after each application. This technique is
commonly used to treat dilute wastes which can be applied by spray
application.

B.    **Emission Sources and Models**

    1.    General

Emissions to the atmosphere at a land treatment facility arise
from two primary sources. The first is pools of liquid wastes which
form after application on the surface. These pools remain until the
liquids seep into the underlying soils or are incorporated by tilling.
Volatilization directly into the atmosphere can occur as long as the
liquid wastes are exposed at the surface.

The second source is wastes which have been incorporated into the
soil. Volatile constituents in the wastes can enter the interstices
and eventually diffuse to the soil surface. At the surface the
emissions mix into the atmosphere and are swept away from the site.

    2.    Emission Models

    2.1    Surface Emissions

Emissions from a surface layer of liquids behave in the same
general manner as emissions from a floating layer of organic compounds
on a surface impoundment. For pure or highly concentrated organic
mixes, the following equation describes the emission rate:

$$Q = Ak'_g C$$

where    $Q$    =    emission rate (lb/hr)

         $A$    =    surface area ($ft^2$)

         $k'_g$    =    mass transfer coefficient (ft/hr)

$$C \quad = \quad \text{vapor concentrations of diffusing component in equilibrium with liquid, } lb/ft^3$$

$$\text{and} \quad k_g' \quad = \quad k_g \frac{RT}{P_T}$$

where
$$
\begin{aligned}
k_g &= \text{mass transfer coefficient } (lb\text{-}mol/ft^2 \text{ hr}) \\
R &= \text{gas constant } (atm\text{-}ft^3/lb\text{-}mol - {}^\circ R) \\
T &= \text{temperature } ({}^\circ R) \\
P_T &= \text{total pressure } (atm)
\end{aligned}
$$

This equation is appropriate immediately after application of wastes and as long as a liquid pool remains on the surface. It is also appropriate for use in spills generally. The simple form assumes gas film control and is not appropriate for dilute wastes in which diffusion to the liquid surface from the bulk of the pool is significant in determining the overall mass transfer coefficient.

In the discussions below, it is assumed that the hazardous wastes would be tilled into the soil immediately upon application, or would be injected by a subsurface technique. Emissions from surface pools can be quite substantial; they should be avoided wherever possible.

### 2.2 Emissions from Incorporated Wastes

The model used to estimate emissions from land treatment after the wastes have been incorporated was based on work by Thibodeaux as reported by Hwang, (1982). The equation; for the ith contaminant species, is:

$$Q = AD_{eff} \, C_g \, [h_s^2 + (2 \, D_e \, t \, A \, (h_p - h_s) \, C_g) \, M_o]^{-\frac{1}{2}}$$

where:
$$
\begin{aligned}
Q \quad &= \quad \text{flux rate } (lb/ft^2\text{-}hr) \\[6pt]
D_{eff} \quad &= \quad \text{diffusivity of air filled pore space } (ft^2/hr) \\[6pt]
&= \quad D_{air} \, \varepsilon_T/\tau \\[6pt]
D_{air} \quad &= \quad \text{molecular diffusivity of i in air } (ft^2/hr)
\end{aligned}
$$

$e_T$    =    total porosity

$\tau$    =    tortuosity factor ($\tau \approx 4$)

$h_s$    =    depth of surface injection (ft)

$t$    =    time after application (hr)

$A$    =    surface area of application ($ft^2$)

$h_p$    =    depth of penetration of plow slice (ft)

$M_o$    =    initial mass of component i (lb)

$C_g$    =    concentration of i on gas side of interface ($lb/ft^3$)

This model is appropriate once the waste has been incorporated into the soil. It assumes:

o    The soil column is isothermal
o    No capilliary action
o    No adsorption on soil particles
o    No biochemical degradation

This rather complex equation becomes much simpler under two separate assumptions. The first assumption is that the waste is placed on the surface and immediately tilled. In this case, $h_s$ equals zero. This equation, rearranged in terms of emissions per unit area is:

$$\frac{Q}{A} = \left[ \frac{C_g \, D_{eff} \, (M_o/A)}{2 \, h_p \, t} \right]^{\frac{1}{2}} = D_{eff}/f$$

where:

$$f = \left[ \frac{2t \, D_{eff} \, h_p}{C_g (M_o/A)} \right]^{\frac{1}{2}}$$

As long as the assumption of no biodegradation is valid, i.e., for a short time following applications, the Thibodeau-Hwang equation indicates that emissions would decrease proportional to the square root of time expired after application. Once biodegradation becomes important, so that vapor concentration, Cg, decreases with time, the relationship becomes more complex. Emission rate would decrease more rapidly than the inverse square root form indicates. The exact form depends on the nature of the behavior of concentration with time.

The second simplifying assumption is that the wastes are injected at a depth equal to the depth of penetration of the plow or cultivator. In this case, $h_s$ equals $h_p$ and the equation becomes simply:

$$\frac{Q}{A} = \frac{D_{eff}}{h_s} C_g$$

Similarly in this case, the emission rate would remain constant at first and then would decrease according to the functional form of the relationship between concentrations and time.

### 3.   Controlling Parameters

#### 3.1 Liquid Pools

There is virtually no parameter that can be controlled as long as a liquid pool is present. To minimize volatile emission, the time that the liquid are exposed should be kept as short as possible.

#### 3.2 Incorporated Wastes

Emissions, once the wastes have been incorporated, depend on a number of parameters. In all cases, emissions depend in different functional forms on the concentration of the component in the waste. For the case where wastes are injected at the same depth as the plow cut, the initial emissions depends additionally only on the effective diffusivity and depth of injection. The deeper the injection, the lower would be the emission rate. The rate depends directly on effective diffusivity. Diffusivity in the form developed by Thibodeaux and reported by Hwang (1982) is related to the porosity of the soil and a tortuosity factor. Porosity as in the case of covers for landfills depends on the degree to which the soil is compacted and on the moisture content. Unlike the case for covers, it is important to maintain a relatively high soil air-filled porosity in order to provide a means to carry oxygen to the microorganisms that degrade the organic wastes. Porosity is therefore not considered to be a controllable parameter for land treatment.

For surface injections, the initial emission rate depends on the depth of the plow cut, waste application rate, and time (explicitly) as well as effective diffusivity, and concentration (implicit function of time). Emissions decrease inversely with the square root of time following application. In this configuration the rate depends inversely on the square root of plow depth rather than the first power as in the alternative form. Finally, the initial rate depends directly on the square root of the application rate.

The total emissions for a fixed area over a period of time, say a year, depend on the number of applications. A given quantity of waste can be applied as fewer and larger portions or more and smaller portions. The total emissions over the period will depend on the number of applications. The form of the exact dependence requires that the rate equations above be integrated. The integration cannot be carried out analytically, without an explicit form of the relationship between concentrations and time, which depends on the nature of the biodegradation process and on its coupling to the diffusive mechanism. This relationship is complex and poorly understood at present.

The equations were integrated for several possible forms for $C_g$ as a function of time to determine the general shape of the dependency on application rate. The results indicate that total emissions increase with increasing frequency of applications; thus, to control emissions for a fixed quantity of wastes applied over an intended period, the maximum quantity consistent with the biodegradative behavior should be applied each time.

## C.  Controls

### 1.  Introduction

Emissions from land treatment can be reduced by removal of volatile components of wastes prior to application. Pretreatment will have the same effectiveness for this type of facility as those discussed previously. Post-treatment techniques involving collection systems in conjunction with a treatment process can also be applied. The equations indicate that the rate of emission is quite sensitive to a number of the operating design parameters. There are, as a result, several control methods that involve design and operating practices. There are no available control approaches considered as in-situ methods for this type of facility.

### 2.  Pretreatment

Pretreatment may not be appropriate or effective for land treatment. If the biodegradation processes that occur in the soil produce volatile compounds from the breakdown of heavier molecules, then pretreatment will not be so effective. Pretreatment may be more an alternative to land treatment not an adjunct, since both are designed to handle the same kind of organics.

### 3.  Operating Practices

Volatile emission rate can be controlled through the means by which the wastes are injected and by adjusting the amount applied per unit area per pass. Subsurface injection is considerably more effective in controlling emissions than is surface injection immediately followed by tilling. Wastes should be injected at the maximum depth consistent with the parameters determine the waste degradation rates. Typical common depths of application range from 4 to 8 inches (10-20 cm) below the soil surface (Wooding and Shipp, 1979). A variety of types of equipment have been developed for subsurface injection (EPA, 1983, Overcash and Pal, 1979). In some of these devices, the waste is simultaneously injected below the surface and mixed into the top 6 to 10 inches (15 to 25 cm) of soil. The initial injection may be followed closely by a second pass with a cultivator to distribute the waste uniformly across the treatment area. Monographs discussing land treatment recommend subsurface injection for wastes with volatile components and odorous constituents. Subsurface injection minimizes exposure of the operators during application.

Emission rate is a weak function of the application rate per unit area. For economic reasons, land treatment facility operators apply the maximum waste per unit area consistent with the degradation capabilities and carrying or assimilative capacity of the soils. In practice, application rates often exceed values recommended in conventional design and guidance sources. Emissions increase when the design capacity is overloaded because degradation will be impeded and the volatile materials remain in the soil for longer periods of time. Typical application rates range from about one-half a pound per square foot to about three times that rate. (EPA, 1983)

As noted earlier soil porosity is probably not controllable but care should be taken to prevent excessive drying. If the soil becomes too dry, the effective diffusivity can increase by several orders of magnitude. (See the discussion of landfills for the particulars of the relationship between diffusivity and soil porosity). Soil moisture content should be maintained at the maximum level consistent with the parameters controlling degradation in order to keep atmospheric emissions to a minimum.

4.   In-Situ Controls

No in-situ controls are available for land treatment.

5.   Post-Treatment

It is conceptually possible to consider placing air inflated structures over small land treatment facilities or over sections of larger facilities. In a manner similar to that described above, air containing the volatile emissions can be vented through a treatment unit prior to being released into the atmosphere. Use of such covers would also give the facility operator some degree of control over temperature and moisture.

D.   **Effectiveness**

1.   Introduction

The effectiveness of controls at land treatment focusses on tilling practices. Pretreatment and post-treatment perform essentially the same as described in the preceding sections. Emissions are referred to surface applications as the uncontrolled case.

2.   Emissions Reduction and Effectiveness

Figure VI-1 indicates the percent of volatiles applied in land treatment that would be lost to the atmosphere through volatilization. These curves are based on the equations developed above. The emission loss relationships and the data generated using them assume no biological degradation. This assumption is obviously unrealistic. The main function of land treatment for organic compounds is to provide degradation. The actual emissions and percentage lost to the atmosphere will be less than that depicted herein for this reason. Nevertheless, the equations and the data are useful in describing the relative effectiveness of a variety of alternate controls. The results shown in Figure VI-1 are consistent with data taken in the field. (Minear et al., 1981). As reported in this reference, oil wastes spread and tilled into soil resulted in losses of 0.6% and 2% in two separate field tests (Francke and Clarke, 1974, and Suntech, undated). In another test reported in this same reference, 11% of the wastes from an API separator were lost to the atmosphere (Suntech, undated). In this last test the wastes were spread on the surface but not tilled into the soil. Vapor concentration over the types of wastes treated were not reported. Based on the types of hydrocarbons generally present in the wastes treated, the vapor concentration can be estimated. A value of around 1,000 ppm appears reasonable.

The same kind of data are shown in a somewhat different format in Table VI-1. In this table the quantity of volatile losses per acre per year is shown for compounds of increasing vapor pressure. Comparison of the columns for surface injection versus those for subsurface injection illustrate the reduction that is theoretically possible. Figure VI-2 illustrates the reduction efficiency directly. The potential reductions shown in Table VI-1 and Figure VI-2 may not be achieved in practice, if the rate of biodegradation is reduced sufficiently by subsurface injection. Stated alternatively, if the rate of degradation is slowed down, then there is more time and more unreacted materials available for volatilization. The efficiency depends on the relative volatility of the constituent. Efficiency for low volatility compounds is quite high exceeding 90% for constituents with vapor concentrations below about 1,000 ppm. The efficiency falls off for the more volatile compounds. For reference, the vapor concentration in equilibrium with benzene at normal temperatures is about 100,000 ppm. The corresponding equilibrium concentration for xylene is about 10,000 ppm. The overall effectiveness will depend on the particular mix of compounds in the waste.

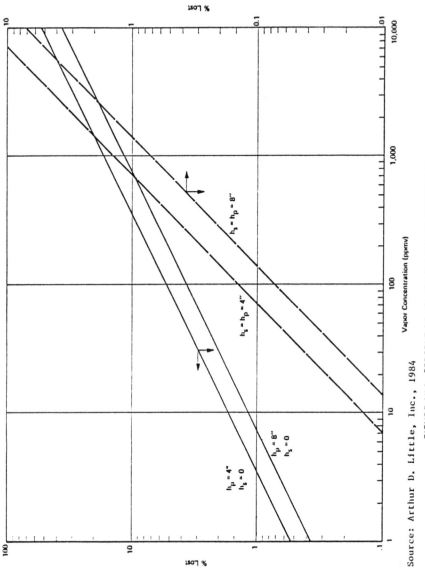

Source: Arthur D. Little, Inc., 1984

FIGURE VI-1   PERCENT OF VOLATILES LOST TO ATMOSPHERE

TABLE VI-1

ANNUAL LOSSES FROM 1 ACRE OF LAND TREATMENT[1]

| | Injection ($h_s^\cdot$) and Plow Depth ($h_p$) | | | |
|---|---|---|---|---|
| | $h_s = 0$ | | $h_s = h_p$ | |
| Equilibrium Vapor Concentration (PPMV)[3] | $h_p = 4''$ | 8" | 4" | 8" |
| 10 | 4.8 | 3.4 | .1 | .05 |
| 50 | 10.7 | 7.6 | .2 | .1 |
| 100 | 15.2 | 10.7 | .4 | .2 |
| 500 | 34 | 24 | 2 | 1 |
| 1000 | 48 | 34 | 4 | 2 |
| 5000 | 107 | 76 | 20 | 10 |
| 10000 | 152 | 107 | 40 | 20 |
| 50000 | 287[2] | 246 | 200 | 100 |
| 100000 | 287[2] | 287[2] | 287[2] | 200 |

Notes:

[1] Units are metric tons per year; 287 metric tons applied per year

[2] Complete (100%) loss

[3] Parts per million by volume

Source:  Arthur D. Little, Inc.

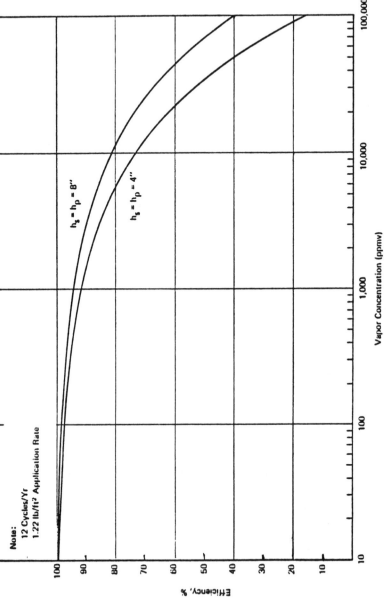

Source: Arthur D. Little, Inc., 1984

FIGURE VI-2    REDUCTION EFFICIENCY FOR SUBSURFACE INJECTION RELATIVE TO SURFACE INJECTION

3.   Costs

3.1  Pretreatment

Pretreatment costs are assumed, as above to be of the order of $1-2 per pound ($2,200-$4,400/Mg) of wastes recovered.

3.2  Operating Practices

The more effective subsurface injection technique is only incrementally more expensive than surface application. The addition of injection equipment to a tank truck spreader adds about 7% (about $5,000) to the cost of the system (Overcash and Pal, 1979). The labor cost should be about the same for either applications technique.

Lower application rates according, to the Thibodeaux relationship, would reduce emissions. This control would be costly as the land required to treat a given quantity of waste would increase proportional to the decrease in application rate. The added cost depends on the cost of land.

3.3  Post-Treatment

Post-treatment using an air-supported cover in combination with carbon adsorption or incineration costs the same as these systems described in the previous chapters.

4.   Cost Effectiveness

Based on the model used herein, subsurface injection is the most cost-effective emission control for land treatment facilities. Predicted efficiency of reduction ranges from about 20-40% for highly volatile components to better than 95% for low vapor pressure constituents. The cost-effective curves, if plotted on the same type of figure used in the discussion of surface impoundments and tanks would fall below pretreatment at about 100 pounds per year, a figure well below the capability of any practical land treatment site. No curves have been prepared, however, given the uncertainty in the performance and lack of cost data.

Improved injection (and reduced application rate) cannot, however, achieve high levels of control for the more volatile materials. Loss of in excess of 30-40% of some constituents may be unacceptable. In this case, pretreatment or post-treatment can be used to reach control levels of 90% or better. If pretreatment is feasible, then as in the other cases above, it is considerably cheaper than post-treatment. Some wastes currently being managed at land treatment are not well suited for the pretreatment techniques described by Spivey et al. (1984). For example, some refinery wastes are quite sludge-like, contain a large fraction of solids and, indeed, arise as residues from typical process recovery operations such as distillation.

In such a case, post-treatment using air-inflated structures plus incineration or adsorption would be the choice. If adsorption is used, the recovered wastes can be reapplied. In this mode, performance would be similar to that of the system described earlier where recovered wastes are reinjected in the aerated impoundment over which the cover is installed.

## E.    References

EPA, 1983, Hazardous Waste Land Treatment, (Prepared by K.W. Brown and Associates) - SW-874.

Francke, H.C. and F.E. Clark, Disposal of Oil Waste by Microbial Assimilation, Oak Ridge National Laboratory, Report No. VC-11/Y-1934, May 1974.

Hwang, S.T., 1982, Toxic Emissions from Land Disposal Facilities. Envir. Progr., Vol. 1, No. 1, (Feb. 1982).

Minear, R.A. et al., 1981, Atmospheric Hydrocarbon Emissions from Land Treatment of Refinery Oil Wastes, American Petroleum Institute, Washington, D.C. Report No. DCN 81-219-060-06.

Overcash, M.R. and D. Pal, 1979, Design of Land Treatment for Industrial Wastes, Ann Arbor Science Publishers, Inc., Ann Arbor, Michigan.

Spivey, J.J., C.C. Allen, D.A. Green, J.P. Wood, and R.L. Stallings, 1984. Preliminary Assessment of Hazardous Waste Pretreatment as an Air Pollution Control Technique. Draft Final Report. Research Triangle Institute, Research Triangle Park, NC. For the U.S. Environmental Protection Agency, IERL, Cincinnati, OH.

Suntech Group, Summary of Results from Toledo, Ohio. Refinery Landfarm Tests, Suntech Environmental Group, Marcus Hook, PA, undated.

Wooding, H.N. and R.F. Shipp, 1979. Agricultural Use and Disposal of Septic Tank Sludge. In Pennsylvania information and recommendations for farmers, septage haulers, municipal officials and regulatory agencies. Pennsylvania State University Coop. Ext. Serv. Spec. Circ. 257.

# Part B

# Evaluation of Models
# for Estimating Air Emissions

The information in Part B is from *Evaluation and Selection of Models for Estimating Air Emissions from Hazardous Waste Treatment, Storage, and Disposal Facilities,* prepared by William Farino, Peter Spawn, Michael Jasinski, and Brian Murphy of GCA Corporation for the U.S. Environmental Protection Agency, December 1984.

# Acknowledgment

This Final Report incorporates comments received from the technical peer review relative to GCA's selection and judgement of models most appropriate for evaluating AERR from TSDFs. Dr. Seong T. Hwang, the EPA/OSW Task Officer, performed technical review of the first and second peer review drafts and provided valuable insights which are incorporated herein. Additional technical review and assistance were received from the following: Dr. Charles Springer, University of Arkansas; Dr. Thomas R. Marrero, University of Missouri; Dr. John Williams, Northeastern University, Dr. C. Kleinstreuer, Rensselear Polytechnic Institute; Dr. Donald Mackay, University of Toronto; and Chemical Manufacturers Association Secondary Emissions Work Group. Comments have also been received from representatives of other EPA offices: IERL, MERL, and OAQPS.

The authors wish to thank the following contributors from GCA: Dr. David Cogley who provided valuable insight relative to vapor diffusion through soils; Mr. Ron Bell who provided the storage tank summary and other research efforts; and Mr. Thomas Nunno who provided assistance in review of AERR models applicable to wastewater treatment systems.

## SPECIAL NOTE

A Draft Final Report (October 1982) and a Revised Draft Final Report (May 1983) previously developed for this project were prepared and furnished to U.S. Environmental Protection Agency, Office of Solid Waste and Emergency Response (OSWER), Land Disposal Branch. Ms. Alice C. Gagnon served as EPA Project Officer and Dr. Seong T. Hwang of EPA/OSWER served as Task Officer for these efforts. On December 23, 1983, hazardous waste treatment, storage, and disposal facility (TSDF) area source emissions regulatory development was transferred from OSWER to the Office of Air Quality Planning and Standards (OAQPS). This Final Report was prepared under the direction of EPA/OAQPS Task Officer Kent C. Hustvedt.

# 1. Introduction

BACKGROUND AND PURPOSE

Under contract with EPA's Office of Solid Waste (OSW), GCA reviewed and evaluated available mathematical models describing the release rate (i.e. mass flux) of volatile air emissions from hazardous waste treatment, storage and disposal facilities (TSDFs). Air emission release rate (AERR) models judged most suitable for estimating air emissions from TSDFs are identified herein along with the rationale for each model selection. Other models available in the technical literature are also described with an indication of their limitations.

The purpose of this report is to provide a source of information for all air emissions release rate models available in the literature. Information contained in this report incorporates comments made by the peer review group in January 1983, and the EPA Task Officer. Since AERR models for TSDFs are continually undergoing development and refinement, and since little field validation data are available, this document presents the best available models as of early 1983.

PROJECT SCOPE AND TECHNICAL APPROACH

Technical efforts were initially limited to an evaluation of existing AERR models reported in the technical literature. A computer-assisted literature search generated 30 references that appeared suitable for review. Additional references became evident as the initial 30 technical articles were evaluated. Telephone surveys were performed to solicit additional information from EPA personnel, active researchers, other EPA contractors, and trade associations such as the American Petroleum Institute and the Chemical Manufacturers Association. A list of the technical references reviewed for this report appears in Appendix A.

On OSW's request, AERR models were investigated for the following TSDF categories:

- Landfills;

- Landfarms (land treatment);

- Surface impoundments;

- Storage tanks;

- Drum handling and storage;

- Treatment units;

- Waste piles (particulate emissions only).*

Treatment units were segregated at the suggestion of OSW into three generic types: (1) open tanks with quiescent surface conditions; (2) open tanks with mixing; and (3) closed-loop treatment units. This approach was developed in order to reduce the large number of treatment unit processes into a manageable size for analysis.

No models were found to describe air emissions from drum handling and storage facilities, although estimates of air emissions can be performed for known spill quantities using appropriate techniques described later in this report. GCA developed a general approach to describe the potential air emissions based on estimates of the fractional loss due to accidental spills.

Some effort was devoted to developing criteria for selecting AERR models most suitable to fulfilling EPA objectives for AERR estimations, and to provide a uniform basis for comparing different models. These efforts were terminated when it became apparent that only a limited number of models were available, and many models had serious limitations. In some cases, only one model was available to describe a particular type of TSDF. Consequently, the following basic criteria were used for model selection:

- Emission release phenomenon actually occurring at a TSDF must be accurately described by the models;

- Input data must be readily obtainable by an Agency engineer either through published literature or calculation techniques easily understood;

- Models must be suitable for use by entry level regulatory engineers with appropriate guidance through a manual.

Section 2 presents the AERR modeling approaches recommended by GCA based on the review and evaluation of current research. Rationale for each selection are also identified. Sections 4 through 8 discuss all AERR models available for each TSDF-type, indicating deficiencies and the rationale for not recommending certain models. The status of AERR Model validation efforts via laboratory and field measurements is provided in Section 3.

---

*All other models address volatile air emissions.

Section 9 presents the general approach developed by GCA to describe AERR for drum storage and handling facilities.  Section 10 describes methods for quantifying particulate emissions from storage piles.  The remainder of this introductory section provides a cursory review of the technical background and discusses the mass transfer principles that form the basis for virtually all AERR models.

TECHNICAL INTRODUCTION

Background

Air emissions of volatile hazardous compounds from TSDFs can enter the atmosphere in two ways:  (1) as the result of forced bulk motion on a macroscopic scale; or (2) as the result of diffusion on a molecular or turbulent scale.  Examples of forced motion include:

● vapor losses during filling of a storage tank;

● vapor movement in a tank or landfill cell due to pressure gradients created by changes in atmospheric pressure;

● convection due to biogenic gas production in landfills.

Air emissions resulting from forced motions described above do not depend on the volatility of the compound; all molecules simply partake in the macroscopic scale gas motion.

Diffusion of volatile air emissions from a TSDF can generally be thought of in terms of escape of material through multiple layers, with the last layer being the lower atmosphere.  For example, as illustrated in Figure 1, nonturbulent diffusive transport in a lagoon proceeds through four stages starting with the bulk liquid, to a laminar liquid layer at the liquid surface, and then through a similar laminar air layer, and finally, into the atmosphere.  The rate of transfer (diffusion) through all layers, combined, determines the release rate to the atmosphere.  This transfer rate is expressed in terms of the mass transfer coefficient, as will be described later.  A compound's volatility, characterized by its solubility, partial pressure, Henry's law constant and diffusivity, provides an indication of the quantity of material in the gas phase which will be subjected to the transfer process.  Additionally, turbulence in any of the layers can significantly increase the transfer rate.

Generally, transfer rates in one or perhaps two of several layers will be so low (due to the high resistance in these layers) that these layers will control the overall atmospheric emission release rate.  In the example given for a lagoon, the laminar air or liquid phase (or both for some compounds) will control the overall transfer rate.

Similarly, in discussing volatile liquids in landfills, it is possible to describe the process as one of diffusion from bulk waste, through a wetted soil region, through dry soil and into a laminar air layer.  However, for any

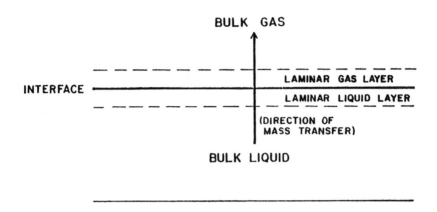

Figure 1.    Nonturbulent two-layer model of mass transfer.

significant burial depth, diffusion of vapors through the overburden of dry soil will be the most significant factor because the transfer coefficient in soil will be small. Figure 2 shows the probable controlling layer or layers for different TSDFs.

For an open landfill, precisely which layer will control the overall release rate will depend on the specific compound of interest and whether, for example, compounds placed initially on the surface have seeped deeply into the soil.

| | SOIL | LIQUID | AIR |
|---|---|---|---|
| LAGOON | | X | X |
| BURIAL (LANDFILL) | X | X | |
| LAND FARM | X | X | X |
| SPILL ON IMPERMEABLE SURFACE | | X[a] | X |
| SURFACE DUMP | X | | X |
| STORAGE TANKS | | X[a] | X |

[a]Liquid phase controlling for dilute aqueous or hydrocarbon mixtures.

Figure 2.  Controlling layer for diffusion losses from different TSDFs.

All state-of-the-art AERR models reviewed were based on mass transfer principles. Whenever the concentration of a chemical or compound varies between two regions there is a natural tendency for mass to be transferred across the interface, and approach equilibrium within the system. Each AERR model reviewed was based on the following relationship which describes diffusion transport across a series of different layers:

Mass flux = (overall mass transfer coefficient) (driving force);

where:

Mass flux = net quantity of material transferred across a unit area normal to the surface in a given unit of time (i.e., grams of a chemical transferred across one square centimeter of an air-liquid interface in one second);

Overall mass transfer coefficient (K) = overall rate of chemical transfer. The reciprocal of the overall K value* is equal to the sum of the reciprocals of individual k-values for each region. Changes in state (liquid-gas) modify this relationship slightly as described below. The reciprocal of the K-value is thus defined as the overall resistance to diffusion. The overall mass transfer coefficient represents the extent of contribution to mass transfer in each region which the chemical must pass through;

Driving force = the differences in chemical potential (i.e., chemical concentration) of the compounds of interest on each side of the boundary between two regions.

The choice of the most accurate model to represent AERR at TSDFs reduces to the problem of finding the "best" individual mass transfer coefficient, k, in each of the regions (soil, liquid, air). The overall mass transfer will, in turn, have a coefficient in the form of a series resistance, $\frac{1}{K} = \sum_i \frac{1}{\alpha_i k_i}$ where K is the overall transfer rate, and $\alpha_i$ is the equilibrium concentration ratio in each boundary layer relative to the layer below. For example, at a liquid-gas interface, $\alpha_i$ is essentially the Henry's Law constant ($H_i$) divided by the gas constant R and absolute temperature T. For many compounds, $H_i$ has been experimentally determined, or can be calculated from other chemical properties (vapor pressure and solubility).

## General Discussion of Available Models

Figure 3 shows for each major reference article investigated during this study the interface or layer which was considered by the respective researcher(s). In Figure 3, note that an "x" to one side of a specific interface indicates that side was considered implicitly to be controlling the overall transfer rate. Figure 3 also shows which forced motions, if any, were treated by various authors.

The substantial difference between each available model is the method employed to calculate the individual mass transfer coefficients (k-values). However, there are also differences in basic assumptions between models. In some cases, assumptions inherent to certain models do not accurately reflect the physical phenomenon responsible for air emissions from TSDFs.

Some authors make assumptions which are inappropriate for our purposes as to which region controls. For example, McCord (1981) bases his approach to non-aerated lagoons on the theory of water evaporation, where the evaporation rate of water is gas phase controlled. However, for most volatile compounds,

---

*In this report, a capital letter K refers to the overall mass transfer coefficient and a lower case k refers to the individual phase mass transfer coefficients.

| Reference | Forced Emissions | Interface Problems[a]: Soil - Air | Liquid - Soil | Liquid - Air | Other Interface |
|---|---|---|---|---|---|
| 1. Mackay & Leinonen ES&T'75 | | | | x | |
| 2. Thibodeaux Parker & Heck AIChE'81 | | | | x | |
| 3. Smith, Bomberger & Haynes EST'80 | | | | x | Liquid - Surfactant - Air |
| 4. Smith, Bomberger & Haynes, Chemosphere'81 | | | | x | |
| 5. Shen, JAPCA, '82 | | | | x | |
| 6. Hwang Env. Prog.'82 | | x | x | x | |
| 7. Thibodeaux, Hwang Env. Prog.'82 | | x | x | | |
| 8. McCord Haz. Waste Conf.'81 | | | | x | |
| 9. Hartley Pesticide, '69 | | | | x | |
| 10. Shen, APCA'80 | | x | | | |
| 11. Steenhuis, Bubenger & Converse, ASAE'79 | | x | | | |
| 12. Thibodeaux, J. Haz. Mat.'81 | | x | | | |
| 13. Farmer, Yang, Letey, 1978, 1980 | | x | | | |

Figure 3. Interface problems investigated by respective researchers for treatment and disposal AERR models. (A complete list of references is provided in Appendix A.)

| Reference | Forced Emissions | Interface Problems[a] | | | |
|---|---|---|---|---|---|
| | | Soil – Air | Liquid – Soil | Liquid – Air | Other Interface |
| 14. Thibodeaux, Springer & Riley, 1981 | x | x | | | |
| 15. Liss & Slater, Nature'74 | | | | x | |
| 16. Mackay, Haz. Assess. of Chem.'81 | | | | x | |
| 17. Owens, Edwards, Gibbs, Int. J. Air, Water Poll.'64 | | | | x | |
| 18. Neely, Dow Chemical, 1976 | | | | x | |
| 19. Dilling, Tefertiller, Kallos EST'75 | | | | x | |
| 20. Dilling EST'77 | | | | x | |
| 21. Mackay & Yuen, Water Pol Journal Canada'80 | | | | x | |
| 22. Arnold Trans of Am. Inst. Chem., 1944 | | | | | x |
| 23. Cohen, Cocchio, Mackay EST'78 | | | x | | |
| 24. Mackay, Matsugu Canadian Journal'73 | | | | | x |
| 25. Mackay, Wolkoff, '73 | | | | x | |
| 26. Thibodeaux, Springer Hedden & Lunney, 1982 | | | | x | |

[a]x denotes the authors interpretation of which phase(s) control the overall transfer rate.

Figure 3 (continued)

especially low solubility organics, the liquid phase controls the overall diffusion rate. As another example, Hartley's (1969) land disposal model is actually most appropriate for lagoons or compounds spread on an impermeable surface. Since the Hartley model does not address diffusion in soil, but only in air, it is not appropriate when wastes are plowed or percolate into the ground.

The principle requirement in selecting the most appropriate k-value (and model) is that it embody the correct mass transfer principles. This does not necessarily assure accuracy, but it does provide a good foundation for model refinements. The mass transfer coefficient in air, for example, is determined by the vertical wind speed profile and turbulence structure in the earth's boundary layer. All references examined which recognize this fact follow the original formulation by Sutton (1953). Diffusion in air-filled spaces within the ground, however, occurs on a molecular level, thus the soil k-value depends primarily on the soil porosity and geometry. The problem is thus one of describing the available "space" within which diffusion occurs. Finally, at the liquid side of an interface, k-value determinations have been obtained mainly in the laboratory. Although laboratory-developed correlations describing k-values are available for both aerated and unaerated surfaces, no extensive data presently exists that places adequate confidence on extrapolating these correlations to field conditions.

The next section summarizes GCA's preliminary model selections for each TSDF category. Later sections provide a more complete literature review describing each available model in more detail.

# 2. Summary and Conclusions

INTRODUCTION

The mechanisms causing air emissions of volatile compounds from TSDFs were briefly described in the introduction. The basis for quantifying air emission release rates (AERR) begins with an accurate description of diffusion across different regions in a TSDF; i.e., the diffusive transport of air emissions across soil, liquid and/or air layers.

Diffusive transport can be quantified by determining the mass transfer coefficient ("k-value") for each region encountered by the potential air emission. Selecting the most appropriate AERR model for TSDFs reduces to a problem of selecting the most accurate and manageable k-value. Once the best technique for determining k-values is established for each type of diffusive transport occurring, the model that most accurately describes the physical situation at a TSDF can be selected.

This section summarizes techniques available in the literature for determining individual k-values*, and presents GCA's recommendations for AERR models based upon our selected k-values. When applicable, recommended model approaches are presented. Later sections of the report discuss models in more detail.

It should be noted that the majority of the k-value correlations presented in this report have not undergone substantial field test validation or verification. For the most part, k-values were developed from laboratory experimentation and some limited field studies. Therefore, caution is advised when applying these correlations to calculate expected emission release rates. Although limited field validation has been performed recently, the accuracy of all emission release rate models is still unknown.

---

*In this report, a lower-case letter k refers to the individual mass transfer coefficients for each region; an upper-case K refers to the overall mass transfer coefficient obtained from the individual k-values.

168

AVAILABILITY AND SELECTION OF MASS TRANSFER COEFFICIENTS (k-VALUES)

Several researchers have developed empirical relationships to describe the mass transfer of contaminants from <u>individual</u> <u>regions</u> or phases, i.e., through soil, liquid and air. These mass transfer coefficients can be categorized according to their applicability to TSDFs as follows:

- <u>Soil ($k_{soil}$)</u>--describes molecular diffusion in porous media; i.e., landfarms and landfills;

- <u>Liquid ($k_{liquid}$)</u>--(a) molecular mass transfer (diffusion) within a stagnant liquid; (b) turbulent diffusion from the surface of a liquid (impoundment) affected by wind; and (c) turbulent diffusion enhanced by mechanical aeration;

- <u>Gas ($k_{gas}$)</u>--(a) turbulent diffusion influenced by wind, and (b) turbulent diffusion enhanced by mechanical aeration.

Table 1 presents the individual mass transfer coefficients available from the literature reviewed by GCA for each region. When more than one technique for determining the k-value was available in the literature, GCA's recommendation of the most appropriate k-value is noted with an asterisk (*). Only one technique was available for quantifying the individual liquid and gas phase k-values for agitated conditions. Likewise, simplification techniques found in the literature are not shown in Table 1, since these techniques provide methods to simplify the calculational effort involved, and do not represent different approaches to determining k-values. The simplification techniques are presented later in this report.

The following discussions review in detail all available k-value relationships shown in Table 1, and describe the theoretical basis used by GCA in selecting the most appropriate equation for AERR estimates.

## Soil-Phase Mass Transfer Coefficient ($k_{soil}$, $k_s$)

Vapor diffusion through air filled pore spaces in soil is a complex phenomenon which is not amenable to a purely theoretical treatment. To obtain an effective diffusion coefficient for a particular soil at a known moisture content, the problem requires that empirical correlations for vapor diffusion in porous media be evaluated to see if they adequately reflect the many experimental observations which are available. GCA has reviewed these correlations for simple systems such as dry sands and for some more complex systems such as platey minerals and soils of known moisture content. With presently available data, it should be possible to provide first order estimates of effective diffusion coefficients. GCA's evaluation of these data is summarized below.

In order to model AERR from landfills and landfarms, vapor movement within the soil must be accurately defined. Farmer et al. (1978) and Thibodeaux (1981a) have both defined the "effective" diffusion coefficient within soil as a function of the volatile components' molecular diffusivity and the physical characteristics of the respective soil.

TABLE 1. MASS TRANSFER COEFFICIENTS (k's) AVAILABLE IN THE LITERATURE APPLICABLE TO MODELING TSDF AIR EMISSION RATES (SEE TEXT FOR TERM DEFINITIONS)

| Transfer Phase | Transfer Coefficient (k) | Author (date) | Comments |
|---|---|---|---|
| 1) Soil ($k_{soil}$, $k_s$) | $D_o = \dfrac{\epsilon_a^{10/3}/\epsilon_T}{h}$ | Farmer (1978) | Based on work by Millington and Quirk describing the effects of soil water content on vapor movement. |
| | $\dfrac{D_o \epsilon}{\lambda h}$ | Thibodeaux (1981a) | Based on the theory of molecular movement in porous media but defines only dry-soil phase. See text. |
| | More general $\epsilon$ power law models | Currie (1960, 1961)[a] and others | Based on Fick's first law; of limited usefulness in TSDF problems. |
| 2) Liquid ($k_{liquid}$, $k_L$) -- a) Molecular Diffusion-- Quiescent State | $\dfrac{D}{h}$ | - - - - | |
| b) Turbulent Diffusion-- Wind Effects | $(11.4 R_e^{*0.195} - 5.0)$<br>$\sim D_L^{0.5}$; from $k_l$ vs wind speed | Cohen, et al. (1978) | Based on laboratory wind wave tank studies for hydrocarbon volatilization from aqueous solutions. Lab studies may not fully represent actual environmental conditions. |
| | $3.12 (1.024)^{\theta} - 20\, U_o^{0.67} H_o^{-0.85} \left(\dfrac{D_i:H_2O}{D_{O_2}:H_2O}\right)$ | Mackay (1981)[a] | Based on solution of Fick's second law and critical review of existing data (postulating reasonable values for $k_L$ as function of wind speed). |
| c) Turbulent Diffusion-- Aeration Effects | $\dfrac{J(POWR)\,(1.024)^{\theta} - 20\,(\alpha)\,(10^6)}{165.04(a_v)(V)}$ | Hwang (1982) from Owens, et al. (1964)[a] | Developed from theoretical, experimental and field measurements of oxygen reaeration rates from flowing streams. |
| | $20 \left(\dfrac{D_i:H_2O}{D_{O_2}:H_2O}\right)^{0.5}$ | Thibodeaux (1978)[a] | Developed from research on the absorption of oxygen into water during agitation. |
| 3) Gas ($k_{gas}$, $k_G$) a) Turbulent Diffusion-- Wind Effects | $0.0292\, U_{air}^{0.78}\, x^{-0.11} Sc^{-0.67}$ | Mackay and Matsugu (1973)[a] | Based on work of Sutton, and field measurements on the evaporation organic compounds and water into air. |
| | $0.03 \left(\dfrac{D_{i,air}}{D_{H_2O,air}}\right) V_{8A}^{-0.05}$ | Thibodeaux and Parker (1974) from Harbeck (1963) | Based on field measurements from the evaporation of water into the air. |
| b) Turbulent Diffusion-- Aeration Effects | $0.00039\, \dfrac{\rho_g D_{i,air}}{d} (N_{Re})^{1.42} (N_{FR})^{-0.21} (N_p)^{0.4} (N_{Sc})^{0.5}$ | Reinhardt (1977), as presented by Hwang (1982)[a] | Based on preliminary experimental investigations for liquid spherical droplets ejected into a gas phase. |

[a]Denotes GCA's recommendation.

Pore diffusion, as related to the soil matrix, occurs by the Fick diffusion mechanism. Within the soil matrix, the effective area for transport is less than that for uniform pore structures because the diffusion paths are irregularly shaped channels. Increases in soil moisture will reduce the air-filled porosity and at the same time increase the diffusion path length. Therefore, the mass flux for soil-phase diffusion needs to be described in terms of an "effective" diffusion coefficient which is less than the molecular diffusion coefficient and dependent upon variables influencing the diffusing vapor; i.e., moisture content.

The effective diffusion coefficient presented by Farmer (1978) was formulated from theoretical soil hydraulic conductivity studies of Millington and Quirk (1961). Based on a theoretical derivation, Millington and Quirk presented the following correlation for the effective vapor diffusion coefficient:

$$\frac{D_e}{D_o} = \frac{\varepsilon_a^{10/3}}{\varepsilon_T^2}$$

where: $D_e$ = effective diffusion coefficient in soil;

   $D_o$ = air diffusion coefficient;

   $\varepsilon_a$ = air-filled porosity;

   $\varepsilon_T$ = total porosity.

For dry soil, when $\varepsilon_a = \varepsilon_T$, this expression reduces to the following:

$$\frac{D_e}{D_o} = \varepsilon_T^{1.33}$$

However, an additional literature review of vapor diffusion in porous media conducted by GCA indicates that the Millington and Quirk correlation is limited to a specific soil type. Further, the Millington and Quirk analysis cannot be of general applicability to the diffusion problem when pores of different sizes are present. In diffusion, flux is proportional to pore radius squared. In hydraulic flow, flux is proportional to pore radius to the fourth power.

In one particularly relevant study, Currie (1960) conducted hydrogen diffusion experiments with 15 different types of dry porous media and concluded that the primary influences on the $D_e/D_o$ ratio were porosity, and particle shape. Differences in particle shape affect the diffusion path. Therefore, the effective diffusion coefficient in porous media is a function of internal geometry and porosity.

The empirical equation proposed by Currie was of the form:

$$\frac{D_e}{D_o} = \gamma \varepsilon^{\mu}$$

where $\gamma$ and $\mu$ are constants for a specific type of porous material. Figure 4 is a log-log plot of $D_e/D_o$ versus $\epsilon$ developed by Currie. Figure 4 shows that the set of data points for a given material lie on a straight line, but the slope of the line (value of $\mu$) varies with each type of material. The following values of $\mu$ were determined from the data on dry soils:

| Soil Type | $\mu$ |
|---|---|
| Solid grains (i.e., sand) | 1.4 |
| Soil crumbs | 1.7 |
| Kaolin (a type of clay) | 2.6 |
| Platey minerals (mica, vermiculite) | 11.0 |

The value of $\gamma$ has been found to range from 0.8 to 1.0 for the materials tested. Thus, an equation with two empirical constants ($\gamma$ and $\mu$) is available to describe the affect of soil porosity ($\epsilon$) on the effective diffusion coefficient ($D_e$) for vapors in dry soil.

Limited data are available for describing the effect of moisture content on the effective diffusion coefficient in wet soils. Currie (1961) examined the effect of moisture on some granular materials, and his data fit the following empirical equation:

$$\frac{D_e}{D_o} = \gamma \ (\epsilon_T)^{\mu - \sigma} \ (\epsilon_a)^{\sigma}$$

where: $\epsilon_T$ = total porosity;

$\epsilon_a$ = air-filled porosity;

$\sigma \cong 4$ for granular materials subjected to moisture testing.

This correlation is obtained by combining Currie's dry soil and wet soil correlations. The complete derivation is provided in Appendix B.

Currie's empirical formula relating $D_e$ to soil parameters accounts for experimental observations for all soils tested. The formula allows one to define $D_e$ at any specified moisture content provided that the required parameters can be experimentally evaluated. Taylor (1949), Rolston and Brown (1976), Lai et al. (1976), and Ball et al. (1981), also present data for wetted granular soils. Unfortunately clay materials (clayey soils) of the sort expected to be employed for landfill caps have not been tested.

Although it appears that Currie's work provides a solid empirical base and adequate general theoretical interpretation for the effect of particle shape and moisture content on $D_e$, there are questions concerning the direct application of Currie's results to estimation of volatile organic emissions from landfills. The unsteady state hydrogen diffusion experimentations may

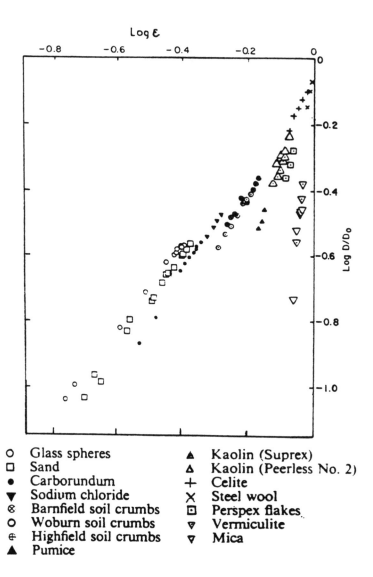

| O | Glass spheres | ▲ | Kaolin (Suprex) |
|---|---|---|---|
| □ | Sand | △ | Kaolin (Peerless No. 2) |
| • | Carborundum | + | Celite |
| ▼ | Sodium chloride | × | Steel wool |
| ⊗ | Barnfield soil crumbs | ⊡ | Perspex flakes |
| ⊙ | Woburn soil crumbs | ▽ | Vermiculite |
| ⊕ | Highfield soil crumbs | ▽ | Mica |
| ▲ | Pumice | | |

Figure 4.    Dependence of coefficient of diffusion on porosity
log $(D/D_0)$ versus log $\varepsilon$ (from Currie (1960)).

not fully represent a hazardous waste landfill situation since hydrogen is a small, non-polar molecular structure compared to organics which are more bulky, and in some cases are polar. In addition, organic vapors are likely to sorb to the organic material present in most soils.

However, data on vapor diffusion in porous media presented in the literature indicates that determination of effective diffusion coefficient is soil-type dependent. At present, GCA suggests that field validation efforts include an analysis of landfill cap material (i.e., soil type, moisture) such that $k_{soil}$ may be properly quantified. Additional laboratory experimentation for wet soils, especially clays, may provide an improved data base.

## Liquid-Phase Mass Transfer Coefficient ($k_{liquid}$, $k_L$)

The transfer of a chemical compound within a liquid medium was identified in Table 1 for three specific cases; i.e., (1) molecular diffusion within a stagnant liquid with no wind and no flow effects; (2) diffusion within the liquid-phase enhanced by the turbulence caused by wind; and (3) diffusion within the liquid-phase enhanced by the turbulence created by mechanical aeration. Figure 5 shows the relationship of the six liquid-phase mass transfer coefficient correlations presented in the literature as a function of wind speed. Liquid phase k-values applicable to TSDFs are highlighted.

Molecular Diffusion--$k_L$ for No Wind Conditions--
The first case of liquid diffusion has limited applicability to TSDFs, but is provided for illustrative purposes in Figure 5. This mass transfer coefficient becomes necessary for the case of open tank storage of a multicomponent waste, where wind effects are eliminated by a large freeboard.

The situation of no wind is not too unusual for open areas during evening hours, typically when water temperatures are higher then the surrounding air. Most mass transfer correlations reviewed incorporate wind speed as a parameter, however, it is likely that emissions release will continue under low or no wind conditions due to turbulence created by thermal agitation. Thibodeaux, et al. (1982b) present a method for determining the mass transfer coefficients in both phases (liquid and gas) by use of the Chilton-Colburn analogy. This analogy permits the evaluation of the mass transfer coefficient through the information obtained from the heat transfer phenomenon.

Figure 5 shows that the mass transfer coefficient calculated from this analogy is greater than that predicted by correlations using wind as a parameter, specifically for the case of low wind speeds (i.e., less than 3 meters per second).

Turbulent Diffusion from Wind and Flow Effects--
In the case of turbulent diffusion, where turbulence is influenced by variations in wind speed, several researchers have conducted experiments in order to develop empirical expressions for determining the liquid-phase mass transfer coefficient. The results of these correlations are shown in Table 2.

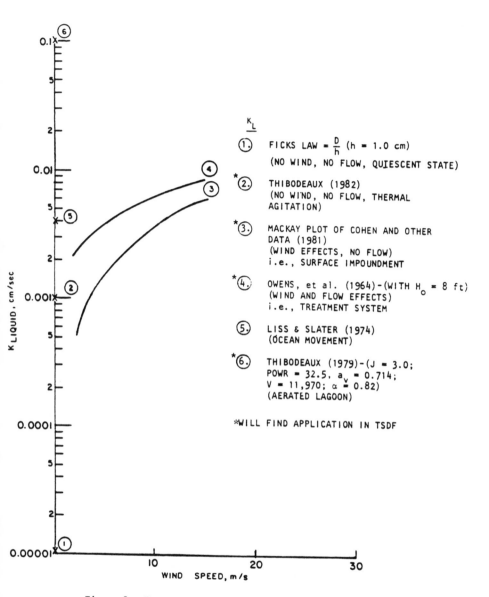

Figure 5. $K_L$ correlations versus wind speed for benzene at 25°C.

TABLE 2.  LIQUID-PHASE MASS TRANSFER COEFFICIENT CORRELATIONS FOR WIND AND FLOW TURBULENCE

[1]  $k_L = (11.4\ Re^{*^{0.195}} - 5)$                    (Cohen, Cocchio, Mackay, 1978)

for 0.11 $\leq Re^* \leq 102$, where $k_L$ is in cm/hr and

$$Re^* = \frac{7.07 \times 10^{-3}(Z_{10})(U_{10})^{1.25}}{\nu_a \exp(56.6/U_{10}^{0.25})}$$

If $Re^* < 0.11$, then $k_L = 2.4$ cm/hr

where:

$Re^*$ = roughness Reynolds number (dimensionless)

$U_{10}$ = wind velocity (cm/s) measured at height $Z_{10}$ above the water surface (cm)

$\nu_a$ = air kinematic viscosity (cm²/sec)

[2]  $k_L = (1.3\ Re^{*^{0.195}} - 0.57)\left(\dfrac{D_{i,H_2O}}{D_{TOL,H_2O}}\right)$                    (Hwang, 1982)

where:

$k_L$ is in lb-mol/ft²-hr

$Re^*$ = roughness Reynolds number (determined as above)

$D_{i,H_2O}$ = diffusion coefficient of compound i in water (cm²/sec)

$D_{TOL,H_2O}$ = diffusion coefficient of toluene in water (cm²/sec)

(continued)

TABLE 2 (continued)

[3a]    $k_L = 27.5 (1.024)^{\xi-20} U^{0.67} H^{-0.85}$    (Owens, Edwards, Gibbs, 1964)

where:

$k_L$ is in cm/hr

$\xi$ = temperature, °C

U = mean velocity of stream (ft/sec)

H = mean depth of stream (ft)

(valid for the following experimentally observed ranges: U = 0.1-5.0; H = 0.4-11.0)

[3b]    $k_L = 29.5 (1.024)^{\theta-20} U^{0.73} H^{-0.75}$    (Owens, Edwards, Gibbs, 1964)

(for fast flowing or shallow streams where U = 0.1-1.8; and H = 0.4-2.4)

[4]    $k_L = 3.12(1.024)^{\theta-20} U_o^{0.67} H_o^{-0.85} \left( \dfrac{D_{i,H_2O}}{D_{O_2,H_2O}} \right)$    (Hwang, 1982)

where:

$k_L$ is in lb-mol/ft$^2$-hr

$\theta$ = temperature (°C)

$U_o$ = surface velocity, ft/sec, normally 0.035 x wind speed (ft/sec) for natural surface, and 0.1 ft/sec for outside region of effect of aerators in biological treatment.

$H_o$ = one-half the effective depth of surface impoundment (ft)

$D_{O_2,H_2O}$ = diffusion coefficient of oxygen in water (cm$^2$/sec)

$D_{i,H_2O}$ = diffusion coefficient of compound i in water (cm$^2$/sec)

(continued)

## TABLE 2 (continued)

[5]    $k_L \alpha D^{0.5}$                                    (Mackay, 1981)

from literature data:

| Wind speed (m/s) | $k_L$ (m/sec) - for Benzene @ 20°C | |
|---|---|---|
| | Environment | Tank |
| 2 | $5 \times 10^{-6}$ | $10 \times 10^{-6}$ |
| 5 | $15 \times 10^{-6}$ | $25 \times 10^{-6}$ |
| 10 | $35 \times 10^{-6}$ | $65 \times 10^{-6}$ |
| 15 | $60 \times 10^{-6}$ | $100 \times 10^{-6}$ |

The first correlation presented in Table 2 was developed from laboratory wind wave tank studies by Cohen, et al. (1978). Cohen et al. developed their equation for benzene and toluene and indicated $k_L$ dependence on wind speed for velocities between 3 and 10 m/s (6.7 to 22.4 mph). Below 3 m/s, $k_L$ was influenced by subsurface agitation, while at wind speeds above 10 m/s, $k_L$ increased was due to the presence of spray, bubble entrainment and white capping. Cohen expressed concern over the direct application of this correlation to environmental conditions because laboratory tests cannot accurately predict conditions of fully developed flow. That is, the width of Cohen's experimental unit may not simulate a larger surface impoundment at a TSDF.

In applying this correlation to different chemical species, one can refer to three mass transfer theories stated in the literature:

● Stagnant film theory;

● Penetration or surface renewal theories; and

● Boundary layer theory.

The first case identifies $k_L$ as a linear function of diffusivity. However, other theories of mass transfer, namely, those mentioned above in the second case, identify this dependency to the 0.5 power. The third theory is generally considered a compromised approach and identifies the diffusivity dependency to the two-thirds power. The exact dependence of $k_L$ on diffusivity remains to be established.

The second equation shown in Table 2 is a reworking of Cohen's correlation presented by Hwang (1982). The remodeled correlation simply expresses $k_L$ in different units (lb-mol/ft$^2$-hr). Hwang also suggests a linear dependency to diffusivity.

Equations (3a) and (3b) presented in Table 2 are based on reaeration stream studies discussed by Owens et al. (1964). Unlike the wind velocity dependency analyzed by Cohen, Owens correlations show $k_L$ dependent on stream velocity and water depth.

Although Owens expression suggests hydraulic effects influence $k_L$, which is in agreement with theoretical models, the direct application of stream coefficients to surface impoundments is questionable since the degree of turbulence offered by a flowing stream is much greater than wind induced turbulence. Thus, Hwang (1982) provides a modification to Owen's correlation to define the surface velocity of water as a function of wind speed (Equation (4) in Table 2). This modification is a more accurate representation of surface impoundment or treatment basin surface actions. Both correlations also show that $k_L$ is dependent upon water depth for a flowing system. Note that Hwang's modification also reflects a change in $k_L$ units (lb-mol/ft$^2$-hr).

Mackay (1981) critically reviewed the existing data base for $k_L$ determinations, which included laboratory tank experiments and field measurements in lakes. Mackay's data base included the laboratory studies of Cohen, et al. (1978), Liss and Slater (1974), Kanwisher (1963), and Deacon (1977). Field studies of Broecker and Peng (1974), Emerson (1975) and Schwartzenbach, et al. (1974) were also mentioned. Mackay presented what he considered as reasonable values of $k_L$ for environmental and laboratory conditions as a function of wind speed (correlation (5) in Table 2). The data presented in Table 2 are for benzene at 20°C. As stated by Mackay, it appears the best correlating approach for estimating $k_L$ for different chemical species under similar conditions is to use the square root of the diffusivity ratio. Mackay also mentions that there are theoretical reasons suggesting that the ratio of Schmidt numbers to the power -0.5 is the best correcting ratio for both solute diffusivity and temperature. However, he grants no judgment on this statement.

The turbulent $k_L$ correlation selected should be based on the degree to which water is flowing through the system. For nonflow systems, i.e., wind induced flow, $k_L$ as presented by Mackay (1981) appears most appropriate since he compiled data from several researchers. Selecting the Mackay data, or rather the plot of Mackay's data, avoids the calculation of a roughness Reynolds number.

For flowing systems, the correlation presented by Hwang (1982) based on the work of Owens (1964) seems most appropriate. Whereas, Owens turbulent $k_L$ is dependent upon the stream velocity, Hwang's approach considers the effect of the water's surface velocity, which is a function of wind speed. For large impoundments of water where the mean water velocity may be quite low, it is the upper surface layer of water directly affected by wind that will affect $k_L$ the most.

It can be concluded from data presented in the literature, and from theories of mass transfer, that the liquid phase mass transfer coefficient depends somewhat on diffusivity. However, the magnitude of this dependency is not precisely defined by existing research. The mass-transfer coefficient will be proportional to a power of the molecular diffusivity between 0.5 to 1.0, where the lower number represents conditions of high turbulence and the higher number represents near stagnant conditions.

Springer (1983) noted that one aspect of ongoing work at the University of Arkansas involves developing a correlation which employs a dimensionless ratio of fetch to depth. The correlation will be shown to fit data from earlier stream reaeration studies as well as recent experimental work, including the "reasonable values" presented by Mackay (1981).

Turbulent Diffusion from Mechanical Mixing--
The third case of liquid phase mass transfer involves turbulent diffusion created by mechanical aeration. Considerable research has been conducted on scale models and prototype mechanically agitated water surfaces to determine the absorption rate of oxygen. Since oxygen absorption is liquid-phase

controlling, the reported values can be transformed to yield a liquid-phase mass transfer coefficient under agitated conditions. The following relationship is the only correlation appearing within the literature:

$$k_L = \frac{J(POWR)(1.024)^{\theta-20}(\alpha)(10^6)}{165.04(a_v)(V)} \left( \frac{D_{i,H_2O}}{D_{O_2,H_2O}} \right)^{0.5}$$    (Thibodeaux, 1978)

where: $k_L$ is in lb-mol/ft$^2$-hr;

        $J$ = oxygen-transfer rating of mechanical aerator (3 lb $O_2$/hr-hp);

    POWR = total power input to aerators (hp)-rated hp x efficiency (0.65-0.9);

        $\theta$ = temperature (°C);

        $\alpha$ = oxygen-transfer correction factor (0.8-0.85);

      $a_v$ = surface area per unit volume of surface impoundment (ft$^{-1}$);

      $V$ = volume of surface impoundment in region of effect of aerators (ft$^3$);

$D_{i,H_2O}$ = diffusion coefficient of compound i in water (cm$^2$/sec);

$D_{O_2,H_2O}$ = diffusion coefficent of oxygen in water (cm$^2$/sec).

## Gas-Phase Mass Transfer Coefficients ($k_{gas}$, $k_G$)

The final mass transfer region that all chemical compounds must encounter is diffusion within the gas phase. As illustrated in Table 1, two cases of gas-phase mass transfer must be considered; i.e., (1) turbulent diffusion influenced by wind speed, and (2) turbulent diffusion created by mechanical aeration.

Turbulent Diffusion from Wind Effects--
A review of the literature revealed two correlations developed from experimental field measurements for determining the influence of wind speed on the gas-phase k-value. These relationships appear in Table 3.

The correlation proposed by Mackay and Matsugu (1973) was developed from experiments on the evaporation of isopropyl benzene (cumene), gasoline and water into air. Their research showed that the work of Sutton (1953), who assumed that the wind velocity profile follows a power law, could be used to quantify the rate of evaporation from a smooth liquid surface, and subsequently obtain an expression for the gas-phase mass transfer coefficient. The final result of Mackay and Matsugu's research provided the correlation shown in Table 3 as a function of wind speed and the effective diameter of the liquid surface.

TABLE 3.   GAS PHASE MASS TRANSFER COEFFICIENT CORRELATIONS

[1]   $k_G = 0.0292 \ U^{0.78} \ X^{-0.11} \ Sc^{-0.67}$   (Mackay, Matsugu, 1973)

where:   $k_G$ is in m/hr

   U = wind speed, m/hr

   X = effective pool diameter, m

   $S_c$ = gas Schmidt number (dimensionless), calculated from:

$$S_c = \mu_g / \rho_g \ D_{i,air}$$

   where $\mu_g$ is the absolute viscosity of the gas (g/cm sec), $\rho_g$ is the density of the gas (g/cm³), and $D_{i,air}$ is the molecular diffusivity of compound i in air (cm²/sec).

[2]   $k_G = 0.03 \left( \dfrac{D_{i,air}}{D_{H_2O,air}} \right) V_8 \ A^{-0.05}$   (Harbeck, 1962; as presented by Thibodeaux and Parker, 1974)

where:   $k_G$ = individual gas phase mass transfer coefficient (lb-mol/ft²-hr)

   $D_{i,air}$ = molecular diffusivity of compound i in air (cm²/sec)

   $D_{H_2O,air}$ = molecular diffusivity of water vapor in air (cm²/sec)

   $V_8$ = wind velocity measured at 8m above the water surface (miles per hour)

   A = surface area of the impoundment (acres)

Hwang (1982) modified the Mackay/Matsugu correlation to reflect a change in $k_G$ units (lb-mol/ft$^2$-hr). The modification alters the equation constant to 0.0958 and includes multiplying by the ratio of air's density to its molecular weight.

The results provided by Harbeck (1962), shown as Equation (2) in Table 3, were developed entirely on water evaporation measurements from reservoirs. Theoretically, since water evaporation is gas-phase controlling, these measurements can provide a measure of the gas-phase mass transfer coefficient. However, the results obtained by Mackay and Matsugu not only describe the evaporation of water, but also provide good correlation to the work of Sutton. Therefore, the selected k-value for the gas-phase (influenced by wind) is that shown by Mackay and Matsugu.

Turbulent Diffusion from Mechanical Mixing--
    The literature provided limited information to describe the turbulent diffusion of a chemical within the gas-phase as a function of mechanical aeration. However, through limited experimental observations and additional investigations by Reinhardt (1977), an empirical expression has been developed to approximate the gas-phase mass transfer coefficient under aerated conditions. This relationship appears below:

$$k_G = 0.00039 \frac{\rho_g D_{i,air}}{d} (N_{Re})^{1.42} (N_{Fr})^{-0.21} (N_P)^{0.4} (N_{Sc})^{0.5}$$

(Reinhardt, 1977, as -given by Hwang, 1982)

where:

$k_G$ is in lb/ft$^2$-hr;

$\rho_g$ = density of the gas (lb/ft$^3$);

$D_{i,air}$ = diffusion coefficient of compound i in air (ft$^2$/hr);

d = diameter of aerator turbine or impeller (ft);

$(N_{Re})$ = gas Reynolds number = $\rho_g d^2 \omega / \mu_g$;

$\omega$ = rotational speed of turbine impeller (rad/sec);

$\mu_g$ = absolute gas viscosity (g/cm-sec);

$(N_{Fr})$ = Froude number = $d^2 \omega / g$;

g = gravitational constant (32.17 ft/sec$^2$);

$(N_P)$ = power number = $Prg / \rho_L d^5 \omega^3$;

Pr = power to the impeller (ft-lb$_f$/sec);

$\rho_L$ = density of the liquid $(lb/ft^3)$;

$(N_{Sc})$ = gas Schmidt number = $\mu_g/\rho_g D_{i,air}$;

Note:  $(N_{Sh})$ = Sherwood number = $(k_G)d/\rho_g D_{i,air}$ and therefore;

$(N_{Sh})$ = 0.00039 $(N_{Re})^{1.42}(N_P)^{0.4}(N_{Sc})^{0.5}/(N_{Fr})^{0.21}$

## Overall Mass Transfer Coefficient (K-Value)

As described previously, correlations for determining the individual mass transfer coefficients for the soil, liquid and/or gas-phases have been developed by several researchers. However, as shown previously in Figure 2, there are seldom instances where only one phase occurs for a particular TSDF. Commonly, the mass transfer coefficients for two or more phases must be calculated to determine the overall mass transfer coefficient (K-value).

The two-phase resistance theory is used to describe transfer within multiphase systems. The theory describes the relationship between an overall mass transfer coefficient (K-value) and the individual soil, liquid and/or gas phase mass transfer coefficients (k-values) identified previously in this section. Specifically, the reciprocals of the individual mass transfer coefficients are summed and combined with the equilibrium concentration (described by the Henry's Law constant--$H_i$) established at either the liquid or gas interphases, thus describing the overall resistance (1/K) for the multiphase case. For example, the overall diffusion of a given contaminant from a lagoon requires the mass transfer of that contaminant through the bulk of the liquid ($k_{liquid}$) and through the air above the liquid ($k_{gas}$). In terms of the two-phase resistance theory, this multiphase phenomenon can be described as follows:

$$\frac{1}{K} = \frac{1}{k_{liquid}} + \frac{RT}{k_{gas} H_i}\left(\frac{C_{gas}}{C_{liquid}}\right) \qquad \text{(Hwang, 1982)}$$

where:   K   = overall liquid-phase mass transfer coefficient;

$k_{liquid}$ = individual liquid-phase mass transfer coefficient;

$k_{gas}$ = individual gas-phase mass transfer coefficient;

$H_i$   = Henry's Law constant for compound i in atm-m$^3$/gmol;

RT   = 0.024 atm-m$^3$/gmol (@ 298°K);

$C_{gas}$, $C_{liquid}$ = molar densities of the gas and liquid, respectively.

However, one phase frequently dominates the overall mass transfer (or resistance, 1/K) and therefore controls the rate of mass transfer. The determining factor for this occurrence has been theoretically and experimentally found to be dependent upon the value of $H_i$. For example, if $H_i$ is large (more than $10^{-3}$), the liquid-phase resistance (1/$k_{liquid}$) often controls, whereas if $H_i$ is small (less than 2 x $10^{-5}$), the gas-phase resistance (1/$k_{gas}$) often controls.

## Special Case for Overall K-Value

For the case of agitated conditions (specifically aerated impoundments), two distinct zones occur at the liquid surface: (1) turbulent, and (2) natural (convective). To describe this phenomenon using the two-phase resistance theory, it is necessary to determine the overall mass transfer coefficient for each distinct zone and sum these values, proportioned to the affected area of each zone. This can be described as follows:

$$\frac{1}{K_L^{(n)}} = \frac{1}{k_{liquid}^{(n)}} + \frac{RT}{(k_{gas}^{(n)})(H_i)} \left( \frac{C_{gas}}{C_{liquid}} \right)$$

and

$$\frac{1}{K_L^{(t)}} = \frac{1}{k_{liquid}^{(t)}} + \frac{RT}{(k_{gas}^{(t)})(H_i)} \left( \frac{C_{gas}}{C_{liquid}} \right)$$

where:

$$K_L = (K_L^{(n)}) \frac{A_n}{A} + (K_L^{(t)}) \frac{A_t}{A}$$

where: $K_L^{(n)}$ = overall liquid-phase mass transfer coefficient for the natural zone;

$K_L^{(t)}$ = overall liquid-phase mass transfer coefficient for the turbulent zone;

$K_L$ = area-averaged overall liquid-phase mass transfer coefficient;

$A_n$ = effective surface area of the natural zone;

$A_t$ = effective surface area of the turbulent zone;

$A$ = total surface area of the surface impoundment.

In order to calculate the overall mass transfer coefficient, the individual k-values for liquid and gas phases must be in similar units. Hwang (1982) has provided the following conversion method:

$$k_L \left( \frac{lb\text{-}mol}{ft^2\text{-}hr} \right) = k_L \left( \frac{cm}{hr} \right) \frac{\rho_m}{MW_m} \times \frac{(30.48)^2}{454}$$

$$k_g \left( \frac{lb\text{-}mol}{ft^2\text{-}hr} \right) = k_g \left( \frac{m}{hr} \right) \frac{\rho_{air}}{MW_{air}} \quad (3.28)$$

where:  $\rho_m$  = density of mixture (g/cm$^3$) (= 1 for water);

   $MW_m$  = molecular weight of mixture (= 18 for water);

   $\rho_{air}$  = density of air (lb/ft$^3$);

   $MW_{air}$  = molecular weight of air.

## SELECTION OF MOST APPROPRIATE AERR MODELS

Once the appropriate mass transfer coefficients are selected, the model which most accurately reflects the actual emission release phenomena at TSDFs can be selected. The following paragraphs identify models selected for each TSDF-type, based on the previous discussion of k-value selection. A more detailed discussion of each model appears later in this report.

### Surface Impoundment (SI) Model Selection

SI AERR models are required for four typical scenarios encountered in the field; i.e., steady-state conditions with and without mechanical aeration, and unsteady-state conditions with and without mechanical aeration. AERR models for the cases of mechanical and diffused aeration with biological activity are treated separately in Section 8. Table 4 summarizes the available models found in the literature and indicates the scenario(s) for which each model is applicable. GCA's recommended models are footnoted.

Several SI models shown in Table 4 were eliminated from consideration because they did not completely describe the theoretical conditions encountered in the field. Specifically, McCord (1981), Smith, et al (1980 and 1981), and Mackay and Wolkoff (1973), were dismissed from further consideration, as described in Table 4.

Shen (1982) modified the original Thibodeaux, Parker and Heck (1981d) nonaerated, steady-state model by incorporating; (1) the $k_L$ relationship presented by Owens, et al (1964) which is more applicable to flowing streams (not stagnant surface impoundments), and (2) the most appropriate $k_G$ relationship (in GCA's opinion as described earlier). Shen also suggests a simplification technique for calculating the Schmidt number based upon the compound's molecular weight. This simplification technique is described further in Section 4 of the report.

The Thibodeaux, Parker and Heck (1981d) approach appears most appropriate for estimating AERR from nonaerated SIs under steady-state conditions i.e., fairly continuous inflow of contaminants. For unsteady state nonaerated SIs, where the contaminant is discharged to the SI as a slug, or pulse injection, the Mackay and Leinonen (1975) approach appears most appropriate. Section 4 of this report elaborates on each SI model summarized herein.

TABLE 4.    AERR MODELS FOR SURFACE IMPOUNDMENTS

| Model | Applicability | Primary reason for recommending or not recommending |
|---|---|---|
| Mackay and Wolkoff (1973) | Nonaerated<br>Unsteady state | Not based on two-film resistance theory |
| Mackay and Leinonen (1975)[a] | Nonaerated<br>Unsteady state | Only unsteady state model based on two-film resistance theory |
| McCord (1981) | Aerated<br>Nonaerated<br>Steady state | Only considers gas-phase resistance. Most chemical compounds are liquid-phase controlled |
| Smith, et al (1980-81) | Nonaerated<br>Steady state | Based on two-film resistance theory, but requires complex laboratory experiments. |
| Thibodeaux, Parker and Heck (1981d)[a] | Aerated<br>Nonaerated<br>Steady state | Based on two-film resistance theory - recommended for steady-state conditions |
| Shen (1982) | Nonaerated<br>Steady state | Modified Thibodeaux, Parker and Heck (1981d) model. Simplification technique for Schmidt number is provided, and $k_G$ value used is recommended |

[a]Denotes recommended models.

For aerated SIs, only two equations are available. McCord (1981) based his model on evaporation of water to address diffusion of volatile compounds. However, the evaporation of water is gas-phase controlling, whereas volatilization of most chemical compounds is liquid-phase controlling. This inaccuracy was also noted for McCord's nonaerated SI model. Only the Thibodeaux, Parker and Heck (1981d) model accurately describes the aerated SI in terms of the two-film resistance theory. Thus, Thibodeaux, Parker and Heck (1981d) is recommended for both aerated and nonaerated SIs under steady state conditions.

The Mackay and Leinonen (1975) unsteady-state model may also be applicable to an aerated impoundment. This statement is justifiable by recalling that the Thibodeaux, Parker and Heck (1981d) model is applicable to both a nonaerated and aerated SI, and is similar in form. The only difference is that the aerated impoundment must also consider liquid and gas-phase mass transfer in the turbulent zone created by mechanical aeration, in addition to the liquid and gas-phase k-values from the convective (natural) zone. These additional considerations can be incorporated into the unsteady-state model, by using the appropriate overall K-value for both the turbulent and convective zones.

Landfill Model Selection

Landfill models are required for two cases: (1) hazardous waste disposal only, and (2) codisposal of hazardous waste with solid waste material. Table 5 summarizes the available models for landfills for no gas generation, effects of biogas generation, effects of barometric fluctuations and also for open dumps. GCA's recommendations are footnoted.

Limited field validation test data makes the selection of landfill AERR model difficult at this time. The key to selecting a landfill model is an accurate description of pore diffusion as it relates to various soil matrices and moisture contents. As described in the $k_{soil}$ subsection, the effective diffusivity for pore diffusion may be best described by a power law relationship to air-filled porosity, which appears to be dependent on soil type. Limited data for gas diffusion in wet soils complicates a similar power law relationship.

For now, GCA believes that power law relationship models deserve field validation. Field testing should include a method of determining the power law relationship of various soil mixtures and moisture contents which may be typical of hazardous waste landfill sites.

Accurately determining the effective diffusivity parameter for the AERR landfill model may only be one of many other potentially major problems to consider. Other problems include quantifying lateral gas migration, and adsorption of organic vapors onto soil. Like the effective diffusivity determination, each of these factors may also be soil specific.

Shen (1980) modified Farmers' original equation, thus introduced some simplification errors. Shen defined the air emissions of a specific component

TABLE 5.   AERR MODELS FOR LANDFILLS

| Model | Applicability | Primary reasons for recommending or not recommending |
|---|---|---|
| Farmer, et al (1978)[a] | Covered landfills | • Experimental data limited to pure component hexachlorobenzene |
| Shen (1980) | Covered landfills | • Simplification of Farmer's model and contains over-simplifications (see text) |
| Thibodeaux (1981a)[a] | Covered landfills-with no gas generation | • Defines pore diffusion in terms of soil porosity and tortuosity, but limited for dry, granular soils |
| Thibodeaux (1981a)[a] | Covered landfills-with biogas generation | • Describes porous media diffusion with convective effects due to biogas generation<br>• No experimental data |
| Thibodeaux, Springer, Riley (1981b) | Covered landfills-with barometric fluctuations | |
| Shen (1981) | Open dump | • Does not take into account vertical structure of the earth's boundary layer (Sutton)<br>• Limited to gas phase diffusion only |

[a]Denotes recommended models.

based on its weight percent in the waste mixture. For waste mixtures, consideration of waste composition on a mole basis and the activity coefficient is important.

Thibodeaux (1981a) also provides a model for codisposal of hazardous waste with solid waste material to account for the air emissions resulting from the ascending gases. One may need to correlate the quantity of gases generated with the type and quantity of material buried and respective soil properties in order to determine the expected gas velocity. Sufficient data may exist such that a general correlation can be developed.

Thibodeaux et al. (1981b) also identify air emissions release rate for landfills which includes the effect of barometric pressure fluctuations. According to Springer (1983), computer simulation studies have shown that the effect of barometric pressure fluctuation (pumping) does not average out over long periods of time, and thus should be considered for annual AERR estimates. The effect of internal gas generation also plays an important role when hazardous waste is codisposed with solid waste. The convective effects created by the anaerobic gas generation can carry volatile vapors to the soil surface, thus increasing the flux rate.

## Open Dump Models

Only one equation appeared in the literature for AERR from open dump sites, proposed by Shen (1981). Shen's equation is not considered accurate for our purposes because it neglects the vertical structure of the earth's boundary layer described by the work of Sutton (1953). The starting point for an appropriate open dump equation should be based on work conducted by Thibodeaux and Hwang (1982) for landfarming emissions. Assuming solid material in the open dump that does not intermix or seep into the soil, the flux rate can be estimated by:

$$q_i = k_{Gi}C_iX_i$$

where; $k_{Gi}$ is the gas-phase mass transfer coefficient for component i calculated by the correlation developed by Sutton (1953) and previously described under $k_G$ subsection; $C_i$ equilibrium vapor concentration of compound i in g/cm$^3$; and $X_i$ is the mole fraction of compound i in the waste. Sutton accurately described the gas-phase mass transfer coefficient as a function of wind speed, surface roughness, and surface length. For liquid material in an open dump, the landfarming model described below would be most appropriate.

## Landfarming Model Selection

The selection of an AERR model for landfarming is relatively straightforward since only two candidates were found in the literature. The Hartley (1969) model describes AERR for landfarming operations only as a surface evaporation problem (i.e., gas phase resistance only). This may be sufficient for short-term emissions determination following initial spreading, but for landfarming, where multicomponent waste oil or sludge is generally incorporated into the soil matrix, one needs to define vapor movement in the porous soil. In addition, Hartley's evaporation model uses water evaporation as the basis (reference compound) in determining the AERR for other compounds. It does not appear correct to compare the rate of evaporation of volatile compounds to that of water.

The Thibodeaux-Hwang (1982) model correctly identifies the flux rate of a chemical compound as vapor diffusion through soil medium. The model also accounts for the soil "dry down", the increased resistance contributed by the soil phase. The effect of liquid-phase resistance is considered based on the assumption that the oil-layer diffusion length is considerably small (i.e., $Z_0$ less than soil particle diameter) after the oil waste is incorporated into the soil. However, one should examine the contribution of liquid-phase resistance for the initial emission release rates immediately after surface application where the oil layer thickness may be considerably larger than a soil particle size.

Although the Thibodeaux-Hwang model most accurately identifies the soil-phase resistance as a factor in air emissions for landfarming operations, it does not identify the liquid-phase resistance as probably being most important, especially for emission rate determination immediately after application where waste oil may be applied several inches thick. Typically, waste oil or sludge applied to the soil may contain many organic volatile compounds. As the liquid diffusion coefficient of a compound is generally four or five magnitudes smaller than its molecular (gas-phase) diffusion coefficient, this suggests that the liquid mass transfer coefficient should be incorporated into the short-term analysis. Assumptions made the Thibodeaux-Hwang landfarm model include:

- isothermal soil column;

- no vertical liquid movement by capillary action;

- no adsorption by soil particles; and

- no biochemical oxidation.

The use of these assumptions may not be entirely valid for landfarming operations. Thermal gradients frequently exist in the upper soil layers, thereby creating the upper movement of heat and water in soil. In addition, the purpose of landfarming is to employ the microbiological actions of the upper soil layer to degrade organic material. Thus, biochemical oxidation and gas generation will influence the air emissions rate.

Very little experimental data are available to verify this model, thus, factors such as solar radiation effects, gas generation due to biological activity, and microbiological destruction rates should be quantified when model verification tests are performed.

Storage Tank Model Selection

Models for storage tanks have long been available through EPA in the standard air emission estimation handbook, AP-42. Section 7 describes the state-of-the-art techniques recommended by EPA for fixed-roof and floating roof tanks. No models were selected for pressure tanks since emissions from such tanks are not regularly occurring events.

Air emissions from fixed roof tanks occur from breathing losses and working losses. The best available model for estimating breathing loss emissions was developed by API (1962) and modified by TRW/EPA (1981). The TRW/EPA modification was necessary since test results showed the API (1962) equation over-estimated breathing losses by approximately a factor of four. The best available model for estimating working loss emissions was developed by API in 1962. Recent test results show that this model accurately predicts emissions.

The models deemed most applicable to estimating standing storage losses and withdrawal losses from external floating roof tanks were developed by API (1980) and modified slightly by EPA (1981). Emissions estimates for internal floating roof tanks are available in API Publication 2519.

Emission factor equations in AP-42 are applicable to pure component liquids. Thus, GCA has provided a method of determining storage tank emissions for mixtures. In addition, EPA requested that GCA identify applicable AERR models for open storage tanks. Details of storage tank AERR models and these two special cases appear in Section 7.

WASTE TREATMENT PROCESSES

As suggested by OSW, the approach used for properly identifying the appropriate AERR model for the variety of treatment systems was to categorize each system into one of three main topics:

- open tank-no mixing - i.e., no disturbance of air-liquid interface;

- open tank-mixing - i.e., obvious disturbance of air-liquid interface;

- closed system - i.e., no air emissions directly from process during normal operations.

Table 6 presents each of these categories and lists typical treatment processes.

TABLE 6.   WASTE TREATMENT PROCESSES

- OPEN TANK-NO MIXING

    - sedimentation
    - chlorination
    - equalization
    - anaerobic treatment

- OPEN TANK-MIXING

    - Low-rate mixing

        - neutralization
        - chemical precipitation
        - flocculation/coagulation

    - high-rate mixing

        - activated sludge process (contact stabilization)
        - aerated lagoons
        - ozonation
        - rotating biological discs
        - trickling filter
        - dissolved air flotation

- CLOSED TANK SYSTEMS - chemical separation techniques

    - steam distillation/stripping
    - extraction
    - decantation
    - ion exchange separations
    - activated carbon systems[a]
    - dialysis and electrodialysis
    - filtration

[a]Powdered activated carbons can be used in existing tanks, filtration, or settling equipment (i.e., open tanks).

For many treatment processes, some form of reaction, chemical or biological, will occur within the tank. Assuming steady-state conditions (no fluctuations in flow or formation of additional chemical compounds) a mass balance for a specific component can determine the concentration of that component in the tank. Once the in-tank concentration of the component is known, air emissions can be estimated by applying one of the surface impoundment models previously recommended, depending upon whether or not the tank is agitated. Determining the in-tank concentration of organic compounds, however, will be related to the following:

- the compound's solubility in water;

- the compound's affinity for the formed sludge, floc, etc., versus water;

- digestion capability of microbial population within the tank.

Organic compounds removal from wastewater streams by one of the closed systems shown in Table 6, is common in the chemical and petrochemical industries. Aside from system leakage or abnormal operation, there should be no air emissions from these closed treatment systems. The effluent streams from these unit operations may go on to further treatment (biological, or cooling towers and lagoons) or disposal (surface impoundment, or landfarming), therefore it may be necessary to determine the concentrations of organics in the respective streams.

Generally, the efficiency of a processing unit (steam stripper, activated carbon bed, etc.) should be known based on laboratory or pilot scale testing. However, in the absence of efficiency data, one can calculate a systems removal efficiency based on chemical engineering principles. For instance, knowing the design of a stripping tower (i.e., tray data and flow rates) one can determine the towers efficiency. Likewise, knowing the absorptive capacity of carbon for a specific organic compound (generally through laboratory analysis) and the bed design data, the effluent concentration of wastewater can be determined.

OPEN TANK WITH MIXING

GCA's evaluation of open tank treatment processes with mixing focused mainly on the activated sludge biological treatment process (high rate mixing). Other processes which might fall into the low rate mixing subcategory would include neutralization or precipitation process involving the addition and mixing of a chemical reagent. The emissions from these processes are best described by application of Thibodeaux's ASI model using very low power input to identify the area of turbulent mixing. This model is discussed in detail in Section 4 of this report.

Biological Treatment Systems

Efforts in selecting models for predicting air emission release rates from activated sludge treatment processes focused on four models given in the literature:

- Thibodeaux (1981d)--Aerated Surface Impoundment (ASI) Model;

- Hwang (1980)--Activated Sludge Surface Aeration (ASSA) Model;

- Freeman (1979)--Activated Sludge Surface Aeration (ASSA) Model;

- Freeman (1980)--Diffused Air Activated Sludge (DAAS) Model.

All of the models presented predict the emissions release rate based on the concentration of compound i $(X_i)$ in the aeration basin of the activated sludge (AS) process. The complexity of the models are generally dependent upon the information assumed to be unknown. The Thibodeaux (1981d) ASI model, which assumes the concentration $X_i$ is known, is by far the simplest model. Thibodeaux applies the two-film resistance theory to predict the air stripping losses due to mass transfer at the air-water interface (aeration basin surface). Thibodeaux's model isolates zones for convective and turbulent mixing and applies appropriate mass transfer expressions for each zone. Thus, having determined an area-averaged overall mass transfer coefficient $(K_L)$, the ASI model applies this to the concentration $(X_i)$ in the basin to predict the air emission release rate.

Both the Hwang (1980) and Freeman (1979) ASSA model employ identical mass transfer expressions to predict emissions from the activated sludge process. These models differ from the Thibodeaux model in that they predict air emissions based on the influent concentration $(S_o)$ of the compound (substrate). The Hwang (1980) ASSA model considers the interaction of; (1) substrate removal by biodegradation; (2) removal by air stripping; and (3) adsorption of substrate to the biomass (sludge). Whereas, the Freeman (1979) ASSA model considers biodegradation and air stripping, but ignores adsorption of substrate to the biomass. The work of Kincannon (1981,1982), discussed briefly in Section 8, also indicated that sludge adsorption was not a major pollutant removal mechanism when biodegradation, air stripping, and adsorption are considered collectively. However, other studies (Hwang, 1981; Patterson, 1981) suggest that biosorption removal may be quite significant for a number of organic priority pollutants. Both ASSA models calculate the effluent concentration (Se) based upon a steady-state equilibrium reached by the above competing removal processes. The air emission release rate is then computed by the ASI mass transfer model.

The different biooxidation rate kinetics equations selected by Hwang and Freeman lead to major differences in the complexity of each model. Hwang's ASSA model employs the relatively simple 1st order Grau kinetics to predict the rate of biooxidation. Freeman selects a biokinetic model developed by Gerber for refinery wastes, which ties the substrate removal rate to the biomass concentration, the oxygen concentration, and the substrate concentration.

Solution of Hwang's model is tedious but workable. In addition, Hwang has provided biokinetics rate data (values of $K_{1(s)}$) for a large number of organic compounds. Thus, data necessary to develop solutions based on Hwang's model are reasonably accessible. Conversely, solution of Freeman's model requires solving three simultaneous nonlinear equations for three unknowns.

Hand calculation of the solution is rather cumbersome, thus the solution is best conducted by a computerized iteration technique. In addition, biokinetic rate constants for the Gerber Model (3rd order Monod) used by Freeman have been compiled only for a limited number of compounds (acrylonitrile, benzene, styrene ). Consequently, application of Freeman's ASSA model using Gerber kinetics will be limited by this data requirement. In short, the solution of the Freeman model is very complex and not well suited to EPA's intended purposes under this program.

Freeman's (1980) diffused air activated sludge (DAAS) model was the only model reviewed which adequately models diffused air systems. The DAAS model is similar to Freeman's ASSA model with the exception of the mass transfer model for air stripping. In the DAAS model, Freeman predicts the air stripping losses based on the mass transfer of the organic compound into the sparger bubbles as opposed to the surface aeration model which predicts mass transfer through the basin surface. Because the DAAS mass transfer model is slightly more sophisticated than the more empirical surface aeration (ASI) mass transfer model, Freeman's DAAS is also slightly more difficult to solve. Freeman's DAAS model assumes that the mass transfer process is liquid phase controlled and assumes that transfer occurring at the surface of the basin is negligible.

The concern in employing these models for the purpose of predicting air emission release rates is accuracy. Hwang's ASSA model is based on limited laboratory and treatment facility data. Efforts to verify the Hwang model were only moderately successful. Freeman's diffused air mass transfer model showed good agreement with verification data based on sterile process experiments. Freeman's DAAS model has been calibrated to laboratory data, however, no additional verification has been reported in the literature. Given the status of model verification presented above, concern over putting untested models into general use is warranted.

GCA recommends that the Agency employ the following models for use in air emission assessments:

- Thibodeaux (1981d) ASI model--for known effluent concentrations;

- Hwang (1980) ASSA model--for unknown effluent concentrations;

- Freeman (1980) DAAS mass transfer model--for known effluent concentrations;

- Hwang (1980) ASSA model modified after Freeman (1980) DAAS-- for diffused air systems with unknown effluent concentrations.

The Hwang ASSA model is selected over Freeman's model because the Hwang model can account for adsorption to wasted sludge, employs simpler bioxidation kinetics for which rate constant data are available, and is capable of modeling adsorption of substrate to the biomass. Freeman's DAAS mass transfer equations are selected for modeling diffused air systems because they are currently the only reasonable choice. GCA recommends that the DAAS mass transfer equations be integrated into Hwang's ASSA model for predicting DAAS air emission release rates where the effluent concentration is unknown. Again, this is recommended to simplify the problem solution and to avoid costly experiments to determine Gerber's biokinetic rate constants.

WASTE PILES (PARTICULATE EMISSIONS)

EPA requested GCA to survey available information for estimating particulate emissions from waste storage piles. (Note that all other information contained in this report pertains to estimating volatile air emissions). The most approriate approach to estimating waste pile particulate emissions is the emission factor equations developed from limited field test data by the Midwest Research Institute (MRI) under contract with EPA. The MRI correlations have been frequently used by regulatory agencies in estimating particulate emissions from industrial storage piles (coal, iron ore, gravel, etc.) and are presented in Section 10 of this report. An emission factor equation in AP-42 provides a more general approach that does not account for site operational procedures.

# 3. AERR Model Validation Efforts Reported in the Literature

INTRODUCTION

The AERR models have generally not been validated or calibrated by comparison with a substantial field data base. Only three test reports or reference articles reported comparisons between predictive AERR models and actual field measurements at the end of 1982. Additional reports provided ambient air monitoring data which cannot be directly applied to the AERR models without back-calculating the AERR via a dispersion model. However, some field work aimed at model validation is currently underway (IERL-Cincinnati with Thibodeaux, et al.) and in the planning stages for 1983 (IERL-Cincinnati with Radian Corporation and OSW - Engineering Science, Inc.).

This section summarizes model validation efforts conducted prior to 1983.

AERR MODEL COMPARISON WITH FIELD DATA

Table 7 lists four reports in which measured AERR were compared with predictive modeling equations. Test data from the most recent sampling of surface impoundment air emissions by Thibodeaux, et al. (1982) appear most promising. Initial comparisons indicate that the predictive model estimates are within a factor of two to three of the measured emission rate.

## Thibodeaux, Parker and Heck (1981d)

The objectives of this report were to develop a measurement technique for volatile organic carbon (VOC) emissions emitted from wastewater treatment basins and to develop a mathematical model that would predict emission rates. The concentration-profile (CP) technique was used for determining the flux rate of methanol, acetone and total hydrocarbon emissions. The following individual mass transfer correlations were used in the predictive model:

- $k_g$ (convective)--Harbeck;

- $k_g$ (turbulent)--Reinhardt;

- $k_l$ (convective)--Cohen;

- $k_l$ (turbulent)--Thibodeaux, Parker.

TABLE 7.   AERR MODEL VALIDATION EFFORTS

| Test report | Authors (date) | Comparisons made |
|---|---|---|
| Measurement of Volatile Chemical Emissions from Wastewater Basins | Thibodeaux, Parker, Heck (1981d) | Development of concentration-profile technique: comparisons of measured methanol flux rates with predictive model |
| Evaluation of VOC Emissions from Waste-water Systems (Secondary Emissions) Draft Report | Cox, Steinmetz, Lewis (1982) | Concentration-profile technique measurements compared to predictive model for aerated and non-aerated conditions |
| Air Emission Monitoring of Hazardous Waste Sites | Thibodeaux, Springer, Lunney, James, Shen (1982) | Concentration-profile technique measurements compared to predictive model for aerated and nonaerated conditions |
| Suggested Control Measures to Reduce Organic Compound Emissions Associated with Volatile Organic Waste Disposal | Ames, Shiroma, Wang, Lam, O'Brien (1982) | Comparisons of measured land-treatment emission rates with Hartley and Thibodeaux-Hwang models |

Although flux rate measurements were made for methanol, acetone and total hydrocarbon, only methanol data were used in the predictive model comparison effort. The authors decided not to use acetone in the comparison because of uncertain water chemistries at the low concentrations encountered. Chemistry data with correlation coefficients of less than 0.8 were excluded. The following flux rate comparisons were reported:

- measured--1.4 to 3.8 $ng/cm^2$-sec (11 to 29 lb/acre-day);

- predicted--0.37 to 5.9 $ng/cm^2$-sec (2.8 to 45 lb/acre-day).

The comparison of all four mass transfer coefficients indicated that methanol volatilization is gas-phase controlling.

### Cox, Steinmetz and Lewis (1982) - Draft Report

This draft report compares the measured air emissions from two wastewater treatment facilities to results obtained by using the predictive model. The CP technique was used to measure chemical flux rates from an aerated and a nonaerated pond. The following individual mass transfer correlations were used in the predictive model:

- $k_g$ (convective)--Mackay, Matsugu;

- $k_g$ (turbulent)--Reinhardt;

- $k_l$ (convective)--Cohen;

- $k_l$ (turbulent)--Thibodeaux.

A Final Report was not available for review, but the following preliminary results were noted. Predictive flux rates initially reported ranged from 15 to 38 percent of measured flux rates for nonaerated conditions, and from 0.1 to 2.5 percent of measured flux rates for the aerated site. However, in applying the predictive model equation, the approach used for calculating the equilibrium constant was incorrect. The use of Raoult's law may be only applicable to hydrocarbon mixtures, not to aqueous solutions. Henry's law, or the use of an activity coefficient are more correct approaches. Field data from this report may still be valuable subsequent to complete peer review and verification of results.

### Thibodeaux, et al. (1982)

The work presented in this paper identified the emission rates of two surface impoundments (one aerated and one non-aerated) at a hazardous waste facility. Measurements at the site by the CP techniques provided data to "field test" the emission release rate models for chemicals in which the mass transport process is liquid-phase controlled. Additionally, the authors identified the capability of the CP technique in detecting and quantifying absorption and vaporization from area source. Comparison of calculated flux rate versus measured flux rate for 1,1-dichloroethane and benzene were found to be within a factor of ±2.

Ames, et al. (1982)

In an attempt to illustrate the difference between the Hartley and
Thibodeaux-Hwang AERR models for landfarming, this report provided sample
calculations based on experimental data.  Figure 6 shows the predictive
results of the two models and also displays the experimental results.  The
experimental data were not available to GCA, thus the variables and
assumptions used in the calculations could not be verified.

One limitation of the Hartley model, clearly illustrated in Figure 6, is
the assumption of a maximum emission rate throughout the exposure period.  The
Thibodeaux-Hwang model only slightly underpredicts the measured results.

ADDITIONAL DATA AVAILABLE

Additional test reports contain ambient air emissions monitoring data for
hazardous waste facilities.  A document prepared by Fred C. Hart Associates,
entitled "Development of a Data Base on Air Emissions from Hazardous Waste
Facilities," provided a cursory review of 21 hazardous waste site reports
received from the 10 EPA regions.  It is not known if data provided in the
original test reports are adequate for application to the various AERR
models.  This data base may need further review.

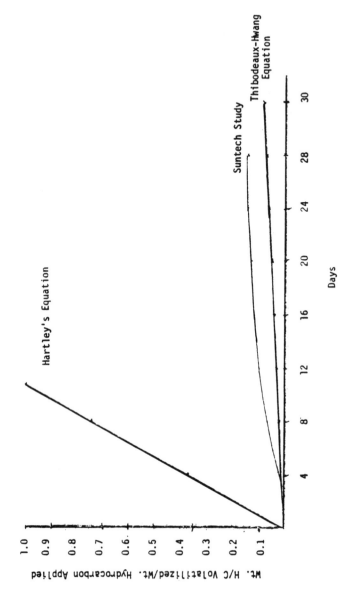

Figure 6.  Comparison of Hartley and Thibodeaux-Hwang landfarming AERR models
with measured air emissions data (Ames, et al., 1982).

# 4. Review of Surface Impoundment AERR Models

INTRODUCTION

Surface impoundments (SIs) are defined in Chapter 40, CFR Part 260, Subpart B, Section 260.10 (amended by 46 FR 27476, May 20, 1981) as "a facility or part of a facility which is a natural topographic depression, man—made excavation, or diked area formed primarily of earthen materials (although it may be lined with man—made materials) which is designed to hold an accumulation of liquid wastes or wastes containing free liquids, and which is not an injection well. Examples of surface impoundments are holding, storage, settling, and aeration pits, ponds, and lagoons."

This section presents all models for SIs found in the literature. GCA's recommendations for SI models and mass transfer coefficients were presented previously in Section 2.

Eight models or variations thereof appear in the literature for estimating the air emission release rates (AERR) from SIs (nonaerated and aerated) as follows:

- Mackay and Wolkoff (1973) unsteady-state, nonaerated;

- Mackay and Leinonen (1975) unsteady-state, two-film theory, nonaerated model;

- Thibodeaux, Parker and Heck (1981d) steady-state, two-film theory one nonaerated and one aerated model;

- Shen (1982) modification of Thibodeaux, Parker and Heck nonaerated model;

- Smith, et al. (1980-81) steady-state, first-order kinetic, nonaerated model;

- McCord (1981) steady-state, modified Nusselt equation, one nonaerated and one aerated model.

In addition, the literature contains a calculational model, proposed by Thibodeaux, Parker, Heck and Dickerson (1981c). This model is not suitable for our purposes since detailed field measurements are required as input data.

The eight predictive models can be categorized according to their
theoretical basis.  Five of the eight models incorporate the two-film
resistance theory.  The three predictive models that are not based on the
two-film resistance theory have been criticized by more recent research.  For
example, the two predictive models presented by McCord (1981) are based on the
Nusselt equation, not the two-film resistance theory.  The theoretical basis
of water evaporation implied by the Nusselt equation is considered a
unrealistic approach to the volatilization process of most sparingly
(slightly) soluble organics.  The Mackay and Wolkoff (1973) predictive model
is not based on the two-film resistance theory and was essentially revised by
Mackay and Leinonen's 1975 research.

Models which incorporate the two-film resistance theory are considered
the most accurate methods of describing the actual volatilization process, and
subsequent flux rate of an individual compound into the atmosphere.  However,
the models based on this theory require determining the individual liquid
($k_L$) and gas-phase ($k_G$) mass transfer coefficients for each compound of
interest.  The availability of liquid and gas-phase k-values is limited, and
are primarily based on laboratory experiments or on field measurements of
lakes, rivers, and the ocean.  Consequently, the accuracy of all available
state-of-the-art models is somewhat in question due to problems involved with
precisely determining k-values for specific situations at TSDFs.

Table 8 summarizes input parameters required for each model reviewed by
GCA.  Following sections describe all models available for aerated and
nonaerated surface impoundments.

## NONAERATED SURFACE IMPOUNDMENTS

AERR models found in the technical literature for nonaerated SIs are
reviewed below.  The summary in Section 2 previously presented GCA's
selections for each SI type.

### Mackay and Wolkoff (1973)

Mackay and Wolkoff (1973) proposed a model to quantify the volatilization
of low solubility compounds (hydrocarbons and chlorinated hydrocarbons) from
rivers, lakes and oceans.  Their approach was based on equilibrium
thermodynamic principles of water evaporation, in contrast with more recent
research which is based on mass transfer principles of a concentration
gradient across an interface.  Mackay and Wolkoff assumed that the AERR could
be calculated on the basis of water evaporation and the ratio of the
contaminant to water in the vapor.  This approach assumes that the diffusion
or mixing in the water phase is sufficiently fast so the concentration of the
contaminant at the water-air interface is close to that in the bulk of the
water body.  However, recent research on mass transfer rates suggests that
diffusion in the water phase is the rate controlling variable for most
low-solubility compounds.  In other words, current mass transfer theory shows
that the basic assumption in the Mackay and Wolkoff model is inappropriate for
modeling AERR from hazardous waste SIs.  Other stated assumptions in the
Mackay and Wolkoff model include:

TABLE 8.  INPUT PARAMETERS REQUIRED FOR EACH SURFACE IMPOUNDMENT MODEL

| Parameter | Mackay (1975)[a] | Thibodeaux (1981d)[b] | Shen (1982)[c] | Smith (1980-81)[d] | McCord (1981)[e] |
|---|---|---|---|---|---|
| N, mass flux rate of compound i across the phase boundary | X | | | | |
| $K_{i,L}$, overall liquid phase mass transfer coefficient of compound i | X | X | X | | |
| $k_{i,L}$, individual liquid phase mass transfer coefficient of compound i | X | X | X | X | |
| $k_{i,G}$, individual gas phase mass transfer coefficient | X | X | X | X | |
| R, ideal gas law constant | X | | | X | |
| T, temperature | X | | X | X | |
| $H_i$, Henry's Law Constant | X | | X | X | |
| $C_i$, concentration of compound i at time t | X | | | | |
| $C_{io}$, concentration of compound i at zero time | X | | | | |
| L, impoundment depth | X | | X | X | |
| $P_i$, equilibrium partial pressure of compound i in the vapor | X | | | | X |
| $P_{i\infty}$, vapor pressure of compound i | X | | | | |
| $C_{i\infty}$, solubility of compound i in water | X | | X | | |
| $Q_i$, rate of air emissions of compound i | | X | X | | |
| K, constant establishing equilibrium between the liquid and air phases | | X | X | | |
| $x_i$, mole fraction of compound i in the liquid phase | | X | | | |
| A, surface area of the disposal facility | | X | X | | |
| $MW_i$, molecular weight of compound i | | X | X | | |

(continued)

**TABLE 8 (continued)**

| Parameter | Mackay (1975)[a] | Thibodeaux (1981d)[b] | Shen (1982)[c] | Smith (1980-81)[d] | McCord (1981)[e] |
|---|---|---|---|---|---|
| $ke^*$, roughness Reynolds number | | x | | | |
| $MW_{benzene}$, molecular weight of benzene | | x | | | |
| $U_{air}$, wind speed | | x | x | | x |
| $N_{Sc}$, gas Schmidt number | | x | x | | |
| $\mu_g$, absolute gas viscosity | | x | | | |
| $\rho_g$, density of gas | | x | | | |
| $D_{i,air}$, diffusion coefficient of compound i in air | | x | | | |
| $\rho_{air}$, density of air | | x | | | |
| $MW_{air}$, molecular weight of air | | x | | | |
| $ERP_i = Q_i$ | | | | | |
| $K_{oL} = K_{iL}$, overall liquid phase mass transfer coefficient | | | | | |
| $H = L$, depth of lagoon | | | | | |
| $P$, total pressure | | | x | | |
| $M$, average molecular weight of liquid | | | x | | |
| $Z$, length of lagoon surface | | | | x | x |
| $R_v$, volatilization rate | | | | x | |
| $k_v c$, volatilization rate of compound c | | | | x | |
| $E_s$, evaporation rate at steady state | | | | | x |
| $E_o$, initial evaporation rate | | | | | x |
| $L_s$, amount of compound in lagoon at steady state | | | | | x |

(continued)

**TABLE 8** (continued)

| Parameter | Mackay (1975)[a] | Thibodeaux (1981d)[b] | Shen (1982)[c] | Smith (1980–81)[d] | McCord (1981)[e] |
|---|---|---|---|---|---|
| $L_o$, initial amount of compound in lagoon | | | | | x |
| $I$, weight of percent of compound in water or lagoon liquid | | | | | x |
| VP = $P_{is}$, vapor pressure of the compound | | | | | |
| $u_o$ = $U_{air}$, wind speed | | | | | |
| $K$, amount of compound in feed stream | | | | | x |
| $F$, feed rate to lagoon | | | | | x |
| $V$, operating volume of lagoon | | | | | x |
| sp. gr., specific gravity of liquid waste | | | | | x |
| $M_B$ = $M_i$, flux rate | | | | | x |
| $D_{Bi}$ = $D_{i,air}$, diffusivity of compound i in air | | | | | |

(continued)

NOTES FOR TABLE 8:

MODEL DEFINITIONS--AS PRESENTED IN THE LITERATURE

[a]Mackay and Leinonen - Unsteady-State Predictive Model for Nonaerated Surface Impoundments.

$$(1) \quad N_i = K_{iL} \ (C_i - P_i/H_i)$$

$$(2) \quad \frac{1}{K_{iL}} = \frac{1}{k_{iL}} + \frac{RT}{H_i \ k_{iG}}$$

$$(3) \quad C_i = C_{io} \ exp \ (-K_{iL} \ t/L)$$

$$(4) \quad P_i = C_i \ P_{is}/C_{is}$$

[b]Thibodeaux, Parker and Heck-Steady-State Predictive Model for Nonaerated and Aerated Surface Impoundments.

$$(1) \quad Q_i = K_L A \ (x_i - x_i^*) \ MW_i$$

$$(2) \quad \frac{1}{K_L} = \frac{1}{k_L} + \frac{1}{K \ k_G}$$

[c]Shen's Simplification of the Steady-State Predictive Model Proposed by Thibodeaux, Parker and Heck.

$$(1) \quad (ERP)_i = (18 \times 10^6) \ K_{oA} \ AC_i$$

$$(2) \quad \frac{1}{K_{oA}} = \frac{1}{k_L} + \frac{1}{K \ k_G}$$

$$(3) \quad k_L = (4.45 \times 10^{-3}) \ M_i^{-0.5} \ (1.024)^{t-20} \ U_o^{0.67} H^{-0.85}$$

$$(4) \quad K = (H_i/PM) \times 10^6$$

$$(5) \quad k_G = (8.05 \times 10^{-4}) \ M_i^{-1} \ U_o^{0.78} \ Z^{-0.11} \ Sc^{-0.67}$$

[d]Smith, Bomberger and Haynes - First Order Kinetic Equation for Nonaerated Surface Impoundments.

$$(1) \quad R_v = \frac{-d(c)}{dt} = k_v^c \ (c)$$

NOTES FOR TABLE 8 (continued):

(2) $\quad k_v{}^c = \dfrac{1}{L} \left[ \dfrac{1}{k_L{}^c} + \dfrac{RT}{H_c \, k_G{}^c} \right]^{-1}$

$^e$ McCord's Steady-State Predictive Model for Nonaerated Surface Impoundments.

(1) $\quad E_s = \dfrac{E_o \, L_s}{L_o}$

(2) $\quad E_o = 0.53425 \, A_m \, \dfrac{D}{L}{}^{0.22} \, (\%) \, (v_p) \, (W_o)^{0.78}$

(3) $\quad L_s = \dfrac{K}{1-X}$

(4) $\quad 1-X = \dfrac{E_o}{L_o} + \dfrac{F}{V}$

(5) $\quad K = 0.0834 \, (F) \, (\%) \, (\text{sp. gr.})$

(6) $\quad L_o = 0.0834 \, (V) \, (\%) \, (\text{sp. gr.})$

$^f$ McCord's Steady-State Predictive Model for Aerated Surface Impoundments

(1) $\quad E_s = \dfrac{E_o \, L_s}{L_o}$

(2) $\quad E_o = \dfrac{150.315}{(460+T)} \, \dfrac{H}{r} \, \dfrac{L_o}{V_L} \, \dfrac{V.P.}{\text{sp.gr.}} \, \dfrac{d}{\pi} \, \dfrac{1}{\sqrt{0.17678R^{1/2}}}$

(3) $\quad L_s = \dfrac{k}{1-x}$

(4) $\quad 1-x = \dfrac{E_o}{L_o} + \dfrac{F}{V}$

(5) $\quad K = 0.0834 (F)(\%)(\text{sp.gr.})$

(6) $\quad L_o = 0.0834 (V)(\%)(\text{sp.gr.})$

- The contaminant is truly in solution, not in suspended, collodial, ionic, complexed, or absorbed form.

- The vapor formed is in equilibrium with the liquid at the interface.

- The water evaporation rate is negligibly affected by the presence of the contaminant.

Dilling's (1977) preliminary work indicated that the Mackay and Wolkoff model was inadequate since neither absolute nor relative predicted rates were in agreement with experimental data.

### Mackay and Leinonen (1975)

Mackay and Leinonen (1975) extended Mackay's 1973 research in order to develop a more realistic AERR estimate for low-solubility compounds from the entire water body, and not just from the water surface. Mackay and Leinonen incorporated the work of Liss and Slater (1974) to develop an unsteady-state model to determine the AERR of a single compound. Liss and Slater applied the two-film resistance theory to estimate the flux of gases across the air-ocean interface. Contrary to Mackay's 1973 thermodynamic model theory, Liss and Slater's 1974 work suggested that for most low-solubility gases, the water (liquid) phase controls; i.e., the liquid phase offers more resistance to contaminant transport than through the gas phase.

The details of the Mackay and Leinonen model appear in Table 9. The model is based on unsteady state conditions; i.e., the chemical contaminant of interest enters the water body in a discrete slug as a pulse injection. All other SI models are based on steady state conditions; i.e., a fairly constant influx of the contaminant into the SI. As in the case of all other predictive models for nonaerated SIs, this model simplifies the actual situation by assuming well-mixed air and water phases separated by an interface with near stagnant films of air and water on either side. Thermoclines; i.e., other rate-limiting diffusion processes at depths in the water body, are not considered in this model. The Mackay and Leinonen model can be simplified by assuming that... "if $P_i$ is negligible; i.e., the background atmospheric level of the contaminant is low compared to the local level..." then the unsteady-state flux equation becomes:

$$N_i = K_{iL}C_i$$

where:  $C_i = C_{io}\exp(-K_{iL}t/L)$

$$\frac{1}{K_{iL}} = \frac{1}{k_{iL}} + \frac{RT}{(H_i k_{iG})}$$

In 1975, the only technique apparently available to Mackay and Leinonen for calculating the mass transfer coefficients (k-values for gas and liquid phases) was the 1974 Liss and Slater work shown below:

$$k_{iL} = k_{L,CO_2} \left(\frac{MW_{CO_2}}{MW_i}\right)^{0.5}$$

TABLE 9.    MACKAY AND LEINONEN UNSTEADY-STATE PREDICTIVE MODEL
FOR NON AERATED SURFACE IMPOUNDMENTS

---

**Model Form:**

$$N_i = K_{iL} (C_i - P_i/H_i)$$

where;

$$\frac{1}{K_{iL}} = \frac{1}{k_{iL}} + \frac{RT}{H_i \, k_{iG}}$$

$$C_i = C_{io} \exp(-K_{iL} \, t/L)$$

$$P_i = C_i P_{is}/C_{is}$$

**Definition of Terms:**

$N_i$ = mass flux rate of compound i across the phase boundary ($mol/m^2$-hr)

$K_{i_L}$ = overall liquid phase mass transfer coefficient of compound i (m/hr)

$k_{iL}$ = individual liquid phase mass transfer coefficient of compound i (m/hr)

$k_{iG}$ = individual gas phase mass transfer coefficient of compound i (m/hr)

$R$ = ideal gas law constant ($8.2 \times 10^{-5}$ atm-$m^3$/mol-°K)

$T$ = temperature (°K)

$H_i$ = Henry's law constant of compound i (atm-$m^3$/mol)

$C_i$ = concentration in $mol/m^3$ of compound i at time t (hr)

$C_{io}$ = concentration in $mol/m^3$ of compound i at zero time

$L$ = impoundment depth (m)

$P_i$ = equilibrium partial pressure of compound i in the vapor (atm)

$P_{is}$ = vapor pressure of compound i (atm)

$C_{is}$ = solubility of compound i in water ($mol/m^3$)

and

$$k_{iG} = k_{G,H_2O} \left( \frac{MW_{H_2O}}{MW_i} \right)^{0.5}$$

where:

$k_{L,CO_2}$ = liquid phase mass transfer coefficient of $CO_2$ = 20 cm/hr (reported to be at the sea surface);

$MW_{CO_2}$ = molecular weight of $CO_2$ (44 g/g mole);

$MW_i$ = molecular weight of compound i;

$k_{G,H_2O}$ = gas phase mass transfer coefficient of $H_2O$ = 3,000 cm/hr (reported to be at the sea surface);

$MW_{H_2O}$ = molecular weight of $H_2O$ (18 g/g mole).

As described in the following sections, one emphasis of more recent research has been refinement of techniques for calculating the individual liquid and gas phase k-values.

### Thibodeaux, Parker and Heck (1981d) Including Hwang (1982) and Shen (1982) Modifications

Thibodeaux, Parker and Heck proposed a steady-state model that assumes a constant influx of contaminant, as opposed to Mackay and Leinonen's unsteady state model. The Thibodeaux, Parker and Heck model, shown in Table 10, is also based on the two-film resistance theory. Empirical relationships for mass transfer coefficients developed from laboratory and field experiments at lakes, rivers and oceans were incorporated (Cohen, Cocchio and Mackay-1978; Mackay and Matsugu-1973; Owens, Edwards and Gibbs-1964). Although the k-values incorporated by Thibodeaux, Parker and Heck represent an improved data base compared to Mackay and Leinonen's approach, the calculation became substantially more complex as shown in Table 11. Some of the input parameters for these k-values are difficult to obtain, calculate or estimate, and in some cases, require field or laboratory measurement.

Hwang (1982) presented a simplified technique for determining the mass transfer coefficients for various compounds by referring to a typical compound whose k-values are known. In his review of several laboratory experiments, Hwang suggests that the k-values for oxygen and water vapor can be used to determine the k-value for the compound of interest. In his examples, the k-values for the reference compounds are based on a temperature of 25°C. The approach for calculating mass transfer coefficients identified in Hwang's paper appear in Table 12.

To apply Hwang's simplified equations, the liquid phase mass transfer coefficient for oxygen and the gas phase mass transfer coefficient for water vapor must be determined either from: (1) published data; or (2) calculating from the empirical relationships shown previously in Table 11 (the latter

TABLE 10.   THIBODEAUX, PARKER AND HECK STEADY-STATE PREDICTIVE MODEL
FOR NON AERATED SURFACE IMPOUNDMENT

**Model Form:**

$$Q_i = K_L A (x_i - x_i^*) MW_i$$

where;

$$\frac{1}{K_L} = \frac{1}{k_L} + \frac{1}{Kk_G}$$

and;

$$x_i^* = 0 \text{ since } y_i \text{ is negligible in comparison to } x_i$$

therefore,

$$Q_i = K_L A (x_i) MW_i$$

**Definition of Terms:**

$Q_i$ = rate of air emissions of compound i (g/sec)

$K_L$ = overall liquid phase mass transfer coefficient of compound i
(g-mol/cm$^2$-sec)

$k_L$ = individual liquid phase mass transfer coefficient of compound i
(g-mol/cm$^2$-sec)

$k_G$ = individual gas phase mass transfer coefficient of compound i
(g-mol/cm$^2$-sec)

$K$ = constant establishing equilibrium between the liquid and air phases,
expressed by $y_i = Kx_i$ (Henry's law constant in mole fraction form).

$x_i$ = mole fraction of compound i in the liquid phase = $C_{iL} (MW_w/MW_i)$

$A$ = surface area of the disposal facility (cm$^2$)

$MW_i$ = molecular weight of compound i (g/g-mol)

$MW_w$ = molecular weight of water (18 g/g-mol)

TABLE 11.    SUMMARY OF EMPIRICAL RELATIONSHIPS TO DETERMINE THE
INDIVIDUAL LIQUID AND GAS PHASE MASS TRANSFER
COEFFICIENTS FOR A NONAERATED IMPOUNDMENT

___

Liquid Phase Coefficient ($k_L$):

[1]    $k_L = \left(11.4 \, Re*^{0.195} - 5\right) \left(\dfrac{MW_{benzene}}{MW_i}\right)^{0.5}$    (Cohen, Cocchio and Mackay, 1978)

for $0.11 < Re* \leq 102$, where $k_L$ is in cm/hr and

$$Re* = \frac{7.07 \times 10^{-3} \, (Z_{10}) \, (U_{10})^{1.25}}{Va \, \exp \, (56.6/U_{10}^{0.25})}$$

if $Re* < 0.11$, then $k_L = 2.4$ cm/hr

where;

$Re*$ = roughness Reynolds number

$MW_{benzene}$ = molecular weight of benzene (78.1 g/g-mole)

$MW_i$ = molecular weight of compound i (g/g-mole)

$U_{10}$ = wind velocity (cm/s) measured at height $Z_{10}$ (10 m) above the water surface (cm)

$Va$ = air kinematic viscosity (cm²/sec)

[2]    $k_L = (1.3 \, Re*^{0.195} - 0.57) \left(\dfrac{D_{i, H_2O}}{D_{TOL, H_2O}}\right)$    (Cohen, et al. as presented by Hwang, 1982)

where;

$k_L$ is in lb-mol/ft²-hr. (Note: This equation is a modified form of equation 1 to obtain the $k_L$ value in units of lb-mol/ft²-hr).

$Re*$ = roughness Reynolds number (determined as above)

$D_{i, H_2O}$ = diffusion coefficient of compound i in water (cm²/sec)

$D_{TOL, H_2O}$ = diffusion coefficient of toluene in water (cm²/sec)

___

(continued)

TABLE 11 (continued)

---

[3]  $k_L = 3.12 \ (1.024)^{\theta-20} \ U_o^{0.67} \ H_o^{-0.85} \left( \dfrac{D_{i,H_2O}}{D_{O_2,H_2O}} \right)$    (Owens, Edwards and Gibbs, 1964, as presented by Hwang, 1982)

where;

$k_L$ is in lb-mol/ft$^2$-hr

$\theta$ = temperature (°C)

$U_o$ = surface velocity, ft/sec, normally 0.035 x wind speed (ft/sec) for natural surface, and 0.1 ft/sec for outside region of effect of aerators in the biological treatment.

$H_o$ = effective depth of surface impoundment (ft)

$D_{O_2,H_2O}$ = diffusion coefficient of oxygen in water (cm$^2$/sec)

Gas Phase Coefficient ($k_G$):

[1]  $k_G = 0.0958 \ U_{air}^{0.78} \ N_{sc}^{-0.67} \ d_e^{-0.11} \ \dfrac{\rho_{air}}{MW_{air}}$    (MacKay and Matsugu, 1973)

where;

$k_G$ is in lb-mol/ft$^2$-hr

$U_{air}$ = wind speed (m/hr)

$N_{Sc}$ = gas Schmidt number = $\mu_g / \rho_g D_{i,air}$

$\mu_g$ = absolute gas viscosity.(g/cm-sec)

$\rho_g$ = density of gas (g/cm$^3$)

$D_{i,air}$ = diffusion coefficient of compound i in air (cm$^2$/sec)

$d_e$ = effective diameter of the quiescent area of the impoundment
(m) = $\left( \dfrac{4Ac}{\pi} \right)^{0.5}$

$A_c$ = area of convective (natural) zone of impoundment surface (m$^2$)

$\rho_{air}$ = density of air (lb/ft$^3$)

$MW_{air}$ = molecular weight of air (28.8 lb/lb-mole)

---

(continued)

TABLE 11 (continued)

---

[2]  $k_G = 0.03 \left(\dfrac{D_{AA}}{D_{WA}}\right) V_8/A^{0.05}$    (Thibodeaux and Parker, 1974)

where;

$k_G$ is in lb-mol/ft$^2$-hr

$D_{WA}$ = diffusivity of water vapor in air (cm$^2$/sec)

$D_{AA}$ = diffusivity of compound A in air (cm$^2$/sec)

$V_3$ = wind speed at 8 meters above the water surface (miles/hr)

A = surface area of water body (acres)

---

TABLE 12.   HWANG's (1982) SIMPLIFICATION OF THE EMPIRICAL LIQUID
AND GAS PHASE MASS TRANSFER COEFFICIENTS FOR A
NONAERATED IMPOUNDMENT

Liquid Phase Coefficient of Compound i ($k_{L,i}$):

$$k_{L,i} = \left(\frac{MW_{O_2}}{MW_i}\right)^{0.5} \left(\frac{273 + \theta}{298}\right) k_{L,O_2}.$$

Gas Phase Coefficient of Compound i ($k_{G,i}$):

$$k_{G,i} = \left(\frac{MW_{H_2O}}{MW_i}\right)^{0.335} \left(\frac{273 + \theta}{298}\right)^{1.005} k_{G,H_2O}$$

where;

$MW_{O_2}$ = molecular weight of oxygen (g/g-mole)

$MW_{H_2O}$ = molecular weight of water vapor (g/g-mole)

$MW_i$ = molecular weight of compound i (g/g-mole)

$\theta$ = temperature of concern (°C)

$k_{L,O_2}$ = liquid-phase mass transfer coefficient of oxygen at 25°C
(g-mol/cm$^2$-sec)

$k_{G,H_2O}$ = gas phase mass transfer coefficient of water vapor at 25°C
(g-mol/cm$^2$-sec)

case is more accurate). Therefore, from these deterministic equations, the complex empirical relationships remain a key variable in obtaining either the liquid and/or gas phase mass transfer coefficients for the compound of interest.

Shen (1982), using the rate expression proposed by Thibodeaux et al. (1981d), described a simplifying method for estimating the volatilization rate of volatile organic compounds from waste lagoons. Shen's modifications included the use of Owen's correlation for $k_L$ determination and the use of Mackay and Matsugu's correlation for $k_G$. Owen's correlation was obtained from reaeration studies of flowing streams, and thus may not be directly applicable to stagnant waste lagoons. For both $k_L$ and $k_G$, Shen made some calculation errors in converting the units of the original correlations. Thus, correlations shown previously in Table 8 should be modified as shown in Table 11.

In applying MacKay and Matsugu's correlation, Shen suggests a molecular weight dependency for Schmidt number determinations, such that the following simplification could be employed:

| Molecular Weight | $Sc^{-0.67}$ |
|---|---|
| Less than 100 | 0.7 |
| 100-200 | 0.6 |
| More than 200 | 0.5 |

## Smith, Bomberger and Haynes (1980-1981)

Smith, Bomberger and Haynes proposed a technique for determining an overall volatilization rate constant based on laboratory measurements. The volatilization rate constant is essentially the overall K-value that represents the individual liquid and gas phase k-values combined. If found adequate, this approach would eliminate the complex empirical relationships for determining the individual values proposed by other researchers.

Smith, Bomberger and Haynes essentially expanded on several concepts presented by Liss and Slater (1974); i.e., (1) mass transfer in the liquid phase controls more than 95 percent of the volatilization rate if $H_c$ (Henry's law constant) is greater than approximately 3500 torr-LM$^{-1}$ (0.0044 atm-m$^3$-mol$^{-1}$); (2) if $H_c$ is less than approximately 10 torr-LM$^{-1}$ (1.2 x 10$^{-5}$ atm-m$^3$-mol$^{-1}$), mass transfer in the gas phase controls more than 95 percent of the volatilization rate; and (3) if $H_c$ is between 10 and 3500 torr-LM$^{-1}$, both gas and liquid phase mass transfer are important. Smith, Bomberger and Haynes categorized chemicals from these three classifications as high, low, and intermediate volatility compounds, respectively. A technique was developed for estimating the volatilization rate constant ($k_v^c$) for each class of volatile compounds. Table 13 shows these three estimation techniques and the model equation presented by Smith, et al. appears in Table 14.

TABLE 13.    SUMMARY OF TECHNIQUES DEVELOPED BY SMITH, BOMBERGER AND HAYNES
TO ESTIMATE THE VOLATILIZATION RATE CONSTANT ($k_v^c$) FOR
HIGH, LOW AND INTERMEDIATE VOLATILITY COMPOUNDS

High-Volatility Compounds ($H_c > 3,500$ torr-LM$^{-1}$):

$$k_v^{\ c} = \left(k_v^{\ c}/k_v^{\ o}\right)_{lab} \times \left(k_v^{\ o}\right)_{env}$$

where;

$k_v^{\ c}$ = volatilization rate constant of compound c (hr$^{-1}$)

$\left(k_v^{\ c}/k_v^{\ o}\right)_{lab}$ = ratio of volatilization rate constants for compound c and
oxygen measured simultaneously in the laboratory (shown to be a
constant for high-volatility compounds

$\left(k_v^{\ o}\right)_{env}$ = oxygen reaeration constant in a real water body $\left(k_v^{\ o}\right)_{env}/L$, since
the second term in the two-film resistance theory model is small
compared to the first for high-volatility compounds

Low-Volatility Compounds ($H_c < 10$ torr-L M$^{-1}$):

$$k_v^{\ c} = \left(k_v^{\ c}/k_v^{\ w}\right)_{lab} \times \left(k_v^{\ w}\right)_{env}$$

where;

$k_v^{\ c}$ = volatilization rate constant of compound c (hr$^{-1}$)

$\left(k_v^{\ c}/k_v^{\ w}\right)_{lab}$ = ratio of volatilization rate constants for compound c and water
vapor measured simultaneously in the laboratory (shown to be a
constant for low-volatility compounds)

$\left(k_v^{\ w}\right)_{env}$ = $k_g^{\ w}H_c/LRT$, since the first term in the two-film resistance theory
is small compared to the second for low-volatility compounds

$k_g^{\ w}$ = gas phase mass transfer coefficient for water, determined from Fick's
law, i.e., $N^w = k_g^{\ w} (P_s^{\ w} - P^w)/RT$

where;

(continued)

TABLE 13 (continued)

$N^w$ = measured water evaporation flux rate ($mol/cm^2$-hr)

$P_s^w$ = saturated partial pressure of water at temperature T

$P^w$ = actual partial pressure of water at temperature T

Intermediate Volatility Compounds ($H_c$ = 10 to 3500 torr-LM$^{-1}$):

Combines the estimation techniques of both high- and low-volatility compounds.

TABLE 14.  SMITH, BOMBERGER AND HAYNES FIRST ORDER KINETIC EQUATION FOR NONAERATED SURFACE IMPOUNDMENTS

**Model Form:**

$$R_v = \frac{-d(c)}{dt} = k_v^c \, (c)$$

$$k_v^c = \frac{1}{L} \left[ \frac{1}{k_L^c} + \frac{RT}{H_c k_G^c} \right]^{-1}$$

**Definition of Terms:**

$R_v$ or $\frac{-d(c)}{dt}$ = volatilization rate of compound c from water (mg/l-hr)

$k_v^c$ = volatilization rate constant of compound c $(hr^{-1})$

L = length of impoundment (cm)

$k_L^c$ = individual liquid phase mass transfer coefficient of compound c (cm/hr)

$k_G^c$ = individual gas phase mass transfer coefficient of compound c (cm/hr)

R = ideal gas constant (L torr/°K mol)

T = temperature (°K)

$H_c$ = Henry's law constant (torr L/mol)

The key limitation to the Smith, Bomberger and Haynes approach is that laboratory determinations of the ratios $(k_v^c/k_v^o)$ and $(k_v^c/k_v^w)$ are compound specific, sophisticated and expensive. Only a very limited number of compounds have been tested to date. Using other methods for determining these ratios; i.e., using values of the diffusion coefficients, increase the overall uncertainty of this model.

## McCord (1981)

A fourth predictive model for estimating the emission rate of volatile compounds from nonaerated lagoons was presented by McCord (1981). This steady-state model, shown in Table 15, was based upon Nusselt's equation which describes water evaporation rates from a lagoon. Consequently, McCord's model is limited to situations where volatilization is controlled by the gas phase mass transfer. This limitation is similar to that of Mackay and Wolkoff's (1973) model. It has been shown by more recent research that most volatilization is liquid phase controlled. Additionally, McCord assumes that equilibrium at the air-water interface follows Raoult's law. However, for dilute aqueous solutions, the equilibrium should actually be determined on the basis of Henry's law.

## AERATED IMPOUNDMENTS

### Thibodeaux, Parker, Heck (1981d)

Thibodeaux, Parker and Heck (1981d) proposed use of their steady state model for nonaerated SIs for use with aerated impoundments with some modifications. The theory of two-film resistance is applicable to an aerated impoundment if one considers that turbulence, caused by aeration, creates two distinct zones at the impoundment surface; i.e., (1) turbulent, and (2) convective or natural. The overall liquid phase mass transfer coefficient for the entire system must be modified to account for each distinct zone, proportional to the affected area of each zone. The resulting overall K-value proposed by Thibodeaux, Parker and Heck for aerated SIs appears in Table 16.

As shown previously in Table 11, empirical relationships were developed by Thibodeaux, et al. to estimate individual liquid and gas phase k-values for a nonaerated (entirely convective zone) SI; i.e., $(k_L)_c$ and $(k_G)_c$, respectively. Similar relationships have also been experimentally developed to account for the turbulent zone caused by mechanical aeration. These empirical equations are shown below:

- Individual liquid phase coefficient:

$$(k_L)_T = \frac{J(POWR)(1.024)^{\theta-20}(\alpha)(10^6)}{165.04(a_v)(v)} \left( \frac{D_{i,H_2O}}{D_{O_2,H_2O}} \right)^{0.5}$$

TABLE 15.    McCORD'S STEADY-STATE PREDICTIVE MODEL FOR
NON AERATED SURFACE IMPOUNDMENTS

Model Form:

$$E_s = \frac{E_o L_s}{L_o}$$

where;

$$E_o = 0.53425 \, A_m \left(\frac{D}{L}\right)^{0.22} (\%) \, (VP) \, (W_o)^{0.78}$$

$$L_s = \frac{K}{1-x}$$

$$1-x = \frac{E_o}{L_o} + \frac{F}{V}$$

$$K = 0.0834(F) \, (\%) \, (sp.gr.)$$

$$L_o = 0.0834(V) \, (\%) \, (sp.gr.)$$

Definition of Terms:

$E_s$ = evaporation rate of a compound at steady state (lb/hr)
$E_o$ = initial evaporation rate of a compound (lb/hr)
$L_s$ = amount of compound in lagoon at steady state (lb)
$L_o$ = initial amount of compound in lagoon (lb)
$A_m$ = surface area of lagoon ($m^2$)
$D$ = diffusivity coefficient of a compound in air ($m^2$/hr)
$L$ = a lagoon dimension (m)
$\%$ = weight percent of compound in water or lagoon liquid
$VP$ = vapor pressure of pure compound. (atm)
$W_o$ = wind velocity (m/sec)
$K$ = amount of compound in feed stream (lb/hr)
$F$ = feed rate (=discharge rate) in lagoon (gal/hr)
$V$ = operating volume of lagoon (gal)
sp.gr. = specific gravity of liquid waste

TABLE 16.    K-VALUES FOR AERATED SIs PROPOSED BY
THIBODEAUX, PARKER AND HECK

$$K_L = (K_L)_c \frac{A_c}{A} + (K_L)_T \frac{A_T}{A}$$

where;

$K_L$ = overall liquid phase mass transfer coefficient for the entire aerated surface impoundment (also termed the "area-averaged" coefficient) to be used in the basic model equation.

$(K_L)_c$ = overall liquid phase mass transfer coefficient

for the convective zone of the impoundment

$$\frac{1}{(K_L)_c} = \frac{1}{(k_L)_c} + \frac{1}{K(k_G)_c}$$

$A_c$ = effective surface area of the convective zone

$A$ = total surface area of the aerated impoundment
$(A=A_T+A_c)$

$(K_L)_T$ = overall liquid phase mass transfer coefficient for the turbulent zone of the impoundment.

$K$ = equilibrium constant

$$\frac{1}{(K_L)_T} = \frac{1}{(k_L)_T} + \frac{1}{K(k_g)_T}$$

$A_T$ = effective surface area of the turbulent zone

where:

$J$ = oxygen-transfer rating of mechanical aerator (3 lb $O_2$/hr-hp);

POWR = total power input to aerators (hp)-rated hp x efficiency (0.65-0.9);

$\theta$ = temperature (°C);

$\alpha$ = oxygen-transfer correction factor (0.8-0.85);

$a_v$ = surface area per unit volume of surface impoundment (ft$^{-1}$);

$V$ = volume of surface impoundment in region of effect of aerators (ft$^3$);

$D_{i,H_2O}$ = diffusion coefficient of compound i in water (cm$^2$/sec);

$D_{O_2,H_2O}$ = diffusion coefficient of oxygen in water (cm$^2$/sec)

- **Individual gas phase coefficient:**

$$(k_g)_T = 0.00039 \frac{\rho_g D_{i,air}}{d} (N_{Re})^{1.42} (N_{FR})^{-0.21} (N_p)^{0.4} (N_{Sc})^{0.5}$$

where:

$\rho_g$ = density of the gas (lb/ft$^3$)

$D_{i,air}$ = diffusion coefficient of compound i in air (ft$^2$/hr)

$d$ = diameter of aerator turbine or impeller (ft)

$N_{Re}$ = gas Reynolds number = $\rho_g d^2 \omega / \mu_g$

$\omega$ = rotational speed of turbine impeller (rad/sec)

$\mu_g$ = absolute gas viscosity (g/cm-sec)

$N_{FR}$ = Froude number = $d\omega^2/g$

$g$ = gravitational constant (32.17 ft/sec$^2$)

$N_p$ = power number = Pr g/$\rho_L d^5 \omega^3$

Pr = power to the impeller (ft-lb$_f$/sec)

$\rho_L$ = density of the liquid (lb/ft$^3$)

$N_{Sc}$ = gas schmidt number = $\mu_g / \rho_g D_{i,air}$

Note: $N_{Sh}$ = Sherwood number = $(k_g)_T \, d/\rho_g D_{i,air}$

and therefore, $N_{Sh} = 0.00039(N_{Re})^{1.42}(N_p)^{0.4}(N_{Sc})^{0.5}/(N_{FR})^{0.21}$

Freeman (1979), working with surface aerated waste treatment basins, has tabulated values published in the literature for estimating; (1) effective surface area of the turbulent zone $(A_T)$; (2) surface area per unit volume of surface impoundment $(a_v)$; and (3) volume of surface impoundment in region of effect of aerators (V), based on the mechanical aerators horsepower, $\omega$. These parameters are shown in Table 17.

As discussed in the preceeding section, several other researchers have shown that the individual mass-transfer coefficients for various compounds in a nonaerated impoundment (convective zone coefficients) can be simplified by comparison to a typical compound. This simplification process has been applied for determining the individual mass transfer coefficients from the turbulent zone of an aerated impoundment as shown below:

- Individual liquid phase coefficient:

$$(k_{L,i})_T = \left(\frac{MW_{O_2}}{MW_i}\right)^{0.25} \left(\frac{1.024^{\theta-20}}{1.024^5}\right) \left(\frac{273+\theta}{298}\right)^{0.5} \left(k_{L,O_2}\right)$$

- Individual gas phase coefficient:

$$(k_{G,i})_T = \left(\frac{MW_{H_2O}}{MW_i}\right)^{0.25} \left(\frac{298}{273+\theta}\right)^{0.92} \left(k_{G,H_2O}\right)$$

However, as noted previously, a key limitation to this simplification procedure is that the values of $k_{L,O_2}$ and $k_{G,H_2O}$ are best determined by using the complex empirical relationships shown earlier.

### McCord (1981)

One additional predictive model was proposed by McCord (1981) to estimate the emission rate of volatile compounds from aerated lagoons. This steady-state model, based upon Arnold's (1944) studies on the diffusion of volatile compounds from a liquid surface into air, appears in Table 18.

Discussions in the preceeding section concerning McCord's steady-state predictive model for nonaerated impoundments showed that the basis for McCord's model is evaporation of water from the impoundment surface; i.e., gas phase controlling. Therefore, in reality, this model does not accurately represent the volatilization rate of most sparingly (slightly) soluble volatile organic compounds.

TABLE 17.  TURBULENT AREAS AND VOLUMES FOR SURFACE AGITATORS[a]

| $\omega$, Motor horsepower, hp | $A_T$, Turbulent area, $ft^2$ | Effective depth, ft | V, Agitated volume, $ft^3$ | $a_V$, Area per volume $ft^2/ft^3$ |
|---|---|---|---|---|
| 5 | 177 | 10 | 1,767 | 0.100 |
| 7.5 | 201 | 10 | 2,010 | 0.100 |
| 10 | 227 | 10.5 | 2,383 | 0.0952 |
| 15 | 284 | 11 | 3,119 | 0.0909 |
| 20 | 346 | 11.5 | 3,983 | 0.0870 |
| 25 | 415 | 12 | 4,986 | 0.0833 |
| 30 | 491 | 12 | 5,890 | 0.0833 |
| 40 | 661 | 13 | 8,587 | 0.0769 |
| 50 | 855 | 14 | 11,970 | 0.0714 |
| 60 | 1,075 | 15 | 16,130 | 0.0666 |
| 75 | 1,452 | 16 | 23,240 | 0.0625 |
| 100 | 2,206 | 18 | 39,710 | 0.0555 |

[a]Data for a high speed (1,200 rpm) aerator with 60 cm propeller diameter (d).

TABLE 18.   McCORD'S STEADY-STATE PREDICTIVE MODEL FOR AERATED SIs

Model Form:

$$E_s = \frac{E_o L_s}{L_o}$$

where;

$$E_o = \frac{150.315}{(460 + T)} \; \frac{H}{r} \; \frac{L_o}{V_L} \; \frac{VP}{sp. \; gr.} \; \frac{d}{\pi} \; \frac{1}{\sqrt{0.17678R}^{1/2}}$$

$$L_s = \frac{K}{1-x}$$

$$1-x = \frac{E_o}{L_o} + \frac{F}{V}$$

$$K = (0.0834)(F)(\%)(sp.gr.)$$

$$L_o = (0.0834) \; (V) \; (\%) \; (sp. \; gr.)$$

Definition of Terms:

$E_s$ = evaporation rate of a compound at steady-state conditions (lb/hr)

$E_o$ = initial evaporation rate of a compound (lb/hr)

$L_s$ = pounds of a compound in the lagoon at steady-state conditions

$L_o$ = initial pounds of a compound in the lagoon

$T$ = temperature (°F)

$H$ = amount of water delivered by the aeration pump (gpm)

$r$ = radius of spherical drops emitted (cm)

$V_L$ = volume of liquid waste in lagoon (U.S. gal)

$VP$ = vapor pressure of the volatile compound at temperature T (atm)

sp. gr. = specific gravity of the liquid waste

$d$ = diffusivity coefficient of a compound in air ($cm^2$/sec)

(continued)

TABLE 18 (continued)

R = maximum diameter of the aeration spray falling back into the lagoon (m)

K = amount of compound in the feed stream (lb/hr)

F = feed rate (= discharge rate) in lagoon (gal/hr)

V = operating volume of lagoon (gal)

% = weight percent of a compound in water or lagoon liquid

# 5. Review of Landfill AERR Models

INTRODUCTION

Six air emission modeling equations were presented in the literature for landfill facilities:

- Three models for covered landfills without internal gas generation (Farmer, Shen, Thibodeaux); and

- One model for each of the following scenarios: (1) covered landfills with internal gas generation (Thibodeaux); (2) covered landfills with gas generation and barometric pumping effects (Thibodeaux); and (3) uncovered landfills (Shen).

This section presents all models for landfills found in the literature. Tables 19 and 20 respectively summarize available model equations for landfills and the necessary input parameters required for each equation. GCA's recommendations for landfill models and mass transfer coefficients were presented previously in Section 2.

Two of the three AERR equations for covered landfills without internal gas generation were based on laboratory studies conducted by Farmer, et al., (1978). Shen (1980) modified Farmer's equation by changing the flux rate to an emission rate by multiplying by the landfill surface area. In addition, to determine the AERR for a specific component of the waste, Shen multiplies the emission rate by the weight fraction of that component in the bulk waste. Details of both models are provided in this section.

Farmer's laboratory experiments assumed the vapor release of hexachloro-benzene (HCB) was diffusion controlled within the soil phase because of HCBs insolubility in water (i.e., not effected by water movement) and HCBs resistance to biological activity. However, Thibodeaux's (1981a) equation takes on a similarity of the two-resistance theory of mass transfer. Thibodeaux not only describes the vapor movement within the soil phase (as does Farmer), but also addresses the vapor movement from air-soil interface to the overlying air. In practice, the air-soil interface transfer should provide negligible resistance, and thus can be disregarded.

TABLE 19.   REVIEW OF AVAILABLE AERR LANDFILL MODELS[a]
(AS PRESENTED IN THE LITERATURE)

A.   Farmer, et al. (1978) for covered landfills:

(1)   $J = D_o \; (P_a^{10/3}/P_T^2) \; (C_s - C_2)/L$

(2)   $D_A = D_B \left(\dfrac{M_B}{M_A}\right)^{1/2}$

(3)   $C_s = P^o \dfrac{M}{RT}$

(4)   $P_a = P_T - \theta$

(5)   $P_T = 1 - \dfrac{\beta}{\rho}$

(6)   $\theta = \dfrac{\omega\beta}{\rho_\omega}$

B.   Shen's (1980) Modification of Farmer's Equation:

(1)   $E_i = D_i \; C_{si} \; A \; P_T^{4/3} \; \left(\dfrac{1}{L}\right)\left(\dfrac{W_i}{W}\right)$

(2)   $D_i = D_A = D_B \left(\dfrac{M_B}{M_A}\right)^{1/2}$

(3)   $C_{si} = C_s = \dfrac{P^o M}{RT}$

(4)   $P_T = 1 - \dfrac{\beta}{\rho}$

C.   Thibodeaux's (1981a) Landfill Equation--Without Internal Gas Generation:

(1)   $N_A = {}^1K_{A3} \; (\rho_{A1}^* - \rho_{A1})$

(2)   $N_A = N_{A \; soil} + N_{A \; air\text{-}soil}$

(continued)

TABLE 19 (continued)

(3) $N_{A_{soil}} = \dfrac{D_{A3}}{h} (\rho^{*}_{Ai} - \rho_{A1i})$

(4) $N_{A\ air-soil} = {}^{3}K_{A1} (\rho_{A1i} - \rho_{A1})$

(5) $D_{A3} = D_{A1} \dfrac{\varepsilon}{\tau}$

(6) $\dfrac{1}{{}^{T}K_{A3}} = \dfrac{1}{{}^{T}K_{A3}} + \dfrac{1}{{}^{3}K_{A1}}$

D.    Thibodeaux's (1981a) Landfill Equation--With Internal Gas Generation

(1) $N_A = V_y \dfrac{(\rho^{*}_{A1} - \rho_{A1i})}{\exp (h\ V_y/D_{A3}) - 1} + V_y\ \rho^{*}_{Ai}$

(2) $\rho_{A1} = \rho^{*}_{A1} - (\rho^{*}_{A1} - \rho_{A1i}) \dfrac{1 - \exp (y\ V_y/D_{A3})}{1 - \exp (h\ V_y/D_{A3})}$

E.    Thibodeaux, et al. (1981b) Landfill Equation With Internal Gas Generation and Barometric Pumping:

(1) $N_A = \dfrac{D_{A3}}{L}\ \rho^{*}_{AL}\ \dfrac{R\ \exp\ R}{\exp\ R-1}$

(2) $R = \dfrac{LV}{D_{A3}}$

(3) $V = \dfrac{K}{\mu L} (P - \pi)$

F.    Shen's (1980) Open Dump Equation:

(1) $\dfrac{dV}{dt} = 2\ C_e\ \omega \sqrt{\dfrac{DLV}{\pi\ Fv}}\ \dfrac{W_i}{W}$

[a]Definition of terms appears in Table 20.

TABLE 20.    INPUT PARAMETERS REQUIRED FOR EACH LANDFILL MODEL IN TABLE 19

| Parameter | A | B | C | D | E | F |
|---|---|---|---|---|---|---|
| J, vapor flux through soil | X | | | | | |
| $D_o$, diffusion coeff. in air | X | X | | | | |
| $C_s$, conc. material in air | X | X | | | | |
| $C_2$, conc. of material at soil surface | X | X | | | | |
| L, soil depth | X | X | | | | |
| $P_a$, soil air filled porosity | X | X | | | | |
| $P_T$, total soil porosity | X | X | X | | | |
| $D_A$, air diffusion coeff. for unknown compound | X | X | | | | |
| $D_B$, air diffusion coeff. for known compound | X | X | | | | |
| $P^o$, vapor pressure | X | X | | | | |
| M, molecular weight | X | X | | | | |
| R, molar gas constant | X | X | | | | |
| T, absolute temperature | X | X | | | | |
| $\beta$, soil bulk density | X | X | | | | |
| $\omega$, gravimetric soil water content | X | X | | | | |
| $\rho_\omega$, density of water | X | X | | | | |
| $\rho$, particle density | X | X | | | | |
| $W_i/W$, weight % of component i in bulk waste | | X | | | | |
| $E_i$, emission rate of vapor i | | X | | | | |
| $N_a$, mass flux rate | | | X | X | X | |

(continued)

TABLE 20 (continued)

| Parameter | A | B | C | D | E | F |
|---|---|---|---|---|---|---|
| $^1K_{A3}$, overall soil phase transfer coeff. | | | X | | | |
| $\rho_{A1}^{*}$, conc. of A in sand chamber filled pore spaces | | | X | X | | |
| $\rho_{A1}$, conc. of A in air at distance from air–soil interface | | | X | X | X | |
| $N_A$, rate of vapor movement within soil phase | | | X | X | X | |
| h, depth of fill cover | | | X | X | | |
| $D_{A3}$, effective diffusivity of A within the air-filed soil pore space | | | X | X | X | |
| $D_{A1}$, molecular diffusivity of A in air | | | X | | | |
| $\tau$, tortuosity | | | X | | | |
| $^3k_{A1}$, gas phase mass transfer coeff. | | | X | | | |
| $V_x$, wind speed at 10 m | | | X | | | |
| L, length of ground emission source | | | X | | | |
| $V_y$, mean gas velocity in pore spaces | | | | X | | |
| $\rho_{A1i}$, conc. of A at air–soil interface | | | | X | | |
| L, cap thickness | | | | | X | |
| V, superficial velocity through cap | | | | | X | |
| K, permeability of cap material | | | | | X | |
| $\mu$, cell gas velocity | | | | | X | |
| P, cell gas pressure | | | | | X | |

(continued)

TABLE 20 (continued)

| Parameter | Model | | | | | |
|-----------|---|---|---|---|---|---|
|           | A | B | C | D | E | F |
| $\pi$, atmospheric pressure | | | | | X | |
| $\frac{dV}{dt}$, volume conc. of vapor | | | | | | X |
| $C_e$, equilibrium vapor conc. | | | | | | X |
| $\omega$, width of open dump | | | | | | X |
| V, wind speed | | | | | | X |
| $F_v$, correction factor | | | | | | X |

Thibodeaux (1981a) presents another model describing vapor movement in soil which is subject to the effects of biogenic processes and related gas generation. This "sweeping" action provided by the upward movement of landfill gases provides a parallel transfer motion to the molecular diffusion of vapor in the soil phase.

Thibodeaux (1981b) further developed his model to include the barometric pumping effect caused by fluctuations in atmospheric pressure and landfill cell pressure. Computer simulation tests indicated that the flux rate of benzene, $N_A$, is influenced only slightly by barometric pressure fluctuations under conditions of co-disposal, and influenced significantly under conditions of no internal gas generation (i.e., no co-disposal with municipal refuse).

This statement suggests that for a co-disposal facility, Thibodeaux's model for landfills with gas generation should be employed. For a landfill handling strictly hazardous waste, with no expected internal gas generation, either of two models (Farmer, or Thibodeaux) may be appropriate. However according to Springer (1983), the effect of barometer pumping is not reversible and will not average out over a long period of time. An increase of 10 to 15 percent to emission release rate could be expected.

To determine AERR from an uncovered landfill, Shen (1981) presents the equation based on Fick's Law. This is the only open dump AERR model presented in the literature and it assumes an insoluble material with vapor movement that is only air-phase controlled. A review of Thibodeaux's two-resistance-landfill equation, specifically the rate of vapor movement from the air-soil interface to overlying air (air-phase resistance) indicates it may be a more accurate description of vapor movement in the gas phase because it accurately accounts for wind effects across the earth's boundary i.e., Sutton (1953). Further development for the open dump case may be warranted.

## FARMER, ET AL. (1978) FOR COVERED LANDFILLS

The equation developed by Farmer was intended as a method of assisting a planner in designing a landfill cover that minimized the escape of hexachlorobenzene (HCB) or other volatile organic vapors. Alternatively, the equation could be used to assess the effectiveness of an existing landfill cover for controlling organic vapor flux to the atmosphere. The equation was experimentally verified by Farmer for HCB-containing waste in a laboratory-simulated landfill.

Using Fick's First Law for steady-state diffusion, Farmer describes the volatilization or vapor loss of HCB, or other compounds, as a diffusion controlled process. With the assumptions of no degradation from biological activity, no adsorption of the compound, no transport in moving water and no landfill gas production, the rate at which a compound will volatize from the soil surface to the atmosphere will be controlled by the diffusion rate through the soil cover. To describe molecular diffusion through a soil surface, Farmer adopted the effective diffusion coefficient suggested by Millington and Quirk (1961) as shown below:

$$D_e = D_o \left( \frac{P_a^{10/3}}{P_T^2} \right)$$

where: $D_e$ = effective diffusion coefficient in soil;

$D_o$ = diffusion coefficient in air;

$P_a$ = soil air-filled porosity;

$P_T$ = total porosity.

This equation describes the effective diffusion coefficient as a power function of soil porosity. This method may be valid, however, as described previously in Section 2, additional research indicated the power function may be soil-type dependent. Thus, this equation appears to be limited to the soil class with which Millington and Quirk experimented with, namely isotropic soil.

The following describes the methods employed by Farmer (1980) to determine equation parameters, and identifies some simplifications to the overall equation.

A.  Diffusion coefficients, $D_o$, are available for many compounds, generally at a specified temperature. The vapor diffusion coefficient for an unknown compound A can be estimated from a known compound (B) by the following equation:

$$D_A = D_B \left( \frac{M_B}{M_A} \right)^{1/2}$$

where    $M_A$, $M_B$ = molecular weights of compounds A and B, respectively.

To determine the diffusion coefficient at a temperature other than a temperature listed, the following equation can be used:

$$D_2 = D_1 \left( T_2 / T_1 \right)^{1.5}$$

where $D_1$ and $D_2$ are the diffusion coefficients at temperatures $T_1$ and $T_2$, respectively.

B.  Saturation vapor density at the bottom of the soil layer, $C_S$, can be obtained based on the ideal gas law:

$$C_S = P^o M / RT$$

where:    $P^o$ = vapor pressure (mm Hg);

$M$ = molecular weight (g/g-mole);

R = molar gas constant (62.36 1-mm Hg/°K-mole)

T = absolute temperature (°K)

Farmer experimented with pure HCB, thus, he calculated the pure component vapor concentration. For mixtures, the vapor concentration of a component in the waste becomes the driving force and this is not equal to the saturation vapor concentration of the pure component. The vapor concentration for a component in a waste mixture is calculated by:

$$C_i^* = \frac{P_i^* M}{RT}$$

where $P_i^*$ = partial pressure of component i in the waste mixture.

The partial pressure in a liquid mixture is determined as follows:

$$P_i^* = \gamma_i P^o X_i$$

where: $\gamma_i$ = activity coefficient of component i; and

$X_i$ = mole fraction of component i in the liquid.

For hydrocarbon mixtures, $\gamma_i$ = 1, but for aqueous mixtures the activity coefficient must be determined.

C.  Soil air-filled porosity, $P_a$, is calculated from the total porosity, $P_T$, and the volumetric soil water content, $\theta$, where

$$P_a = P_T - \theta$$

and                    $$\theta = W\beta/\rho_w$$

where:   W = gravimetric soil water content (g/g);

$\beta$ = soil bulk density (g/cm$^3$);

$\rho_w$ = density of water = 1 g/cm$^3$.

A worst-case assumption would be completely dry soil, thus, $P_a$ = $P_T$, and the flux equation becomes

$$J = D_o P_T^{4/3} \left(C_S - C_2\right)/L$$

The addition of any water to the soil will reduce the air-filled porosity, thus, reduce vapor flux from the soil surface because the diffusion rate through a liquid is generally several magnitudes less than diffusion through air.

D.  The total soil porosity, $P_T$, can be calculated from the soil bulk density, ß, by the following equation:

$$P_T = 1 - ß/\rho$$

where:    ß = soil bulk density (g/cm$^3$);

$\rho$ = particle density (g/cm$^3$).

Particle density is usually taken as 2.65 g/cm$^3$ for most soil mineral material, while soil bulk density usually varies from 1.0-2.0 g/cm$^3$. Therefore, $P_T$ can range from 0.245 to 0.623 cm$^3$/cm$^3$, and $(P_T)^{4/3}$ = 0.153 to 0.532.

Worst-case scenario would be $P_T^{4/3}$ = 0.532 (from ß = 1.0 g/cm$^3$), thus J = 0.532 $D_o$ ($C_S$-$C_2$)/L. This flux rate equation would allow for a worst case estimate of air emissions.

E.  Farmer assumed that the concentration of volatilizing material at the soil surface, $C_2$, is zero. The reasons for this assumption were stated by Farmer as:

1.  the amount of vapor reaching soil surface will be very small; and

2.  the vapor will be rapidly dispersed by wind currents and by diffusion in air. Thus, the simplified version of Farmer's equation becomes:

$$J = D_o \, P_T^{4/3} \, C_S/L$$

The assumption of $C_2$ = 0, implies worst-case situation because any increase in $C_2$ will effectively reduce the concentration gradient (driving force behind vapor flux) and reduce vapor flux from soil surface.

SHEN'S (1980) MODIFICATION OF FARMER'S EQUATION

Basically, Shen took the simplified version of Farmer's equation for vapor flux from a soil surface and converted to an emission rate by multiplying by the exposed area. To determine the emission rate of a specific waste, component i, Shen multiplied by the weight percent of component i in the bulk waste (Wi/W). However, weight percent should be replaced by mole fraction. Additionally, in calculating the emission rate of a specific compound in a waste mixture, consideration of waste composition and the activity coefficient is important. This procedure was previously outlined.

Since both Farmer's and Shen's equations are nearly identical, the parameters $D_i$, $C_{si}$, and $P_T$ can be calculated and simplified in similar manners. For illustrative purposes, Table 21 presents typical parameter values and ranges for the modified landfill equation.

TABLE 21. TYPICAL PARAMETER VALUES AND RANGES FOR THE FARMER LANDFILL MODEL AS MODIFIED BY SHEN

Equation 1: $E = D C_s A P_t^{4/3} \dfrac{1}{L} \times \dfrac{W_i}{W}$  Equation 2: $D_i = D_K \left[\dfrac{M_K}{M_i}\right]^{0.5} \left[\dfrac{T}{T_K}\right]^{1.5}$  Equation 3: $C_s = \dfrac{PM}{RT}$  Equation 4: $P_t = 1 - \dfrac{a}{\rho}$

| Symbol | Units | Definition | Source of input parameter | Typical value | Expected range | Comments |
|---|---|---|---|---|---|---|
| D | cm²/sec | Diffusion coeff. | Calc. from Eq. 2 | NA | $10^{-1}$-$10^{-3}$ | Methanol high at 0.157; trichloroethylene low at 0.0525 @ 20°C |
| $C_s$ | g/m³ | Saturation vapor conc. | Calc. from Eq. 3 | NA | $10^{-3}$-$10^{-5}$ | Methyl chloride high at 2.39 x $10^{-3}$; benzene low at 3.0 x $10^{-5}$ |
| A | cm² | Surface area | Measured or typical value assumed | 100 m² | 25-5,000 m² | Highly variable parameter |
| $P_t$ | Dimensionless | Soil porosity | Calc. from Eq. 4 | 0.43 | 0.25-0.62 | |
| L | cm | Depth of soil cover | Measure or typical value assumed | 30 | 25-35 | |
| $W_i/W$ | g/g | Weight fraction of chem. | Measure or typical value assumed | 2% | 0.5-5% | Highly variable parameter |
| $D_K$ | cm²/sec | Known D of ref. compound | Chemical handbook | | | |
| $M_K$ | g/mole | MW of ref. compound | Chemical handbook | NA | 50-190 | For organic compounds, chloromethane is low @ 51, PCB is high @ 189 |
| $M_i$ | g/mole | MW of compound | Chemical handbook | NA | 50-190 | For organic compounds, chloromethane is low @ 51, PCB is high @ 189 |
| T | °K | Temperature | Measured or typical value assumed | 293 | 250-310 | Highly variable parameter |
| P | mm Hg | Vapor pressure | Chemical handbook | NA | 0.02-750 | PCB low at 0.015, Cyclohexane high at 721 |
| R | $\frac{\text{mm Hg-liter}}{\text{°K-mole}}$ | Gas constant | Chemical handbook | 62.36 | NA | Constant value |
| a | g/cm³ | Soil bulk density | Soil handbook | 1.5 | 1-2 | |
| ρ | g/cm³ | Particle density | Soil handbook | 2.65 | 2.4-2.8 | ρ = 2.65 for most mineral matter |

THIBODEAUX'S (1981a, 1981b) LANDFILL EQUATIONS

Thibodeaux (1981a, 1981b) presented three equations for estimating air emissions from covered landfills; (1) without internal gas generation, (2) with internal gas generation, and (3) with internal gas generation including the effect of barometric pumping pressure.

## Without Internal Gas Generation

Thibodeaux's equation for covered landfills without gas generation appears similar to the two resistance theory of mass transfer: i.e., Flux = (overall transfer coefficient) (concentration gradient). The equation is the algebraic sum of the two individual phase equations; the rate of vapor movement within the soil phase, and rate of vapor movement from air-soil interface to overlying air.

The rate controlling portion of Thibodeaux's flux rate equation is the rate of vapor movement within the soil phase. The mass transfer coefficient for vapor diffusion in soil is described for vapor movement through a porous media. However, other research previously presented in Section 2 indicates that a more correct expression is to show the effective diffusivity dependency upon a power function of porosity. This power is a function of soil type and moisture content. The second portion of Thibodeaux's equation describes the vapor movement through air, based on the work of Sutton (1953).

## With Internal Gas Generation

In addition to molecular diffusion of landfill vapors through soil, there is a "convective sweep" of chemical vapors toward the surface created by the formation of landfill gases ($CO_2$, $H_2$, $CH_4$), especially applicable to co-disposal sites. Therefore, the flux equation contains both a diffusive and convective term.

## Barometric Pumping Effects

Additional work by Thibodeaux, Springer and Riley (1981b) addressed a third vapor phase transport mechanism in addition to the diffusive and convective mechanisms previously discussed; the barometric pumping pressure. Atmospheric pressure fluctuations develop pressure gradients which pump vapors and gases from landfill cells to the air above. This pumping enhances vapor phase mass transfer.

Changes in barometric pressure will correspondingly change R (see Equation E-2, Table 19) to a positive, zero, or negative value (thus effecting flux rate). For instance:

for $< \pi$, R  0 (air inflow):    $N_A$ is suppressed (= 0).

$p = \pi$, R = 0:    $N_A = D_{A3}\rho_{A1}^*/L$ (rate of vapor movement in soil without gas generation--see Thibodeaux's previous equation)

p > π, R > 0 (air outflow):  $N_A$ is increased due to pressure
gradient driving force.

It was also shown that cell gas pressure, p, will vary with time and is
dependant upon gas generation and the cyclic behavior of fluctuating
atmospheric pressures, such that:

$$\frac{dp}{dt} = \frac{r_B \rho_W p}{\varepsilon_c} - \frac{Kp(p - \pi)}{\varepsilon_c h_c L \mu}$$

where:  $r_B$ = biogas generation rate (cm$^3$/g-sec);

$\rho_W$ = bulk density of waste (g/cm$^3$)

p = cell gas pressure (atm);

$\varepsilon_c$ = cell porosity;

K = permeability of cap material (cm$^2$-cp/sec-atm);

π = atmospheric pressure (atm);

$h_c$ = cell depth (cm)

L = cap thickness (cm)

μ = cell gas viscosity (cp)

Computer simulation tests indicated that the flux rate of benzene, $N_A$,
is influenced only slightly by barometric pressure under conditions of
co-disposal (gas generation) and influenced significantly by barometric
pressure under conditions of no internal gas generation.  The air emissions
model equations which incorporate factors for gas generation (and their
related effects on flux rate) should be applied for situations of co-disposal.

SHEN'S (1980) OPEN DUMP EQUATION

The emission rate equation for open dumps presented by Shen is derived
from Fick's Law and Arnold's (1944) equation for a surface exposed to open
air.  A major limitation of this model is that it only considers air-phase
diffusion and it does not accurately describe the effect of boundary layer
formation caused by ambient wind, as shown by Thibodeaux in describing the
gas-phase resistance of vapor movement.  This limitation suggests further
development of this model is appropriate.

# 6. Review of Land Treatment AERR Models

INTRODUCTION

The technical literature contains two models for calculating air emission release rates (AERR) from landfarming (land treatment) operations. Table 22 summarizes the available equations for land treatment facilities, showing the input parameters required for each model. Table 23 presents typical parameter values and ranges for the Thibodeaux-Hwang (1982) AERR model.

The Thibodeaux-Hwang (1982) model most accurately defines the physical situation existing in a landfarm, namely the mass transfer of a chemical species from a soil-waste mixture. One apparent limitation to this model is that it describes vapor diffusion as being soil-phase controlled, based on the assumption that the oil-layer diffusion length is on the order of a soil particle size. This assumption may only be valid after a certain exposure time. For petroleum waste, the sludge material generally subjected to landfarming is a multicomponent viscous material. Molecular diffusion coefficients for this material may be several orders of magnitude less than that for soil-phase diffusion. Thus, the diffusion rate determining factor will be the application thickness of the waste material. Additionally, it may be possible that both phases (soil and oil) will contribute equally to the diffusion rate of a specific component.

THIBODEAUX-HWANG: MODELING AIR EMISSIONS FROM LANDFARMING OF PETROLEUM WASTES (1982)

With the assumptions that the soil column is isothermal, that no vertical movement of waste occurs by capillary action, no adsorption of material occurs on soil particles and that no biochemical oxidation occurs, the Thibodeaux-Hwang model describes the vapor movement of a chemical species from a soil-waste mixture. This model is applicable to either surface application or subsurface injection methods of landfarming.

The one key variable in the Thibodeaux-Hwang derivation appears to be the oil-layer diffusion length, $Z_0$. This term is related to application rate of oil waste and thus its value will determine the extent that oil-phase diffusion will play an important role in the flux rate determination. The simplified version of the Thibodeaux-Hwang model assumes a low value for

243

TABLE 22.   INPUT PARAMETERS REQUIRED FOR EACH LAND TREATMENT AERR MODEL

| Parameter | Thibodeaux – Hwang[a] | Hartley[b] |
|---|:---:|:---:|
| $D_{ai}$, diffusivity of i in air filled pore spaces | X | |
| $D_{ai}$, molecular diffusivity of i in air | X | |
| $D_{wi}$, diffusivity of i in waste oil | X | |
| $Z_o$, oil-layer diffusion length | X | |
| $A_s$, interfacial area per unit volume of soil | X | |
| $\epsilon$, soil porosity | X | |
| $\tau$, tortuosity | X | |
| $h_s$, depth of surface injection | X | |
| A, surface area of application | X | X |
| $h_p$, depth of penetration or plow slice depth | X | |
| $M_{io}$, initial mass of component i | X | |
| t, time | X | |
| $C_{ig}$, concentration of i on gas side of interface | X | |
| $H_c$, Henry's Law constant in concentration form | X | |
| $C_{il}$, concentration of i in oil phase | X | |
| H, Henry's Law constant | X | |
| $C_g$, molar density of vapor | X | |
| P, total pressure | X | |
| $\beta$, soil bulk density | X | |
| $\rho$, particle density | X | |
| $E_w$, water evaporation rate | | X |
| RH, relative humidity | | X |
| $P_a$, vapor pressure of chemical | | X |
| $P_w$, vapor pressure of water | | X |
| $M_a$, molecular weight of chemical | | X |
| $M_o$, modelucar weight of water | | X |
| Wi/W, weight fraction of chemical in the waste | | X |

[a]Thibodeaux - Hwang:

(1) $\quad q_i = D_{ei} C_{ig}/[h_s^2 + (2 D_{ei} t A (h_p - h_s) C_{ig})/M_{io}]^{1/2}$

(2) $\quad D_{ei} = D_{ai} \epsilon/\tau$

(3) $\quad \epsilon = 1 - \beta/\rho$

(4) $\quad C_{ig} = \left[ \cfrac{H_c}{1 + \left( \cfrac{6 D_{ei} Z_o H_c}{D_{wi} a_s (h_p^2 + h_p h_s - 2h_s^2)} \right)} \right] C_{iwo}$

(5) $\quad H_c = HC_g \dfrac{10^6}{P}$

[b]Hartley:

(1) $\quad E_a = \dfrac{E_w}{(1-RH)} \times \dfrac{P_a (M_a)^{1/2}}{P_w (M_w)^{1/2}} \times A \times \dfrac{W_i}{W}$

TABLE 23.  TYPICAL PARAMETER VALUES AND RANGES FOR THE THIBODEAUX-HWANG LAND TREATMENT AERR MODEL

| Symbol | Units | Definition | Source of input parameter | Typical value | Expected range | Comments |
|---|---|---|---|---|---|---|
| $q_i$ | g/cm²-sec | flux rate | calculated from equation 1[a] | ----- | ----- | ----- |
| $D_{ei}$ | cm²/sec | diffusivity of i in air-filled pore spaces | calculated from equation 2[b] | ----- | ----- | ----- |
| $D_{ai}$ | cm²/sec | molecular diffusivity of i in air | chemical handbook or calculated from reference chemical | $10^{-1}$ | $10^{-3}$ | Highly variable input parameter |
| $\epsilon$ | dimensionless | soil porosity | calculated from equation 3[c] | 0.43 | 0.25 - 0.62 | ----- |
| $\bar{\tau}$ | dimensionless | tortuosity | constant value | $\bar{3}$ | N.A. | $\bar{3}$ is a constant value (assumed) |
| $h_s$ | cm | depth of surface injection | measured on typical value assumed | 0 | 0 - 20 cm | hs = 0 for surface application |
| $A$ | cm² | surface area of application | measured | N.A. | $(3-10) \times 10^7$ | highly variable input parameter |
| $h_p$ | cm | depth of penetration or plow slice depth | measured or typical value assumed | 19 | 8 - 30 | value of hp subscript may be low shortly after surface application |
| $M_{io}$ | g | initial mass of component i | measured | N.A. | N.A. | ----- |
| $t$ | sec | time after application | measured | N.A. | $0-5 \times 10^6$ | ----- |
| $C_{ig}$ | g/cm³ | concentration of i on gas side of interface | calculated from equation 4[d] | N.A. | N.A. | ----- |
| $H_c$ | cm³ oil/cm³ air | Henry's Law Constant in concentration form | calculated from equation 5[e] | N.A. | N.A. | ----- |
| $C_{il}$ | g/cm³ | conc. of i in oil phase | measured | N.A. | N.A. | highly variable parameter |
| $H$ | atm-m³/mol | Henry's Law Constant | chemical handbook | N.A. | $10^{-3} - 10^{-5}$ | highly variable parameter |
| $C_g$ | g-mol/cm³ | molar density of vapor | use typical value or calculate (see comments) | $1/2.44 \times 10^4$ | N.A. | $C_g = \frac{P^o}{RT}$ |

(continued)

TABLE 23 (continued)

| Symbol | Units | Definition | Source of input parameter | Typical value | Expected range | Comments |
|---|---|---|---|---|---|---|
| P | atm. | total pressure | measured or typical value assumed | 1 | 0.7 - 1 | standard atm. pressure is 1 atm |
| $\beta$ | $g/cm^3$ | soil bulk density | soil handbook | 1.5 | 1 - 2 | . |
| $\rho$ | $g/cm^3$ | particle density | soil handbook | 2.65 | 2.4 - 2.8 | $\rho = 2.65$ for most mineral matter |

[a]Equation 1. $q_i = D_{ei} C_{ig}/lh_g^2 + (2 D_{ei} tA(h_p - h_g)C_{ig})/M_{io}]^{1/2}$

[b]Equation 2. $D_{ei} = D_{ei} \epsilon/\tau$

[c]Equation 3. $\epsilon = 1 - \beta/\rho$

[d]Equation 4:

$$C_{ig} = \left[ \frac{H_c}{1 + \left( \dfrac{6 D_{ei} Z_o H_c}{D_{vi} a_s (h_p^2 + h_p h_s - 2h_s^2)} \right)} \right] C_{iwo}$$

[e]Equation 5. $H_c = HC_g \left( \dfrac{10^6}{P} \right)$

$z_0$, i.e., soil particle size thickness, thus for all cases that make use of this assumption Equation (4) from Table 22 becomes:

$$C_{ig} = H_c C_{iL}$$

$C_{ig}$    = concentration of component i in gas phase;

$H_c$    = Henry's Law constant (concentration form);

$C_{iL}$    = concentration of component i in oil phase.

Thibodeaux and Hwang also present a flux rate equation which identifies the air emission release rate immediately after waste application:

$$q_i = k_g C_i^*$$

where: $q_i$    = flux rate ($g/cm^2$-s);

$k_g$    = gas-phase mass transfer coefficient (cm/s);

$C_i^*$    = vapor concentration of chemical i ($g/cm^3$).

This equation assumes no mass-transfer resistance in the soil phase and thus is applicable only for short periods of time (i.e., immediately after application, or for spills, and while a liquid pool is still visible). More importantly, the equation is only valid for a pure compound because it provides no description of liquid-phase resistance. Here again the application thickness will be a controlling variable. For use in describing air emission release rates of a waste material, it is more correct to use an overall mass-transfer coefficient. Refinement of this equation may generate a method of quantifying air emissions from drum storage facilities.

This short term emissions estimate model closely correlates to the flux rate model described by Hartley. The Hartley model does not accurately describe the physical situation of land application as described later.

There is little experimental data available which can be used to validate this model in either its complete or simplified forms. The apparent good agreement with dieldrin test data as shown by Thibodeaux and Hwang (1982) may not represent true field conditions.

HARTLEY MODEL (1969)

The Hartley model was developed to determine the evaporative loss of pure volatile compounds. Although it is applicable to land spreading operations, it is not applicable to landfarming methods. The model assumes that the rate of mass transfer is controlled by resistance in the gas phase and is proportional to the saturated vapor concentration. The liquid phase resistance, which plays an important role for multicomponent liquid mixtures containing volatile compounds, is completely ignored in the model development.

Simply stated, the Hartley model calculation of the flux rate of a chemical compound is performed on the basis of a known flux rate of a reference compound. Water is generally taken as the reference compound.

Although the Hartley model represents a simple method of calculating air emission release rate of chemical compounds from a soil surface (or any surface), it does not accurately represent air emission release rates of landfarming application as does the Thibodeaux-Hwang model. Several drawbacks to the Hartley model include:

- predicts a maximum emission rate throughout volatilization period;

- doesn't account for incorporation of volatile material within soil;

- not readily extended to complex, multicomponent mixtures;

- based on a nonvolatile reference compound (water).

# 7. Storage Tank Air Emission Estimation Techniques

INTRODUCTION

Air emissions of volatile compounds from storage tanks at TSDFs are a function of several factors including:

- physical and chemical characteristics of the stored liquid;

- tank design;

- tank condition and site conditions; and

- operational characteristics, especially turnover frequency.

For a given liquid, tank design influences the emission rate potential. The five types of storage tanks, which are described later in this section, are:

- fixed roof;

- external floating roof;

- internal floating roof;

- variable vapor space; and

- pressure tanks.

Of the five designs, variable vapor space and pressure tanks generally produce the least air emissions.

In addition to tank design, the true vapor pressure of the material stored is one of the most significant parameters affecting emissions. Consequently, the type of tank selected during plant design is partially dependent on the vapor pressure of the material being stored. According to Erickson (1980), fixed roof tanks are preferred for storing materials with vapor pressures up to 34.5 kPa (5 psia); floating-roof tanks when vapor pressures are in the range of 6.9 to 34.5 kPa (1 to 5 psia); and pressure tanks when vapor pressures are greater than 51.7 kPa (7.5 psia). Other factors such as material stability, safety hazards, health hazards, and multiple use also influence tank selection for a particular organic liquid.

249

Physical actions on the tank such as changes in temperature or pressure affect the volatilization rate. Temperature increases from direct solar radiation and contact with warm ambient air increase volatilization and emission potential. Danielson (1973) notes that for a free vented tank, winds may entrain or educt some of the saturated vapors into the ambient air.

Operating conditions also affect storage tank emissions; i.e., frequency of filling (turnover rate), vapor tightness of the tank, and volume of the vapor space. Fixed-roof tanks maintained completely full limit the volume of vapor space and, thus, emissions. When the turnover rate is long; i.e., extensive time periods between filling/emptying cycles, the free space in a tank becomes more saturated with vapor from the liquid. Thus, during filling of the tank or during breathing cycles, a larger concentration of vapors exists in the air-vapor mixture vented to the atmosphere. Danielson (1973) states that vapor tightness of the tank can influence the evaporation rate, and a lack of tight vapor seal allows increased emissions. Proper seal maintenance for floating tanks is necessary to limit vapor losses.

The American Petroleum Institute (API), EPA and others have developed empirical equations for fixed and floating roof tank emissions based on field test data. Masser (1981) notes that emissions from pressure tanks occur only when the design pressure is exceeded, when the tank is filled improperly, or when abnormal vapor expansion occurs. Because these are not regularly occurring events, and pressure tanks are not a significant source of emissions under normal operating conditions, no equations were found available for estimating air emissions from pressure tanks.

Table 24 summarizes the recommended equations for fixed and floating roof tanks, showing the input parameters required for each model. Following sections describe these models in more detail, and also describe additional, non-recommended models found in the literature. Special case consideration for open tanks and storage of mixtures is also presented.

FIXED ROOF TANKS

A typical fixed roof tank is shown in Figure 7. Air emissions from fixed roof tanks occur from breathing losses and working losses. Masser (1981) defines breathing loss ($L_B$) as vapor expulsion due to vapor expansion and contraction from changes in tank temperature and ambient barometric pressure. Breathing losses occur in the absence of any liquid level change in the tank.

The combined loss from periodic filling and emptying is called working loss ($L_W$). When a tank is filled, vapors are expelled from the tank when the pressure inside the tank causes opening of the relief valve. Emptying loss occurs when air drawn into the tank during liquid removal becomes saturated with organic vapor, expands, and exceeds the capacity of the vapor space.

In 1962, the American Petroleum Institute published API Bulletin 2518 which contained equations for estimating breathing losses and working losses from fixed roof tanks. These equations were used extensively by regulatory

TABLE 24.    INPUT PARAMETERS REQUIRED FOR RECOMMENDED FIXED AND
FLOATING ROOF MODELS

| Parameter | A | B | C | D |
|---|---|---|---|---|
| M, molecular weight of vapor in storage tank | X | X | X | |
| P, true vapor pressure at bulk liquid conditions | X | X | | |
| D, tank diameter | X | | X | X |
| H, average vapor space height | X | | | |
| ΔT, average ambient diurnal temperature change | X | | | |
| $F_p$, paint factor | X | | | |
| C, adjustment factor for small diameter tanks | X | | | |
| $K_C$, product factor | X | X | X | |
| $K_N$, turnover factor | | X | | |
| $K_S$, seal factor | | | X | |
| V, average wind speed | | | X | |
| N, seal related wind speed exponent | | | X | |
| $P^*$, vapor pressure function | | | X | |
| $E_F$, secondary seal factor | | | X | |
| Q, average throughput | | | | X |
| C, shell clingage factor | | | | X |
| $W_L$, average organic liquid density | | | | X |

MODEL DESIGNATIONS

A.   Fixed Roof Tank Breathing Losses (API-1962, modified by TRW/EPA).

$$L_B = 2.26 \times 10^{-2} M \left(\frac{P}{14.7-P}\right)^{0.68} D^{1.73} H^{0.51} \Delta T^{0.50} F_p CK_C$$

B.   Fixed Roof Tank Working Losses (API-1962).

$$L_w = 2.40 \times 10^{-2} MPK_N K_C$$

C.   External and Internal Floating Roof Tank Standing Storage Loss (API/EPA, 1980).

$$L_s = K_s V^N P^* DMK_C E_F$$

D.   External and Internal Floating Roof Tank Withdrawal Loss.

$$L_w = \frac{(0.943)QCW_L}{D}$$

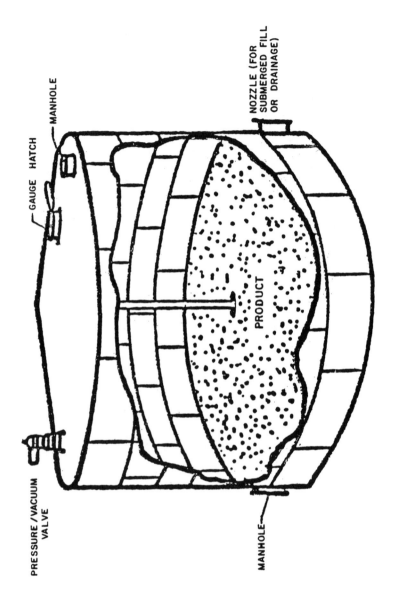

Figure 7.    Typical fixed roof storage tank (Masser – 1981).

agencies in the past to estimate air emissions from fixed roof tanks.
However, test results reported by EPA (1979), the Western Oil and Gas
Association (WOGA) (1977), and the German Society for Petroleum Science and
Carbon Chemistry (DGMK), showed that the 1962 API equation for breathing
losses over-estimated air emissions by roughly a factor of four. However,
working losses estimated by the 1962 API equation were found to be fairly
accurate.

Under EPA contract, TRW (1981), updated the breathing loss equation for
fixed roof tanks based on the more recent test data. The revised equation
provides breathing loss emission estimates which are 72 percent less than
obtained by estimating emissions using the 1962 API equation. EPA's Office of
Emission Standards Engineering Development (ESED) reported to GCA that the
revised equation for breathing losses, prepared by TRW, and the 1962 API
equation for working losses are the best available equations for fixed roof
storage tanks. These equations are incorporated into a revised edition of
EPA's emission factor handbook, AP-42 (April 1981). The revised AP-42
equations appear in Tables 25 and 26 for breathing and working losses,
respectively, from fixed roof tanks.

The fixed roof working loss ($L_W$) is the sum of the loading and
unloading losses. Special tank operating conditions may result in losses
which are significantly greater or lower than the estimates provided by
Table 23 and must be evaluated on a site-specific basis. The total losses
from a fixed roof storage tank are equal to the sum of the breathing losses
($L_B$) and working losses ($L_W$).

Typical values for input parameters to the fixed roof tank equations
appear in Tables 27 and 28, for breathing and working losses, respectively.

EXTERNAL AND INTERNAL FLOATING ROOF TANKS

External Floating Roof Tanks

Standing storage loss, the major element of evaporative loss, results
from wind induced effects acting on the top of an external floating roof
tank. The types of seals used to close the annular vapor space between the
floating roof and the tank wall dictate the nature of wind effects.
Figure 8 depicts a typical external floating roof storage tank.

Standing storage loss emissions from external floating roof tanks are
controlled by either a primary seal, alone, or a primary and a secondary
seal. Three basic types of primary seals used on external floating roofs
are: (1) mechanical (metallic shoe); (2) resilient (nonmetallic); and (3)
flexible wiper. Resilient seals are mounted to eliminate the vapor space
between the seal and liquid surface (liquid mounted), or to allow a vapor
space between the seal and liquid surface (vapor mounted). A primary seal
closes the annular space between the edge of the floating roof and the tank
wall. Some primary seals are protected by a metallic weather shield. Two
configurations of secondary seals currently in use are: (1) shoe mounted; and
(2) rim mounted. Although there are other seal system designs, Masser (1981)
indicates that the systems described here comprise the majority in use today.

TABLE 25.   RECOMMENDED MODEL FOR FIXED ROOF TANK BREATHING LOSSES
(TRW UPDATE OF 1962 API MODEL)

$$L_B = 2.26 \times 10^{-2} M \left[ \frac{P}{14.7-P} \right]^{0.68} D^{1.73} H^{0.51} \Delta T^{0.50} F_p C K_C$$

where:

$L_B$ = fixed roof breathing loss (lb/year)

M = molecular weight of vapor in storage tank (lb/lb mole).
See AP-42, Table 4.3-1.

P = true vapor pressure at bulk liquid conditions (psia).
See note 1.

D = tank diameter (ft).

H = average vapor space height, including roof volume correction (ft).
See Note 2.

$\Delta T$ = average ambient diurnal temperature change (°F).

$F_P$ = paint factor (dimensionless).  See AP-42, Table 4.3-2.

C = adjustment factor for small diameter tanks (dimensionless).
See AP-42, Figure 4.3-4.

$K_C$ = product factor (dimensionless).  See Note 3.

Notes:  (1)  True vapor pressures for organic liquids can be determined from
AP-42, Figures 4.3-5 or 4.3-6, or AP-42, Table 4.3-1.

(2)  The vapor space in a cone roof is equal in volume to a cylinder
which has the same base diameter as the cone and is one third the
height of the cone.

(3)  For crude oil, $K_C$ = 0.65.  For all other organic liquids,
$K_C$=1.0.

TABLE 26.    RECOMMENDED MODEL FOR FIXED ROOF TANK
WORKING LOSSES (1962 API MODEL)

$$L_W = 2.40 \times 10^{-2} \, MPK_N K_C$$

where:

$L_W$ = fixed roof working loss (lb/$10^3$ gal throughput).

M = molecular weight of vapor in storage tank (lb/lb mole).
See AP-42, Table 4.3-1.

P = true vapor pressure at bulk liquid conditions (psia).
See Note 1.

$K_N$ = turnover factor (dimensionless).  See AP-42, Figure 4.3-7.

$K_C$ = product factor (dimensionless).  See Note 2.

Notes:  (1)  True vapor pressures for organic liquids can be determined from
AP-42 Figures 4.3-5 or 4.3-6, or AP-42 Table 4.3-1.

(2)  For crude oil, $K_C$=0.84.  For all other organic liquids,
$K_C$=1.0.

TABLE 27.  TYPICAL RANGES OF INPUT PARAMETERS FOR FIXED ROOF TANK
BREATHING LOSS MODEL:

$$L_B = 2.26 \times 10^{-2} M \left[ \frac{P}{14.7-P} \right]^{0.68} D^{1.73} H^{0.51} \Delta T^{0.50} F_p C K_C$$

| Parameter symbol | Parameter description/units | Typical range of values | |
|---|---|---|---|
| | | For 95% of compounds | For 50% of compounds |
| M | Molecular weight of vapor in storage tank (lb/lb mole) | 32-190 | 70-130[a] |
| P | True vapor pressure at bulk liquid conditions (psia) | 0.00004-6.9@60°F[a] | 1.0 - 4.0[a] |
| D | Tank diameter (feet) | 14.7-98[b] | 16-55[b] |
| H | Average vapor space height (ft) | 7-33[b] | 19-33[b] |
| T | Average ambient diurnal temperature change (°F) | +20° from average daily temp.[c] | +10° from average daily temp.[c] |
| $F_p$ | Paint factor (dimensionless) | 1.0-1.58[a] | 1.2-1.4[a] |
| C | Adjustment factor for small diameter tanks | 0.4-1.0[a] | 0.7-1.0[a] |
| $K_C$ | Product factor (dimensionless) | 0.65-1.0[a] | 0.65-1.0[a] |

[a]Estimated from a review of data presented in AP-42.

[b]Estimated from a review of data presented in Erikson (1980).

[c]Engineering judgement.

TABLE 28.  TYPICAL RANGES OF INPUT PARAMETERS FOR FIXED ROOF
TANK WORKING LOSS MODEL:

$$L_W = 2.40 \times 10^{-2} \, MPK_N K_C$$

| Parameter symbol | Parameter description/units | Typical range of values | |
|---|---|---|---|
| | | For 95% of compounds | For 50% of compounds |
| M | Molecular weight of vapor in storage tank (lb/lb mole) | 32-190[a] | 70-130[a] |
| P | True vapor pressure at bulk liquid conditions (psia) | 0.00004-6.9@60°F[a] | 1.0-4.0[a] |
| $K_N$ | Turnover factor (dimensionless) | 0.22-1.0[a] | 0.3-0.8[a] |
| $K_C$ | Product factor (dimensionless) | 0.84-1.0[b] | N.A. |

[a]Estimated from a review of data presented in AP-42.

[b]$K_C$ for crude oil = 0.84, $K_C$ for all other organic liquids = 1.0.  An
average value of 0.95 is expected to be a good average value with high
precision.

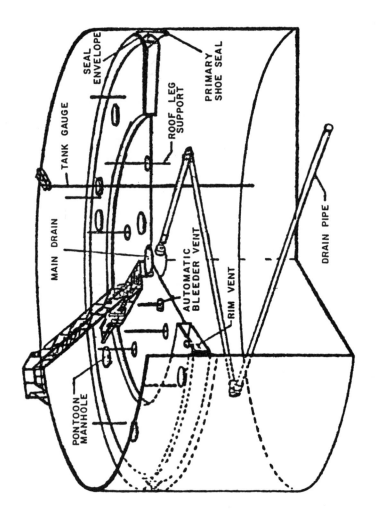

Figure 8. Typical external floating roof storage tank (Masser – 1981).

Withdrawal loss is another source of emissions from external floating roof tanks. This loss is the vaporization of liquids that cling to the tank wall and are exposed to the atmosphere when a floating roof is lowered by withdrawal of liquid.

## Internal Floating Roof Tanks

An internal floating roof storage tank has a permanently fixed roof and a cover inside the tank that either floats on the liquid surface (contact), or rests on pontoons several inches above the liquid (noncontact). Figures 9 and 10 illustrate the contact and noncontact design of internal floating roof storage tanks.

Internal floating roof tanks generally have the same sources of emissions as external floating roof tanks; i.e., standing storage and working losses. Fitting losses through deck fittings in the roof, roof column supports, or other openings, can also account for emissions from internal floating roof tanks.

Typical internal floating roofs incorporate two types of primary seals, resilient foam filled and wiper. Similar to those employed in external floating roof tanks, these seals close the annular vapor space between the edge of the floating roof and the tank wall.

## Recommended Emission Estimation Techniques for Floating Roof Tanks

EPA's Office of ESED recommends in Supplement No. 12 to AP-42 the use of API's February 1980 Bulletin 2517, Evaporation Loss from External Floating Roof Tanks. According to McDonald (1982), EPA currently suggests two changes to the API equations presented in AP-42 for volatile organic liquid storage;

- Change the product factor, $K_c$, from 10 to 1;

- Disregard the secondary seal factor, $E_F$.

Based on the above changes, Table 29 shows EPA's currently recommended equation for standing storage losses from external floating roof tanks.

For estimating standing storage losses from internal floating roof tanks, EPA recommends the approach identified in API's June 1983 Bulletin 2519, Evaporation Loss from Internal Floating-Roof Tanks. Standing losses from internal floating roof tanks are derived by summing the losses estimated from the rim seal area, deck fittings, and deck seams.

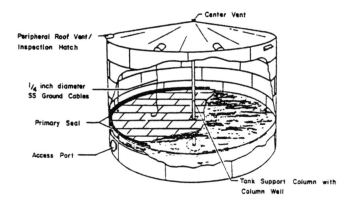

Figure 9.   Noncontact internal floating roof tank.

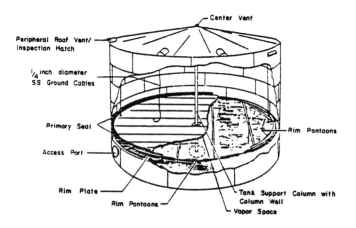

Figure 10.   Contact internal floating roof tank.

TABLE 29.   EPA/API RECOMMENDED TECHNIQUE FOR STANDING STORAGE
LOSSES FROM EXTERNAL FLOATING ROOF TANKS

$$L_S = K_S V^N P^* DMK_C$$

where:  $L_S$ = standing storage loss (lb/yr)

$K_S$ = seal factor (lb-mole/(ft (mi/hr)$^N$yr)).  See Note 1.

$V$ = average wind speed at tank site (mi/hr).  See Note 2.

$N$ = seal related wind speed exponent (dimensionless).  See Note 1.

$P^*$ = vapor pressure function (dimensionless).  See Note 3.

$$P^* = \frac{\left(\frac{P}{P_A}\right)}{\left[1 + \left(1 - \frac{P}{P_A}\right)^{0.5}\right]^2}$$

$P$ = true vapor pressure at average actual organic liquid storage at temperature (psia)

$P_A$ = average atmospheric pressure at tank location (psia)

$D$ = tank diameter (ft)

$M$ = average vapor molecular weight (lb/lb-mole).  See Note 4.

$K_C$ = product factor (dimensionless).  See Note 5.


Notes:  (1)  For petroleum liquid storage:  $K_S$ and $N$ for both primary only and primary/secondary seal systems are found in AP-42, Table 4.3-3.

(2)  If the wind speed at the tank site is not available, wind speed data from the nearest local weather station may be used as an approximation.

(continued)

TABLE 29 (continued).

Notes (continued).

(3)  $P^*$ can be calculated or read directly from AP-42 Figure 4.3-8.
     True vapor pressures for organic liquids can be determined from
     AP-42 Figures 4.3-5 or 4.3-6, or AP-42 Table 4.3-1. If average
     actual organic liquid storage temperature, $T_S$, is unknown, the
     average storage temperature can be estimated from the average
     ambient temperature $T_A(F)$ (available from local weather
     service data), adjusted by the tank paint color factor. See AP-42
     Table 4.3-4.

(4)  The molecular weight of the vapor, M, can be determined by AP-42
     Table 4.3-1, analysis of vapor samples, or by calculation from the
     liquid composition. A typical value of 64 lb/lb-mole can be
     assumed for gasoline, and a value of 50 lb/lb-mole can be assumed
     for U.S. midcontinental crude oils.

(5)  For all petroleum liquids except crude oil:  $K_C = 1.0$

     For crude oil:  $K_C = 0.4$

     For all volatile organic liquids:  $K_C = 1.0$

Withdrawal losses from external floating roof tanks are determined by the EPA/API equation shown in Table 30. The total loss from floating roof tanks in lb/yr is estimated by the following equation:

$$L_T(lb/yr) = L_S(lb/yr) + L_W(lb/yr) \qquad (3)$$

where:  $L_T$ = total loss;

$L_S$ = standing storage loss;

$L_W$ = withdrawal loss.

The equation presented in Table 30 for external floating roof tanks was modified in API Publication 2519 for use in estimating withdrawal losses from internal floating roof tanks. The improvement in withdrawal loss estimate is obtained if the type and number of columns are known.

Tables 31 and 32 show typical ranges of values for input parameters to external floating roof model recommended by EPA.

OTHER STORAGE TANK MODELS

Several other models for storage tanks were found in GCA's literature search and through discussions with EPA and API. However, these approaches, as described below, are not considered state-of-the-art techniques.

Storage Tank Model Developed in the USSR

Moryakov, et al. (1979) presents a model developed in the USSR for estimating storage tanks emissions, but it appears to have two major drawbacks. First, the model is based on tests conducted in a northern climate and has not been thoroughly validated through field measurements. Moryakov reported that further experimental verification is required before the model can be used directly for southern climate zones.

The second major drawback is that the model does not account for what the authors termed small breathing losses. These losses are the equivalent of API's breathing losses for fixed roof tanks and standing storage losses for floating roof tanks. In the U.S., breathing losses are considered significant sources of emissions, and in some cases may exceed working losses or withdrawal losses.

In addition to the above drawbacks, the USSR model was developed to estimate emissions of petroleum products only. It is not known if the model is applicable for estimating emissions of volatile organic liquids.

TABLE 30.  EPA/API RECOMMENDED TECHNIQUE FOR WITHDRAWAL LOSS FROM
EXTERNAL FLOATING ROOF TANKS

$$L_W = \frac{(0.943)\ QCW_L}{D}$$

where:  $L_W$ = withdrawal loss (lb/yr).

Q = average throughput (barrel (bbl)/yr; 1 bbl = 42 U.S. gallons).

C = shell clingage factor (bbl/1000 ft$^2$).  See AP-42, Table 4.3-5.

$W_L$ = average organic liquid density (lb/gal).  See Note 1.

D = tank diameter (ft).

Notes:  (1)  If $W_L$ is not known, an average value of 6.1 lbs/gallon can be
assumed for gasoline.  An average value cannot be assumed for
crude oil, since densities are highly variable.

(2)  The constant, 0.943, has dimensions of (1000 ft$^3$ x gal/bbl$^2$).

TABLE 31.  TYPICAL RANGE OF INPUT PARAMETERS FOR WITHDRAWAL LOSS FROM EXTERNAL FLOATING ROOF STORAGE TANK MODEL:

$$L_W = \frac{(0.943)QCW_L}{D}$$

| Parameter model | Parameter description/units | Typical range of values | | Comments |
|---|---|---|---|---|
| | | For 95% of compounds | For 50% of compounds | |
| Q | Average throughput (bbl/yr) | $6.0 \times 10^5 - 5.0 \times 10^4$ | ? | Average throughput varies greatly. |
| C | Shell clingage factor (bbl/1000 ft²) | 0.0015 - 0.6 | 0.0015 - 0.6 | Clingage factor varies significantly dependent on the product stored and the shell condition. |
| $W_L$ | Average organic liquid density (lb/gal) | N.A. | N.A. | An average value of 6.1 lb/gal can be assumed for gasoline.  An average value cannot be assumed for crude oil or volatile organic liquids, since densities are highly variable (Hasser – 1981). |
| D | Tank diameter (ft) | 30 - 100 | 45 - 85 | Variance between tank sizes is too great to select a constant. |

[a]Based on a 40 m³ tank turned over an average of 200 times per year and an 8,330 m³ tank turned over an average 12 times per year.  Average turnover rates obtained from Erickson (1980).

TABLE 32. TYPICAL RANGES OF INPUT PARAMETERS FOR EXTERNAL FLOATING ROOF TANK STANDING STORAGE LOSS MODEL:

$$L_S = K_S V^{N_P} * DMK_C$$

| Parameter symbol | Parameter description/units | Typical range of values | | Comments |
|---|---|---|---|---|
| | | For 95% of compounds | For 50% of compounds | |
| $K_S$ | Seal factor | 0.2 – 1.2[a] | 0.2 – 1.2[a] | The variance for $K_S$ is great depending on the type of seal. |
| V | Average wind speed at tank site (mi/hr) | 6.2 – 12.4[b] | 7.2 – 10.6[b] | An average value of 8.9 has fairly high precision. |
| N | Seal related wind speed exponent (dimensionless) | 0.4 – 2.6[b] | 0.9 – 2.2[a] | The variance for N is great depending on the type of seal. |
| P* | Vapor pressure function (dimensionless) | 0.05 – 0.7[b] | 0.08 – 0.4[a] | The variance of P* is great depending on the true vapor pressure. |
| D | Tank diameter | 30 – 100[c] | 45 – 85[c] | Variance between tank sizes too great to select a constant. |
| M | Average vapor molecular weight (lb/lb-mole) | 190 – 32[a] | 130 – 70[a] | Variance too great between molecular weights of organic liquid vapors to select a constant. |
| $K_C$ | Product factor | 0.4 – 1.0[a] | 0.4 – 1.0[a] | $K_C$ = 0.4 for crude oil and 1.0 for all other organic liquids. |

[a] Estimated from a review of data presented in AP-42.

[b] Estimated from a review of average wind speeds at 262 U.S. cities published by National Climate Center, Asheville, NC.

[c] Engineering estimate.

## Outdated/Nonrecommended Models for Floating Roof Tanks

In 1962, API Bulletin 2517 presented equations for estimation standing storage loss and withdrawal loss for floating roof tanks. The equations did not differentiate between internal and external floating roof tanks. Input parameters for the API Bulletin 2517 equations appear in Table 33.

In 1978, the Chicago Bridge and Iron Company performed field measurements of emissions from a pilot test tank for EPA. Table 34 presents the equations developed from the pilot tests for both internal and open-top floating roof tanks equipped with primary seals alone, or both a primary and secondary seal. EPA's Office of ESED does not presently recommend use of these equations.

## SPECIAL CONSIDERATIONS

### Storage of Mixtures

The methods thus far reported for estimating storage tank air emissions are for pure compounds. However, to determine the air emission rate for a particular volatile component of a mixture, the emission factor equations presented are still applicable by using the true vapor pressure of the mixture instead of the pure compound vapor pressure.

The true vapor pressure of a mixture is the summation of partial pressures of each component in the stored liquid:

$$P_m^o = \sum P_i = \sum x_i \, \gamma_i \, P_i^o$$

where:  $P_m^o$ = vapor pressure of mixture;

$P_i$ = partial pressure of component i;

$x_i$ = mole fraction of component i in the liquid;

$\gamma_i$ = activity coefficient of component i in the liquid;

$P_i^o$ = vapor pressure of pure component i.

The activity coefficient corrects for deviations from Raoult's Law to account for liquid interactions. For hydrocarbon mixtures, it is generally assumed that Raoult's Law holds, thus the activity coefficient is taken as unity. This activity coefficient is a function of mixture composition and relative concentration. Although the activity coefficients for a particular mixture can be determined experimentally, empirical correlations are provided by Lyman, et al. (1982).

It should be noted that by using the vapor pressure of the mixture in the storage tank models, the emissions estimate will be as total volatile organic carbon (VOC). The concentration of component i in the vapor phase is the

TABLE 33.   INPUT DATA FOR 1962 API FLOATING ROOF MODEL

| Standing storage loss | Withdrawal loss |
|---|---|
| Molecular weight of vapor in storage tank | Density of stored liquid |
| True vapor pressure at bulk conditions | Tank construction factor |
| Tank diameter | Tank diameter |
| Average wind velocity | |
| Tank type factor | |
| Seal factor | |
| Paint factor | |
| Crude oil factor | |

TABLE 34.  FLOATING ROOF EQUATIONS BASED ON 1978 PILOT TESTS
(NOT RECOMMENDED FOR USE)

$$L_T = L_{WD} + L_S + L_F$$

$$L_S = K_S V^n M D \left[\frac{\frac{P}{(14.7)}}{1 + (1 - \frac{P}{14.7})^{0.5}}\right]^2 \frac{1}{2205}$$

$$L_F = NK_F V^n M \left[\frac{\frac{P}{(14.7)}}{1 + (1 - \frac{P}{14.7})^{0.5}}\right]^2 \frac{1}{2205}$$

$$L_{WD} = 0.000198 DTHd \ 1/2205$$

where:  $L_T$ = total loss (mg/yr);

$L_{WD}$ = withdrawal loss (mg/yr);

$L_S$ = seal loss (mg/yr);

$L_F$ = fitting loss (mg/yr);

M = molecular weight of product vapor (lb/lb-mole); 78.1 lb/lb-mole
   for VOC;

P = true vapor pressure of product (psia); 2 psia assumed;

D = tank diameter (ft);

V = average wind speed for the tank site (mph); 10 mph assumed
   average wind speed;

$K_s$ = seal factor

$K_F$ = fitting factor

n = seal wind speed exponent

m = fitting wind speed exponent

C = product withdrawal shell clingage factor

N = fitting multiplier

T = turnovers (per yr),

H = tank height (ft),

d = density of stored liquid at bulk liquid conditions (lb/gal;
   use 8.0).

partial pressure of component i ($P_i$) divided by the vapor pressure of the mixture ($P_m^0$). Thus, the emission rate contribution for component i is determined by multiplying the calculated emission rate for the mixture by the vapor phase mole fraction of component i.

## Open Storage Tanks

Generally, it is unusual to find pure volatile organic liquids stored in open tanks. However, it may be possible that certain wastes such as sludges are stored in open tanks prior to treatment or disposal. The mechanisms of molecular and turbulent diffusion will control the process of evaporation. Thus, the methods for determining an emission rate for this type of storage becomes similar to that developed for surface impoundments.

The major difference between a surface impoundment and an open storage tanks is the potential for a greater freeboard effect. For a substantial freeboard (i.e., greater than the fetch), the wind effect on the liquid surface becomes negligible, thus the emission rate controlling step becomes the molecular diffusion of component i in the waste sludge.

# 8. Air Emission Estimation Techniques for Wastewater Treatment Processes

## INTRODUCTION

The purpose of this section is to identify AERR models which can be used for quantifying air emissions of volatile compounds from hazardous waste treatment systems. The approach used for identifying the appropriate AERR model was to categorize each treatment system into one of these main categories:

- open tanks with no mixing;

- open tanks with mixing; and

- closed systems.

In the case where all volatile species in the waste stream have been identified, and their effluent concentrations measured, AERR models for aerated or nonaerated surface impoundments could be applied to open tank processes depending upon system dynamics (i.e., aerated or nonaerated). For closed system wastewater treatment processes, it can be assumed that aside from system leakage or operational abnormalities, there are no air emissions. The concentration of effluent contaminants can be calculated based on process unit efficiency. Such calculations are provided in the literature but will not be discussed in this report.

Limited data are available regarding the removal efficiency and effluent concentrations for organic pollutants subjected to various biological treatment systems. If outlet concentrations are not available for the subject facility, these limited data may provide a rough cut estimate of the in-tank concentration for a specific compound. However, caution is advised since it is likely such data were developed for single component waste streams. Interactions of multicomponent waste streams will affect the system removal efficiency.

Emissions estimation techniques have been developed for biological treatment processes when only inlet concentrations are known. These models take into consideration pollutant removal by degradation, adsorption and air stripping. A formal discussion of each model is provided in this section.

OPEN TANK SYSTEM-NO MIXING

Typical wastewater treatment systems that fall into the category of open tank-no mixing are:

- sedimentation;

- chlorination;

- equalization.

No AERR Models were located that specifically described air emissions from these treatment systems. However, it can be assumed that these systems fall into a broader category of nonaerated surface impoundments. Treatment systems that fall into the open tank-no mixing category represent plug flow systems. Therefore, applying the nonaerated surface impoundment model one needs to accurately define the in-tank concentration of the specific compound in question.

As described in Section 4, Mackay and Leinonen (1975) presented a surface impoundment model for plug flow, where the in-tank concentration was defined by:

$$C_i = C_{io} \exp \left( -K_{iL} \, t/L \right)$$

where:

$C_i$ = concentration of compound i at time t $(mol/m^3)$;

$C_{io}$ = initial concentration of compound i $(mol/m^3)$;

$K_{iL}$ = overall liquid phase mass transfer coefficient $(m/hr)$;

$t$ = residence time $(hr)$;

$L$ = impoundment depth $(m)$.

This is a simple, first-order decay method for calculating the in-tank concentration, however, the term $K_{iL}$ only accounts for a compound's loss by volatilization. A more correct application would be to define the effects of sorption and probably chemical conversion.

OPEN TANK-MIXING:  BIOLOGICAL TREATMENT SYSTEMS

GCA's evaluation of open tank treatment processes with mixing focused mainly on the activated sludge biological treatment process (high rate mixing). Other processes, which might fall into the low rate mixing subcategory, include neutralization or precipitation involving the addition and rapid mixing of a chemical reagent. The emissions from these processes are best described by application of Thibodeaux's Aerated Surface Impoundment (ASI) model using very low power input to identify the area of turbulent mixing. This model was previously discussed in detail in Section 4.

Major research and development efforts of models to predict hazardous chemical air emissions from biological treatment processes have been conducted by Hwang and Freeman. The following discussion of AERR models developed by these two researchers identifies basic similarities in approach and differences in model details. In addition, Thibodeaux's ASI model is discussed in the context to which it can be applied to the activated sludge (AS) biological treatment process. In summary, four basic models to predict air emissions release rates from the AS process are discussed as follows:

- Thibodeaux (1981d)--ASI model;

- Hwang (1980)--activated sludge surface aeration (ASSA) model;

- Freeman (1979)--ASSA model;

- Freeman (1980)--diffused air activated sludge (DAAS) model.

Table 35 summarizes input parameters required for each model reviewed by GCA. The following sections discuss these models in detail.

All of the models presented predict the air emissions rate based on the concentration of compound i in the aeration basin of the activated sludge (AS) process. For a well mixed system, the basin concentration of a specific compound is assumed equal to the effluent concentration. Models which predict emissions based on the influent concentration ($S_o$ or $C_i$) of the compound tend to be very complex compared to models which assume a known effluent concentration ($S_e$). Three of the four models reviewed can be applied to surface aeration AS systems, whereas only the Freeman DAAS is directly applicable to diffused air systems. In general, the Hwang ASSA model tends to be both adequate and simple to apply compared to Freeman's model.

## Hwang (1980) Activated Sludge Surface Aeration

Hwang's Activated Sludge (AS) treatment AERR model employs techniques presented earlier for predicting emissions from aerated surface impoundments under steady state conditions (Thibodeaux, Parker, and Heck, 1981). The model by Thibodeaux et al., predicts the air emission rate of a compound based on the concentration of the compound (substrate concentration, $S_e$) within the impoundment. Hwang notes that for AS treatment systems, AERR models should consider removal of the compound by: (1) biological oxidation; (2) air stripping; and (3) adsorption to wasted sludge. Thus, Hwang's model predicts the effluent concentrations that result when the mass balance for the compound across the AS process is satisfied. The emission rate from the process is then predicted by the aerated surface impoundment equation. Note that if the pollutant's effluent concentration for an AS process is known, the emission release rate for that pollutant can be calculated directly by the aerated surface impoundment equation. This is based on the complete mix assumption that the effluent concentration adequately represents the concentration of the compound in the aeration tank.

TABLE 35.    SIMPLIFIED LIST OF INPUT PARAMETERS REQUIRED FOR EACH ACTIVATED SLUDGE BIOLOGICAL TREATMENT MODEL[a]

| Parameter | Thibodeaux (1981d)[b] ASI model | Hwang (1980)[c] ASSA model | Freeman (1979)[d] ASSA model | Freeman (1980)[e] diffused air mass transfer model |
|---|---|---|---|---|
| Air emission rate of hazardous substance i | $Q_i$ | $Q_i$ | $N_a$ | |
| Overall mass transfer coefficient | $K_L$ | $K_a$ | $K_L$ | $K_l$ |
| Total surface area of basin | $A_s$ | $A_s$ | $A_s$ | |
| Mole fraction of compound i in the liquid | $X_i$ | $X_i$ | $X_i$ | $X_i$ |
| Mole fraction of compound i at equilibrium | $X_i^*$ | $X_i^*$ | $X_i^*$ | $X_i^*$ |
| Molecular weight of compound i | $MW_i$ | $MW_i$ | $MW_i$ | $MW_i$ |
| Effluent concentration of compound i | | $S_e$ | $C_o$ | |
| Influent concentration of compound i | | $S_o$ | $C_I$ | |
| Recycle ratio, flow rate recycle/flow rate feed | | $r$ | | |
| Hydraulic residence time | | $t_c$ | | |
| Concentration of micro-organisms in basin | | $X$ | $B_O$ | |
| Grau's Biokinetic rate constant | | $k_{1(s)}$ | | |
| Concentration of total substrates | | $S_T$ | | |
| Ratio waste sludge flow/feed flow | | $W$ | | |
| Concentration of substrate (compound i in recycle) | | $S_r$ | $C_R$ | |
| Adsorption constants | | $K_1, K_T$ | | |
| Maximum concentration of substrate on sludge | | $X'$ | | |
| Stripping rate constant | | $k_a$ | | |
| Concentration of micro-organisms in recycle | | | $B_R$ | |
| Concentration of micro-organisms in influent | | | $B_i$ | |
| Concentration of oxygen in influent | | | $O_I$ | |
| Concentration of oxygen in effluent (basin) | | | $O_O$ | |
| Concentration of oxygen in recycle | | | $O_R$ | |
| Feed flow rate | | | $F_I$ | |
| Effluent flow rate | | | $F_O$ | |
| Recycle flow rate | | | $F_R$ | |
| Diameter of region of effect for mass transfer in the turbulent zone | | | $D$ | |

(continued)

TABLE 35 (continued)

| Parameter | Thibodeaux (1981d)[b] ASI model | Hwang (1980)[c] ASSA model | Freeman (1979)[d] ASSA model | Freeman (1980)[e] diffused air mass transfer model |
|---|---|---|---|---|
| Mass transfer coefficient for organic compound i turbulent zone, convective zone | | | $K_a^T$, $K_A^C$ | |
| Mass transfer coefficient for oxygen turbulent zone, convective zone | | | $K_b^T$, $K_b^C$ | |
| Gerber's biokinetic rate constants | | | $K_s$, $K_{O_2}$, $K_1$, $K_2$, $K_3$ | |
| Number of aerators in basin | | | N | |
| Oxygen transferred to basin | | | $N_{O_2}$ | |
| Rate of organic disappearance | | | $r_a$ | |
| Rate of micro-organism growth | | | $r_b$ | |
| Rate of oxygen use | | | $r_{O_2}$ | |
| Substrate use factor, lb mole micro-organism produced/lb mole substrate consumed | | | s | |
| Oxygen use factor, lb mole micro-organism produced/lb mole oxygen consumed | | | t | |
| Basin volume | | | v | |
| Interfacial area per unit volume of basin | | | | $a_v$ |
| Effective mean bubble diameter | | | | $D_B$ |
| Diffusivity of compound i in water | | | | $D_{i,H_2O}$ |
| Acceleration of gravity | | | | g |
| Overall mass transfer coefficient to bubble from liquid | | | | $K_x$ |
| Molecular weight of air | | | | $MW_{air}$ |
| Molecular weight of water | | | | $MW_{H_2O}$ |
| Distribution coefficient, $m_i = \dfrac{y_i}{x_i}$ | | | | $m_i$ |
| Power input to liquid per unit volume of basin | | | | (P/V) |
| Liquid phase Schmidt number = $\dfrac{\mu_1}{\rho_1 D_{i,H_2O}}$ | | | | Sc |
| Volume of bubble | | | | $V_B$ |
| Superficial gas velocity | | | | $V_G$ |

(continued)

**TABLE 35 (continued)**

| Parameter | Thibodeaux (1981d)[b] ASI model | Hwang (1980)[c] ASSA model | Freeman (1979)[d] ASSA model | Freeman (1980)[e] diffused air mass transfer model |
|---|---|---|---|---|
| Superficial liquid velocity | | | | $v_l$ |
| Rise velocity of a bubble swarm | | | | $v_s$ |
| Terminal velocity of a single rising bubble | | | | $v_t$ |
| Mole fractions of compound in gas | | | | $y$ |
| Mole fraction of compound in gas as it enters bottom of basin | | | | $y_1$ |
| Mole fraction of compound in gas as it breaks the surface of the basin | | | | $y_2$ |
| Total rise time of bubble in basin, hours | | | | $\partial$ |
| Viscosity of liquid | | | | $\mu_l$ |
| Density of gas | | | | $\rho_g$ |
| Density of liquid | | | | $\rho_l$ |
| Surface tension of liquid | | | | $\sigma$ |
| Time | | | | $\tau$ |
| Gas holdup fraction in basin | | | | $\phi$ |

[a] Parameters used in computing mass transfer coefficients not included (See Section 4 for details)

[b] Thibodeaux (1981d) ASI model    $Q_i = K_L A_s (X_i - X_i^*) MW_i$

[c] Hwang (1980) ASSA model

(1) without adsorption to sludge    $$S_e = \frac{S_o}{1 + \frac{k_{l(s)}X}{S_o}(1+r)t_c + (1+r)k_a t_c}$$

(2) with adsorption to sludge    $$S_e = \frac{-b + \sqrt{h^2 + 4 \, a \, S_o(1+K_T S_r)}}{2 \, a}$$

where $b = 1 - K_l S_o + K_T S_r + \frac{k_{l(s)}X}{S_o}(1+r)t_c(1+K_T S_r)$

$+ k_a(1+r)t_c(1+K_T S_r) + K_l X'(1+r)W$

$a = K_l(1 + (1+r)\frac{k_{l(s)}X}{S_o}t_c + (1+r)k_a t_c)$

(continued)

TABLE 35 (continued)

[d]Freeman (1979) ASSA model

(1) For substance balance

$$C_I F_I + F_r C_r = C_0 F_0 + \overbrace{\left| N \frac{\pi D^2}{4} K_a^T (X_a) + \left(A_s - \frac{N D^2}{4}\right) K_a^c X_a \right| MW_a}^{N_a} + \overbrace{\left| t \left( \frac{k_5 B_0 C_0 O_0}{\frac{k_5}{k_1} O_0 + K_{O_2} K_s + C_0 O_0 + K_{O_2} C_0} \right) \left(\frac{MW_a}{MW_{O_2}}\right) \right|}^{r_a} v$$

where $X_a = C_0 \left(\dfrac{MW_{H_2O}}{MW_a}\right) \dfrac{1}{\rho_{H_2O}}$

(2) For micro-organism balance

$$B_I F_I + F_r B_r = B_0 F_0 - \overbrace{\left| t \left( \frac{k_5 B_0 C_0 O_0}{\frac{k_5}{k_1} O_0 + K_{O_2} K_s + C_0 O_0 + K_{O_2} C_0} \right) \frac{MW_B}{MW_{O_2}} \right|}^{r_b} v$$

(3) For oxygen balance

$$O_I F_I + F_r O_r = O_0 F_0 + \overbrace{\left( \frac{k_5 B_0 C_0 O_0}{\frac{k_5}{k_1} O_0 + K_{O_2} K_s + C_0 O_0 + K_{O_2} C_0} \right)}^{r_{O_2}} v - \overbrace{\left| N \frac{\pi D^2}{4} K_b T\left(X_{O_2}^* - X_{O_2}\right) + \left(A_s - \frac{N D^2}{4} K_b^c\right)\left(X_{O_2}^* - X_{O_2}\right) \right|}^{N_{O_2}} MW_{O_2}$$

where $X_{O_2} = O_0 \left(\dfrac{MW_{H_2O}}{MW_{O_2}}\right) \dfrac{1}{\rho_{H_2O}}$

[e]Freeman (1980) DAAS model mass transfer equations

$$y_2 = m X_1 \left| 1 - EXP\left(\frac{-6 K_x \theta MW_{air}}{D_B \rho_g m}\right) \right| + y_1 EXP\left(\frac{-6 K_x MW_{air}}{D_B \rho_g m}\right)$$

where:    $D_B = 4.15 \left(\dfrac{\sigma^{0.6}}{(P/V)^{0.4} \rho_1^{0.2}}\right) * ^{1/2} + 0.09 \text{ cm}$

$(P/V) = V_G \rho_1 g + V_1 \rho_1 R$

$V_S = 27.43 \text{ cm/sec} + 2 V_G$

$f = \dfrac{V_G}{0.9 \text{ ft/sec} + 2 V_G}$

$a_v = 1.44 \left| \dfrac{(P/V)^{0.4} \rho_1^{0.2}}{\sigma^{0.6}} \right| \left(\dfrac{V_G}{V_t}\right)^{1/2}$

$k_1 (Sc)^{1/2} = 0.42 \left| \dfrac{(\rho_1 - \rho_g) \mu_1 g}{\rho_1^2} \right|^{1/3} (3600) \dfrac{1}{MW_{H_2O}}$

Figure 11 presents a schematic of the continuous AS process and defines the terms used by Hwang and others. Hwang describes the dynamics of substrate removal in the AS process by employing the following material balance:

$$\text{Input in Influent} - \left[ \underbrace{\frac{(1)}{\text{Biodegradation}}}_{} + \underbrace{\frac{(2)\ \text{Air}}{\text{stripping}}}_{} + \underbrace{\frac{(3)\ \text{Sludge}}{\text{wastage}}}_{} \right] = \text{Output in Effluent}$$

$$Q_F\ S_o - \left[ \frac{K_{1(s)}\ X\ S_e}{S_o}\ V + (K_a\ S_e\ V) + Q_w\ \frac{K_1\ X'\ S_e}{1 + K_T\ S_r + K_1\ S_e} \right] = (Q_F - Q_w)\ S_e$$

where:

X = sludge concentration in reactor (µg/l);

X' = maximum concentration of substrate on sludge (µg/l);

$K_{1(s)}$ = Grau equation rate constant, n=1 (1/day).

Upon simplification and rearrangement;

$$(S_o - S_e) = \left[ \frac{K_{1(s)}\ X\ S_e}{S_o} \right](1+r)\ t_c + K_a\ (1+r)\ S_e\ t_c + W(1+r)\left[ \frac{K_1\ X'\ S_e}{1 + K_T\ S_r + K_1\ S_e} \right]$$

Further rearrangement of the above equation yields the two solution methods presented in Table 36. The first method presented in Table 36 predicts air emissions from the AS process by requiring the solution of a quadratic equation in order to obtain $S_e$. The second solution method is based on the assumption that adsorption to the waste sludge is negligible which is true in many instances.

Biodegradation Kinetics—

Hwang evaluated the four biodegradation kinetic models listed below with respect to the AS process:

1.  First order kinetics:  $-\dfrac{dS}{dt} = k_1\ X\ S$

2.  Monod (1942) kinetics:  $-\dfrac{dS}{dt} = \dfrac{\alpha\ X}{Y}\ \dfrac{S}{k_s + S}$

3.  Grau (1975) kinetics:  $-\dfrac{dS}{dt} = k_{n(s)}\ X\left(\dfrac{S}{S_o}\right)^n$; where n = 1

4.  Hwang (1980) kinetics:  $-\dfrac{dS}{dt} = \dfrac{k_3\ X\ S}{(1 + k_4\ S)\ S_o}$

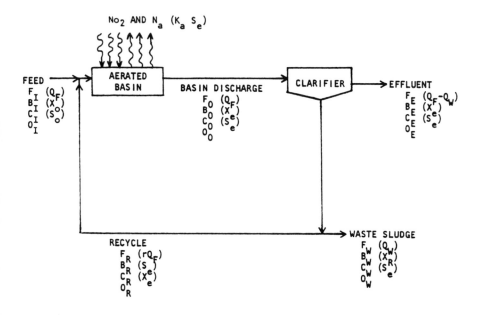

Figure 11.   Model of activated sludge system.

TABLE 36.  HWANG (1980) ACTIVATED SLUDGE TREATMENT PROCESS AERR MODEL

Model Form:[a]

$$Q_i = 4047 \times 10^4 \; MW_i \; K_{Li} \; \bar{A} \; x_i$$

where  $x_i = (Se) \; MW_{H_2O} / 10^6 \; MW_i$

$$K_{Li} = (K_{Li})_c \frac{\bar{A}_c}{\bar{A}} + (K_{Li})_T \frac{\bar{A}_T}{\bar{A}}$$

for significant absorption to waste sludge:

$$S_e = \frac{-b + \sqrt{b^2 + 4 \, a \, S_o (1 + K_T S_r)}}{2 \, a}$$

where  $b = 1 - K_1 S_o + K_T S_r + \frac{k_{1(s)} X}{S_o} (1+r) t_c (1+K_T S_r) + k_a (1+r) t_c (1+K_T S_r) + K_1 X' (1+r) W$

$a = K_1 (1 + (1+r) \frac{k_{1(s)} X}{S_o} t_c + (1+r) k_a t_c)$

excluding adsorption to waste sludge (X' small):

$$S_e = \frac{S_o}{1 + \frac{k_{1(s)} X}{S_o} (1+r) \; t_c + (1+r) k_a \; t_c}$$

[a]Definitions:

$Q_i$ = emission rate of chemical i (g/s)

$MW_i$ = molecular weight of chemical i (g/g mole)

$K_{Li}$ = overall mass transfer coefficient of chemical i expressed in liquid phase concentration (gmol/cm² - s)

$\bar{A}_c$ = surface area of the impoundment expressed in acres (Subscript c denotes the convective area of the surface impoundment; subscript T denotes the turbulent area)

$x_i$ = mole fraction of component i in the liquid phase

$4047 \times 10^4$ = denotes the factor needed to convert the surface area expressed in acres to cm²

X = sludge concentration in reactor (µg/l)

X' = maximum concentration of substrate (compound i) on sludge (µg/l)

$k_{1(s)}$ = Grau equation rate constant n=1 (1/day)

$S_o$ = initial substrate concentration (µg/l)

$k_a$ = area averaged overall air stripping rate constant (1/day)
Note: $K_{Li}$ may be converted to $k_a$ as described earlier under surface impoundments

The suitability of each biokinetic model was tested by Hwang with data from
the following sources:

- EPA, Cincinnati, Ohio; Batch experiments using over 50 pure priority
  pollutants at initial concentrations of 5 and 10 mg/L.

- Union Carbide Corporation; Continuous flow reactors, two in series,
  for BOD degradation of wastewater.

- Catalytic, Inc.; Continuous flow reactor using synthetic mixtures of
  pure components; BOD analyzed.

- J. WPCF; Continuous flow reactor for biodegradation of benzidine in
  wastewater.

Hwang noted that increased initial concentrations resulted in decreased
removal efficiencies, thus eliminating the first order kinetics model from
consideration. Grau kinetics and Hwang kinetics proved superior to Monod or
first order kinetics in fitting the data. The difference between the Grau and
Hwang kinetics was determining whether the plot of $Xt/S_o$ $(S_o-S_e)$ versus
$1/S_e$ yields a straight line through the origin of the plot. Hwang found
that in most cases the Grau kinetics model (n=1) adequately described
biodegration kinetics. Thus, the rate of organic (substrate) removal is
expressed as:

$$- \frac{dS}{dt} = \frac{k_{1(s)} \; X \; S}{S_o}$$

where:

$dS/dt$ = rate of substrate removal ($\mu g/l/day$);

$X$ = biomass concentration (ppm);

$S$ = substrate concentration (usually $S_e$) ($\mu g/l$);

$S_o$ = initial substrate concentration ($\mu g/l$);

$k_{1(s)}$ = Grau kinetics rate constant (n=1) (1/day).

Determination of Biokinetics Rate Constants--
In order to apply the above biokinetics model to an AS system one must
know the Grau kinetics rate constant $k_{1(s)}$ specific to the substrate
compound of concern. Determination of the biokinetic rate constant $k_{1(s)}$
can be accomplished by fitting laboratory or field data to the Grau kinetics
material balance;

$$\frac{Xt}{S_o \; (S_o-S_e)} = \left| \frac{1}{k_{1(s)}} \right| \frac{1}{S_e}$$

Plotting $Xt/S_o$ $(S_o-S_e)$ versus $1/S_e$, should yield a straight line through the origin at a slope of $1/k_{1(s)}$. Hwang has developed values for Grau's $k_{1(s)}$ for most of the 129 priority pollutants based on; (1) laboratory and field data cited above; and (2) estimates based on observed similarities between compounds extended to include untested compounds.

Air Stripping Kinetics--
    The expression for air stripping of volatile compounds (substrate) from the AS treatment basin is given as:

$$-\frac{dS}{dt} = k_a\ S$$

(for single component substrate system)

where:

    $k_a$ = air stripping rate constant, 1/day.

    Hwang employs the area-averaged mass transfer model used by Thibodeaux (1981d), where the overall mass transfer coefficient $k_a$ is based on the area-averaged k-values for zones of turbulence $(k_a)_T$ and convection $(k_a)_c$ as shown below:

$$k_a = (k_a)_T\ \frac{A_T}{A} + (k_a)_c\ \frac{A_c}{A}.$$

Key assumptions of this model are that the AS system can be approximated as two distinct zones where mass transfer takes place at the surface of the reactor only and is only slightly influenced by subsurface phenomena.

    Discussions presented earlier on methods for calculating turbulent and convective zone mass transfer coefficients for use in the above expression are applicable here and will not be repeated. Note that in many cases mass transfer is liquid phase controlled, but, for some low volatility compounds the gas phase mass transfer may be equally controlling. In many cases the turbulent zone will be the major contributor of volatile air emissions, however, Hwang notes that for large lagoons with poor aeration this may not be so.

Adsorption on Sludge--
    Hwang contends that adsorption onto waste activated sludge may be a significant removal mechanism for some organic priority pollutants and cyanides. Hwang assumes a Langmuir type adsorption equation to describe the equilibrium concentration of the substrate (for a single component system) on the sludge as:

$$[B.S] = \frac{K_1\ X'\ S}{1 + K_1\ S + K_T\ S_r}$$

where:

    $[B.S]$ = concentration of a substrate on sludge, $\mu g/l$;

$S$ = substrate concentration in the liquid, $\mu g/l$;

$S_r$ = concentration of total substrates minus substrate S under consideration, $\mu g/l$;

$K_1$, $K_T$ = adsorption constants;

$X'$ = the maximum amount of the substrate adsorbed on sludge, $\mu g/l$.

Similarly, for a multicomponent substrate system the above equation becomes:

$$[B.S_T] = \frac{K_T\,S_T\,X'_T}{1 + K_T\,S_T}$$

(for multicomponent substrate system).

Note that the AS treatment system AERR model presented in Table 36 includes the sludge adsorption term for single component systems.

Kincannon et al. (1981,1982) investigated the importance of each pollutant mechanism (air stripping, sorption and biodegradation) from lab-scale activated sludge treatment systems and found that sorption was not an important mechanism for any of the priority pollutants analyzed. Although some exceptions applied, the most predominate removal mechanism was found to be consistent with chemical classification. For instance, biodegradation removal was predominate for nitrogen compounds, phenols and oxygenated compounds. Aromatics were removed from wastewater by a combination of biodegradation and air stripping, while halogenated hydrocarbons were removed by air stripping.

Determination of Adsorption Rate Constants--
Hwang also presents a method for determining the adsorption rate constants. The above two equations can be rearranged and linearized as follows:

$$\frac{1}{[B.S]} = \frac{1 + K_T\,S_r}{K_1\,X'}\left(\frac{1}{S_e}\right) + \frac{1}{X'}$$

(single component).

$$[B.S] = \frac{1}{K_T\,X_T}\,\frac{1}{S_T} + \frac{1}{X'_T}$$

(multicomponent).

The values of $K_T$ and $X_T'$ are determined by plotting $1/[B.S_T]$ versus $1/S_T$ for single component substrates. Similarly, a plot of $1/[B.S]$ versus $1/S_e$ can be prepared for multicomponent substrate systems. Data presented by Hwang may be used to develop these rate constants. In addition, Hwang has provided a listing of these rate constants for the priority pollutants based on work sponsored by EPA.

### Freeman (1979) Activated Sludge Surface Aeration Model

Freeman's original AS treatment AERR model appears very similar to the work of Thibodeaux (1981d) and Hwang (1980). The major difference between the Freeman and Hwang AS models are the biodegradation kinetics used. In addition, Freeman does not address adsorption of substrate onto the wasted sludge, assuming that adsorption is not a major removal mechanism for organic pollutants. This is a reasonable assumption based on the most recent work of Kincannon mentioned previously.

Freeman uses Thibodeaux's aerated surface impoundment AERR model, based on the two-film resistance theory, to predict the emissions from the AS basin knowing the concentration of substrate in the basin. However, in order to predict the concentration of substrate, Freeman writes the material balance around the basin for the substrate, C, the micro-organisms, B, and oxygen, O, as follows:

$$C_I F_I + F_R C_R = C_O F_O + N_a + r_a V \quad \text{(for compound A)}$$

$$B_I F_I + F_R B_R = B_O F_O - r_b V \quad \text{(for biomass B)}$$

$$O_I F_I + F_R B_R = O_O F_O + r_{O_2} V - N_{O_2} \quad \text{(for oxygen } O_2)$$

Figure 11, shown previously, illustrates the material balance equations presented above and provides a comparison with the Hwang (1980) model. Note that the substrate material balance given above does not consider substrate lost to wasted sludge.

Mass Transfer--
Freeman also employs Thibodeaux's aerated surface impoundment model for predicting the air stripping losses based on mass transfer at the air-water interface. Like Hwang, Freeman isolates zones for convective and turbulent mixing and applies appropriate mass transfer expressions for each. The final form of Freeman's mass transfer expression describing the stripping of a substrate $X_a$ from an AS reactor is shown below:

$$N_a = \left[ \frac{N \pi D^2}{4} K_a^T (X_a) + \left( A_s - \frac{N D^2}{4} \right) K_a^c X_a \right] MW_a$$

Freeman used similar analogies to predict the rate of oxygen transfer into the AS basin as:

$$N_{O_2} = \left[ \frac{N \pi D^2}{4} K_b^T \left( X_{O_2}^* - X_{O_2} \right) + \left( A_s - \frac{N \pi D^2}{4} \right) K_b^c \left( X_{O_2}^* - X_{O_2} \right) \right] MW_{O_2}$$

Note that in Freeman's formula the turbulent area is given as $\pi D^2/4$ where D is the diameter of the region of effect for a given surface aerator. Estimates on regions of turbulence for various sized surface aerators were presented earlier in Section 4. Essentially, all aspects of the mass transfer expression given above are identical to that in the Hwang ASSA model.

Biokinetics--

To represent the biological oxidation process, Freeman selected a model developed by Gerber et al. (1976) as shown below:

$$B + A \underset{k_2}{\overset{k_1}{\rightleftharpoons}} BA$$

$$BA + O_2 \underset{k_4}{\overset{k_3}{\rightleftharpoons}} BAO_2$$

$$BAO_2 \overset{k_5}{\longrightarrow} B + P$$

where:

B = micro-organisms;

A = substrate;

$O_2$ = dissolved oxygen;

P = products.

Assuming that the last step is rate determining, and the basin is completely mixed, Freeman predicts the rate of oxygen uptake as:

$$r_{O_2} = \frac{k_5 \, B_o \, C_o \, O_o}{\frac{k_5}{k_1} O_o + K_{O_2} K_S + C_o \, O_o + K_{O_2} C_o}$$

where:

$K_{O_2} = (k_4 + k_5)/k_3$

$K_S = k_2/k_1$ .

In relating the oxidation of compound A to oxygen uptake and biomass growth, Freeman writes the following:

$$t \, O_2 + s \, A \longrightarrow B + H_2O + CO_2 + \text{products}$$

where:

t and s are stoichiometric constants based on assumptions made about the molecular structure of the biomass in the above equation.

Based on the above equation, the rate of biomass growth ($r_B$) and the rate of substrate (compound A) removal ($r_A$) can be shown as follows:

$$r_B = t \; r_{O_2}\left(\frac{MW_B}{MW_{O_2}}\right)$$

$$r_A = \frac{t}{s} \; r_{O_2}\left(\frac{MW_A}{MW_{O_2}}\right).$$

Thus, the final expression for the rate of substrate utilization would be:

$$r_A = \frac{t \; k_5 \; B_o \; C_o \; O_o}{\left(s \; \dfrac{k_5}{k_1} O_o + K_{O_2} K_S + C_o \; O_o + K_{O_2} \; C_o\right)}\left(\frac{MW_A}{MW_{O_2}}\right).$$

Similarly, Freeman's expression for the value of biomass growth would be:

$$r_B = \frac{t \; k_5 \; B_o \; C_o \; O_o}{\left(\dfrac{k_5}{k_1} O_o + K_{O_2} K_S + C_o \; O_o + K_{O_2} \; C_o\right)}\left(\frac{MW_B}{MW_{O_2}}\right).$$

Model Solution

Solution of the above system of equations requires the following data:

- process flow rates (feed, recycle, effluent) and basin dimensions;

- physical data and energy input data for calculating mass transfer coefficients;

- biokinetic rate constants for the compound being considered (which can be obtained by methods discussed later).

One key difference between the solution method for Freeman's ASSA model and the Hwang ASSA model presented earlier is the extent of data assumed to be known at the start of the problem.  The Hwang model works with a known biomass concentration and assumes that sufficient oxygen is maintained in the aeration basin to support biological growth.  On the other hand, Freeman's model considers the interaction of the substrate (the compound being removed), biomass, and oxygen material balances, thus Freeman's model can be used to predict emissions when less information is known.  However, the solution procedure for Freeman's model is inherently more difficult, as it requires the solution of three simultaneous nonlinear equations to determine three unknowns ($B_o$, $C_o$, and $O_o$) by iterative method.  The solution method is briefly outlined in Table 37.  Freeman recommends that the solution be conducted by application of a small personal computer.  Computerized solution will tend to save time and calculational errors will be avoided.

TABLE 37.   SOLUTION METHOD FOR FREEMAN'S ACTIVATED SLUDGE SURFACE AERATION AERR MODEL

1.  Determine mass transfer coefficients and effective areas of turbulence and convection.  (Discussed in detail in Section 4, Surface Impoundments).

2.  Estimate Stoichiometry of Biological Decomposition of Compound A to products to determine values of stoichiometric constants t and s in the following equation:

$$t\ O_2 + s\ A \rightarrow B + H_2O + CO_2 + Products$$

3.  Obtain biokinetic rate constants through laboratory tests (see text).

4.  Write Material Balance Equation for substrate, micro-organisms, and oxygen as below:

For substrate balance
$$C_I F_I + F_r C_r = C_0 F_0 + \underbrace{\left[N \frac{\pi D^2}{4} k_a^I (X_a) + \left(A_s - \frac{N \pi D^2}{4}\right) k_a^c X_a\right] MW_a}_{N_a} + \underbrace{\left|\frac{t}{s}\left[\frac{k_5 B_0 C_0 O_0}{\left(\frac{k_5}{k_1} O_0 + K_{O_2} K_S + C_0 O_0 + K_{O_2} C_0\right)}\right] \frac{MW_a}{MW_{O_2}}\right| V}_{r_a}$$

where $X_a = C_0 \left(\dfrac{MW_{H_2O}}{MW_a}\right) \dfrac{1}{\rho_{H_2O}}$

For micro-organism balance
$$B_I F_I + F_r B_r = B_0 F_0 - \underbrace{\left|\left[\frac{k_5 B_0 C_0 O_0}{\left(\frac{k_5}{k_1} O_0 + K_{O_2} K_S + C_0 O_0 + K_{O_2} C_0\right)}\right] \frac{MW_B}{MW_{O_2}}\right| V}_{r_b}$$

For oxygen balance
$$O_I F_I + F_r O_r = O_0 F_0 + \underbrace{\left|\left[\frac{k_5 B_0 C_0 O_0}{\frac{k_5}{k_1} O_0 + K_{O_2} K_S + C_0 O_0 + K_{O_2} C_0}\right] MW_{O_2}\right| V}_{r_{O_2}} - \underbrace{V \left[N \frac{\pi D^2}{4} k_b^T (X_{O_2}^* - X_{O_2}) + \left(A_s - \frac{N \pi D^2}{4}\right) k_b^c (X_{O_2}^* - X_{O_2})\right] MW_{O_2}}_{N_{O_2}}$$

where $X_{O_2} = O_0 \left(\dfrac{MW_{H_2O}}{MW_{O_2}}\right) \dfrac{1}{\rho_{H_2O}}$

5.  Select values for $B_0$, $C_0$, and $O_0$ and solve the material balance equations.  Repeat as necessary until the solution converges.

Biokinetic Rate Constant Determination--
According to Freeman, determination of biokinetic rate constants to fit the Gerber kinetics model can be conducted in one of two ways. By following the work of Gerber, laboratory experiments can be conducted at considerable cost and the data may be plotted to directly determine Gerber's rate constants. Freeman estimates that this method could cost in the range of $50,000 to $100,000 per compound to develop the appropriate kinetics data. The cost would tend to be dependent on the compounds characteristics (biodegradability, volatility, etc.).

Another method used by Freeman in experiments with acrylonitrile (AN) involved calibration of the entire model based on laboratory experiments measuring feed AN and AN emissions. In order to apply this calibration method Freeman measured or calculated all variables in the equations shown in Table 37 except the biokinetic rate constants. The rate constants were then determined by "fitting" the model predictions to the measured results under varying conditions. One drawback to this method was that Freeman apparently assumed Gerber's values for $k_{O_2}$ and $k_B$ were applicable to AN biokinetics. While Freeman's assumption may or may not be correct, the model should be accurate at least within the range of testing done by Freeman.

### Freeman (1980) Diffused Air (Subsurface) Activated Sludge Model

In later work Freeman (1980) developed a significant modification to the initial AS model to more adequately describe the mass transfer phenomena taking place in diffused air AS systems. Diffused air systems are commonly employed in laboratory bench testing and occasionally used in actual field applications. Freeman's Diffused Air Activated Sludge (DAAS) model, also called the subsurface aeration model, focuses on the mass transfer of the compound into the air bubbles released by the spargers as the bubbles rise to the surface. This is quite different from the models discussed earlier, since the area across which mass transfer takes place is the surface area of the bubbles, not the basin surface area. Note that this model is better suited to the activated sludge diffused air application than other models discussed. Aside from the variation in mass transfer expressions the DAAS model should be considered identical to Freeman's earlier AS model, thus only mass transfer aspects of the model will be discussed.

Mass Transfer--
If it is assumed that the concentration of compound i in the air diffuser source is negligible, than Freeman's expression for the concentration of compound i in an air bubble as it reaches the surface is:

$$y_2 = m\, X_i \left[ 1 - \exp\left( \frac{-6\, k_x\, \theta\, MW_{air}}{D_B\, \rho_g\, m} \right) \right]$$

where:

m = distribution coefficient;

$X_i$ = mole fraction of compound i in liquid (ppm);

$k_x$ = overall mass transfer coefficient to the bubble from the liquid ($g$-mole/hr-cm$^2$);

$\theta$ = rise time of bubbles (hours);

$MW_{air}$ = molecular weight of air (28.8 g/g-mol);

$D_B$ = mean bubble diameter (cm);

$\rho_g$ = density of air (g/cm$^3$);

$y_2$ = mole fraction of compound i in the gas bubble as it breaks the surface.

The mean bubble diameter required above may be estimated according to the following equation by Calderbank (1967):

$$D_B = 4.15 \left[ \frac{\sigma^{0.6}}{(P/V)^{0.4} \, \rho_1^{0.2}} \right]^{1/2} \psi + 0.09 \text{ cm}$$

where:

$\sigma$ = liquid surface tension (g/s$^2$);

$\rho_1$ = liquid density (g/cm$^3$);

$P/V$ = power input to liquid per unit volume of basin (g-cm$^2$/s$^3$-cm$^3$);

$\psi$ = gas holdup fraction in basin.

The power input per unit volume ($P/V$) to the basin could be estimated from diffuser system power use or from the superficial gas velocity as follows:

$$P/V = V_G \, \rho_1 \, g$$

where:

$V_G$ = superficial gas velocity (cm/s);

$g$ = acceleration due to gravity (980.6 cm/s$^2$).

The gas holdup fraction is estimated by Towell (1965) as:

$$\psi = \frac{V_G}{0.9 + 2 \, V_G}$$

Note that $V_G$ is simply the gas volumetric flow rate divided by the surface area of the tank and 0.9 is an approximation of the terminal gas velocity ($V_t$) in feet per second.

In the application of the above described diffusional air mass transfer model, Freeman assumes the overall transfer is controlled by the liquid film. Thus:

$$\frac{1}{K_x} = \frac{1}{mk_g} + \frac{1}{k_l} = \frac{1}{k_l} \text{ for large value of } k_g \ .$$

Freeman uses Calderbank's (1967) relationship for the liquid film mass transfer coefficient in the following form:

$$k_l \ (Sc)^{1/2} = 0.42 \left[ \frac{(\rho_l - \rho_g) \ \mu_l g}{\rho_l^2} \right]^{1/3} (3600) \frac{\rho_l}{MW_{H_2O}} \ .$$

## Other Activated Sludge Models

Among the numerous other efforts in modeling air emissions from the AS process, the work of Wukash et al. (1981), and Brown and Weintraub (1982) should be referenced for completeness.

Publicly Owned Treatment Works (POTWs)--

Work conducted by Wukash et al. (1981), describing the fate of organic compounds in POTWs, presents a systematic approach towards model development. This study found, as did Hwang, that air stripping, biodegradation and adsorption to biomass were mechanisms associated with pentachlorophenol (PCP) removal in the AS process. Air stripping was found to be responsible for removal of less than 0.04 percent of the PCP applied, while adsorption onto biomass accounted for less than 0.3 percent of the PCP. Important conclusions of this study are summarized below:

- At substrate concentrations below the inhibitory threshold, the equilibrium concentration of a growth limiting substrate can be described by Monod kinetics as follows:

$$\mu = \frac{\mu_m S}{K_s + S}$$

where:

$\mu$ = specific cell growth rate (day $^{-1}$);

$\mu_m$ = maximum growth rate achievable;

$K_s$ = saturation constant (mg/l);

S = substrate concentration (mg/l) also $[S_e$ (Hwang)] or $[C_o$ (Freeman)].

- At much lower substrate concentrations ($S << K_s$), as is the case for most POTWs, the Monod equation reduces to simple first order kinetics:

$$\mu \cong \frac{\mu_m S}{K_s}$$

Under steady state conditions $\mu$ is equal to the reciprocal of the sludge retention time (1/SRT) plus decay (d) thus:

$$\mu = 1/SRT + d = \frac{\mu_m S}{K_s}$$

rearranging:

$$S \cong \frac{K_s}{\mu_m} (1/SRT) + \frac{K_s d}{\mu_m}$$

Wukash et al., found very good correlation with first order kinetics shown above for PCP concentrations below inhibitory levels (< 350 µg/l).

Point Process Wastewater Treatment--

A study done by Brown and Weintraub (1982) indicated that the rate of substrate removal could be predicted by pseudo first order kinetics of the following form:

$$\text{removal rate} = \frac{S_o - S_e}{X_v t} = k S_e$$

where:

$S_o$ = influent COD concentration;

$S_e$ = effluent COD concentration;

$X_v$ = mixed liquor volatile suspended solids (MLVSS);

$Q$ = flow rate;

$V$ = reactor volume;

$t$ = residence time = $V/Q$;

$k$ = coefficient of substrate removal.

The laboratory results from testing several paint process water constituents apparently fit the model shown above extremely well. Substrate concentrations ranged up to 600 ppm in some tests.

The results of the above investigations were included to illustrate the point that one single biokinetic model may not be the only model capable of fitting biodegradation rate data. Furthermore, the number of biokinetic rate expressions is almost as vast as the number of investigators conducting research. No single biokinetic model is currently viewed as the best model, although the Monod model and first order models have been in general use for some time.

As noted earlier, Hwang investigated four biokinetic models before selecting Grau kinetics for use in his model. He expressed concern (Hwang 1982) that a model for complete mix systems would be based on the influent concentration. While this is contrary to basic assumptions for complete mixed reactors [Rx is usually $f(S_e)$], Hwang notes that it apparently works better than other commonly accepted models.

Similarly, Freeman (1982) noted that in his modeling work various biokinetic rate expressions are commonly employed to determine "best fit" to the data. Freeman indicates that any reasonable biokinetic model could be employed using his modeling approach. Freeman selected the Gerber model for its ability to consider the interaction of the substrate balance, the biomass growth balance, and the oxygen balance. However, Freeman also notes that Gerber's (1976) model can reduce to Monod kinetics by assuming constant biomass concentration and sufficient oxygen.

### Selection of Biological AERR Model

The selection of an appropriate AS model or models for predicting emissions of organic compounds should be based upon the following criteria:

- model flexibility;

- simplicity;

- model accuracy;

- underlying assumptions.

The following discussion will focus on the above issues, briefly presenting the relative merits of the models presented earlier.

Model Flexibility--
In order to assess the extent to which models are adequately flexible, it is necessary to understand the circumstances for which they were intended and those for which we wish to apply them. Consider the application of; (1) the Thibodeaux aerated surface impoundment (ASI) model; (2) the Hwang activated sludge surface aeration (ASSA) model; (3) the Freeman ASSA model; and (4) the Freeman diffused air activated sludge (DAAS) model. These models run the gambit from the simple ASI model which requires a known concentration of compound i in the aeration basin, to the Freeman DAAS model which can relate emissions to a wide range of plant design and operating parameters.

Engineers working for regulatory agencies or consultants may desire to assess emissions from planned and existing facilities. Conventional surface aerator design lagoons will be prevalent, however, energy efficient DAAS systems may also be encountered. The ability of each of the four models to predict emissions under varying circumstances is shown in Table 38. As shown in Table 38, the Hwang ASSA Model tends to be very flexible except in predicting emissions from oxygen limited systems, or systems operating under unconventional circumstances.  Of course, only the Freeman DAAS model adequately handles circumstances involving diffused air systems.  Note that where sludge adsorption is determined to be substantial only the Hwang ASSA model is adequate.  In addition, the Hwang model also finds greater immediate application to many organic compounds because adequate biokinetics rate data have been developed for the Hwang (Hwang 1980) model as compared to the Freeman ASSA Model.

Model Simplicity--

One underlying problem in predicting AERR from the AS process is the complex nature of the solution method required.  In comparison to other AERR models, the AS models are among the most complex to solve.  The Thibodeaux ASI model is, of course, relatively simple, whereas the Freeman ASSA and DAAS models requires the assistance of a small computer for practical purposes.  A somewhat subjective ranking of the complexity of the models on a scale of 1 to 10 (10 being the most complex solution) is given below:

| Model | Complexity rating |
|---|---|
| Thibodeaux ASI | 1 |
| Hwang ASSA (without sludge adsorption) | 2 |
| Hwang ASSA (with sludge adsorption) | 4 |
| Freeman ASSA | 8 |
| Freeman DAAS | 10 |

Realizing that most of the AS models presented require time consuming solution techniques, the Hwang ASSA model with sludge adsorption is probably the most complex model an evaluator could be expected to employ without the aid of a small computer.

The issue of model simplicity must be viewed with regard to problem application.  For example, given the concentration of compound i ($S_e$) in the aeration basin, the ASI model is the logical, simple choice.  Use of other models under these circumstances would be poor judgement and likely lead to conflicting information.  Similarly, application of the Freeman ASSA model for a compound for which no biodegradation kinetics data have been generated would be difficult.  However, use of any ASSA model to predict diffused aerator emissions would be in error.  Thus, in this case, the additional complexity of Freeman's subsurface diffused air mass transfer expression is justifiable. Also, the increased complexity of Freeman's ASSA model is justifiable for very high substrate loading conditions, or for modeling an oxygen starved reactor.

TABLE 38.  APPLICATION OF VARIOUS ACTIVATED SLUDGE AERR MODELS

| Model | Applications | | | | | |
|---|---|---|---|---|---|---|
| | Given: $S_e$ only | Surface aeration | | | | Diffused aeration |
| | | Given: $S_o$ conventional system | Given: $S_O$ Oxygen limited | Given: $S_O$ Process upset | Given: $S_o$ substrate adsorbtion | All conditions |
| Thibodeaux ASI | X | | | | | |
| Hwang ASSA | X | X | | | X | |
| Freeman ASSA | X | X | | X | | |
| Freeman DAAS | X | | | | | X |

It is doubtful that such model flexibility will be required for modeling air emissions from planned facilities on a regular basis. While Freeman's models do have valuable use under certain circumstances, it is clear that they were developed for research and design purposes rather than fulfilling analytical needs.

Model Accuracy--

The Thibodeaux ASI model was tested by Cox, et al. (1982) and shown to underpredict benzene emissions as indicated in Section 3. However, it was noted by Hwang* that calculational errors were made in assuming application of Raoult's Law. When applying Henry's Law, the corrected model predictions were found to be within a factor of 2 to 3 to the measured results. Additional model verification was also conducted by Thibodeaux (1981) for methanol. These measured results showed reasonable agreement with predicted emissions.

Tests conducted for the Thibodeaux ASI model are largely applicable to the mass transfer portion of the Hwang ASSA models since they are identical in this respect. Freeman's ASSA model also employs the same mass transfer terms and would be similar in accuracy. Biokinetic rate data and sludge adsorption data were collected for input to the Hwang model. No complete testing of the fully assembled model has been conducted to date. However, if the individual elements are based on test data it can be rationalized that the total model should be reliable.

Freeman (1980, 1982) also conducted laboratory experiments to verify the mass transfer portions of his DAAS model separate from the biokinetics model. Using air stripping experiments with sterile reactors, Freeman (1980) found that the subsurface diffused air model predicted the material balance for acrylonitrile to within 3.2 percent. Freeman (1982) also calibrated the Gerber biokinetics model, using the DAAS mass transfer model, for acrylonitrile and for other organic compounds.

Model Assumptions--

Complete Mix Reactor--One basic underlying assumption of all the models investigated is the complete mix reactor assumption. This assumption is basic to most conventional AS process modeling and will not be disputed. However, GCA notes that there may exist some contradiction between the complete mix modeling assumption and the turbulent/convective zone mass transfer concept used in the ASI and ASSA models. In reviewing these models, GCA questioned the necessity of considering the convective zone mass transfer since the transfer from the turbulent zone should be substantially greater. However, Hwang* pointed out that for large area lagoons the area of convective mass transfer could be very large (inferring low mixing might result). Thus, in this application the complete mix assumption may not be completely accurate.

---

*Telecon with Tom Nunno of GCA, 15 September 1982.

Mass Transfer--Assumptions regarding the governing mechanisms of mass transfer are summarized as follows:

- ASI and ASSA models reviewed assume that mass transfer takes place at the surface according to the two-film resistance theory. In addition, these models generally assume that the concentration of compound i ($X_i$) in the air is neglibible to that in the liquid.

- Freemans DAAS model assumes that mass transfer takes place at the diffused air bubble interface and that, again, the initial concentration ($X_i$) is zero and increases until equilibrium is reached. The Freeman DAAS model also assumes that the overall mass transfer rate ($K_L$) is equal to the liquid phase mass transfer coefficient ($k_l$) because the gas phase mass transfer coefficient is assumed to be very large. In addition, the DAAS model assumes that mass transfer which takes place at the surface is small compared to subsurface mass transfer and, thus, is negligible.

Biokinetics Model--The biokinetic rate expressions employed in the Hwang and Freeman models are quite different in structure, complexity, and underlying assumptions. The Hwang model relies on Grau kinetics which assumes adequate oxygen in the basin. The Freeman model uses Gerber kinetics which considers oxygen requirements in the biokinetic expression. As mentioned earlier, it is difficult to judge the relative merits of biokinetic models. However, it is likely that the Gerber model would be valid under a greater range of conditions than the Grau model.

Sludge Adsorption--Only the Hwang ASSA Model considers adsorption of substrate to the sludge competing with other substrate removal mechanisms. Freeman's ASSA and DAAS models both assume that adsorption of substrate is negligible. While this assumption may be valid for many organic compounds based on the work of Kincannon, adsorption has been shown to be significant in several instances (Hwang 1981, Patterson 1981).

Model Selection

The Hwang ASSA model is preferred to Freeman's model because the Hwang model is most easily applied, employs simpler bioxidation kinetics for which rate constant data is available, and is capable of modeling adsorption of substrate to the biomass. Freeman's DAAS mass transfer equations are recommended for modeling diffused air systems because they are currently the only reasonable choice. The DAAS mass transfer equations could be integrated into Hwang's ASSA model for predicting DAAS air emission release rates where the effluent concentration is unknown. This is recommended to simplify the problem solution and to avoid costly data collection required to use Freeman's biokinetic model.

Based on the factors discussed above GCA recommends that EPA select the following models for the applications listed below:

| Application | Selected model |
|---|---|
| Surface aeration--effluent or aeration basin concentration ($S_e$) known | Thibodeaux ASI model |
| Surface aeration--influent concentration ($S_o$) known | Hwang ASSA model |
| Diffused aeration--$S_e$ known | Freeman DAAS model |
| Diffused aeration--$S_o$ known | Hwang (modified after Freeman) |

The basis for selecting the Thibodeaux ASI and Hwang ASSA models for applications to surface aerated treatment systems includes considerations of simplicity, accuracy and ease of application.  It is believed that these models will adequately predict emissions from the majority of most AS treatment systems and are acceptable for use by most permitting engineers.  In addition, the availability of biokinetic rate data for Grau kinetics makes the Hwang ASSA model particularly attractive for general use.  Because the Hwang ASSA model does not apply well to subsurface diffused air systems, it is recommended that for modeling diffused air systems the Hwang model be modified to incorporate diffused air mass transfer mechanisms similar to that described by Freeman (1980).  Freeman's diffused air mass transfer model could be incorporated into the Hwang model without substantial effort.  This modification would permit the application of Freeman's mass transfer expression integrated with the Grau kinetics and sludge adsorption considerations of Hwang's model.

# 9. Air Emissions from Drum Storage and Handling Facilities

INTRODUCTION

Drum storage and handling facilities encompass diverse operations and sources of air emissions. Sources of routine air emissions include the storage of volatiles in lagoons or storage tanks. Air emission release rate estimation techniques for these activities are described in other sections of this report. The other important air emissions category is thought to be accidental spills.

The purpose of this section is to identify data necessary for estimating air emissions attributed to accidental spills at drum storage and handling facilities. This section provides a general description of a drum storage and handling facility and also some spill rate data for hazardous waste treatment facilities and petroleum handling facilities. At present, no data specific to air emissions from drum handling and storage facilities have been found.

DESCRIPTION OF DRUM STORAGE AND HANDLING FACILITIES

At drum storage and handling facilities, waste material, most commonly solvents, may arrive by tank car or tank truck, as well as in drums. Material may then be pumped to storage tanks, lagoons, or other drums for storage. Materials are segregated by type until enough are collected for reprocessing or disposal either onsite or offsite with removal again either by tank car, or tank truck, and less commonly, in drums.

Drums are removed from a tractor trailer on arrival with a fork lift and are conveyed to and from the drum storage building, ideally a locked metal building with louvered gables where storage occurs on concrete pads. Drums which are stored directly on the ground may freeze to the surface, thus, are subject to rupture during removal.

A number of other operations may also occur. Damaged drums will usually be replaced either upon arrival or upon later inspection. Material may have to be resampled to check compatibility parameters such as flash point, acidity or chloride content. Drums may be intentionally burst if this is the most effective way of removing contents of damaged barrels.

SOURCE OF VOLATILE AIR EMISSIONS

It is useful to classify emissions as routine or accidental, and to further classify routine emissions as either continuous or intermittent. Examples of each category are shown in Table 39. Note that routine continuous air emissions are from sources (lagoons and storage tanks) for which air emission release rate models have been previously discussed. It also appears that intermittent routine emissions cannot be very significant since they involve only one drum at a time over limited time periods.

To describe total facility emissions there are two approaches depending on available data:

● Total facility emissions, in barrels* lost per barrel handled;

● With an accurate description of onsite lagoons and storage tanks, air emissions from these sources can be calculated directly and then supplemented by the accidental spill rate, again parameterized by barrels lost per barrel handled.

The remainder of this discussion assumes the second course.

LITERATURE SURVEY

No examples have been found in the literature corresponding directly to spill rate data at drum storage and handling facilities. Nevertheless, we mention some recent work which is relevant.

Hillenbrand, et al. (1982) have surveyed Pollution Incident Reporting System (PIRS) and Spill Prevention, Control, and Countermeasures (SPCC) data for EPA. The study is restricted to tank farms and chemical plant tanks greater than 100,000 gallons. Data consists mainly of gasoline and fuel oil, but also includes spill reports on non-fuel products, organic and inorganic materials. The report categorized over 3,000 spills by cause (pipe rupture, pipe malfunction, etc.) and quantity. Unfortunately, they have not been able to give a spill rate based on the total amount of material handled (i.e., "exposure variable").

Also under EPA contract, ICF, SCS and Clement Associates (1982) have surveyed release rates and costs for waste treatment technologies. As part of this work, estimates were made of routine and accidental spillage based apparently on "best judgment." Some loss fraction estimates for accident spills for various treatment processes are shown in Table 40.

This report also notes that a December 1980 study by F. G. Bercha and Associates estimated a spill loss fraction of $2.45 \times 10^{-4}$ for hazardous waste loading and unloading operations.

---

*The term barrels is used in this section as a unit of volume whether or not material is in drums, tanks, etc.

TABLE 39.  EXAMPLES OF ROUTINE AND ACCIDENTAL SPILL SITUATIONS
AT A DRUM STORAGE AND HANDLING FACILITY

| Routine | Accidental |
|---------|------------|
| **Continuous** | |
| • Lagoon emissions | • Drum rupture by transfer operations |
| | • Dropped drum |
| • Storage tank breathing and working losses | • Tank or drum overflow |
| | • "Spontaneous" drum failure |
| **Intermittent** | • Pipe rupture or pump malfunction |
| • Open pouring | • Faulty hose coupling |
| • Sampling open drums | |
| • Intentional drum bursting | |

TABLE 40.   ACCIDENTAL SPILL FRACTIONS FOR VARIOUS
            TREATMENT TECHNOLOGIES[a]

| Technology | Accidental spill fraction (amount spilled/ amount handled) |
|---|---|
| Chemical stabilization | $1.5 \times 10^{-3}$ |
| Chemical precipitation | $2.6 \times 10^{-4}$ |
| Chemical destruction | $1.1 \times 10^{-5}$ |
| Chemical coagulation | $2.6 \times 10^{-4}$ |
| Filter press | $1.1 \times 10^{-4}$ |
| Centrifuge | $2.2 \times 10^{-4}$ |
| Vacuum filter | $1.1 \times 10^{-4}$ |
| Evaporation | $7.7 \times 10^{-4}$ |
| Air stripping | $1.3 \times 10^{-4}$ |
| Steam stripping | $2.1 \times 10^{-4}$ |
| Solvent extraction | $1.6 \times 10^{-4}$ |
| Leaching | $3.0 \times 10^{-4}$ |
| Distillation | $2.1 \times 10^{-4}$ |
| Electrolytic decomposition | $2.2 \times 10^{-4}$ |
| Reverse osmosis | $1.9 \times 10^{-4}$ |
| Carbon adsorption, PAC | $4.3 \times 10^{-4}$ |
| Ion exchange | $4.3 \times 10^{-9}$ |
| Average | $3.0 \times 10^{-4}$ |

[a]From "RCRA Risk/Cost Policy Model Project;"
ICF, Inc., SCS Engineers, Inc., and Clements
Associates, Inc., 1982.

Murphy, et al. (1981) presented data for oil spills occurring at the Milford Haven Terminal over an 18-year period. The loss fraction attributable to terminal operations (faulty hose couplings, pumps, valves, etc.) are $4 \times 10^{-5}$.

Until we have the pertinent data in hand we cannot confirm that any of the figures above are relevant to drum handling and storage facilities. Yet these figures do represent what is considered acceptable in operations with a number of similar steps. Our best estimate, therefore, is that the spill rate at drum handling and storage facilities is between 1 and 100 barrels per 100,000 barrels handled.

## OTHER CONSIDERATIONS

The amount spilled is not the same as the amount of volatile released to the atmosphere. Absorbent material will be added, at least to significant spills, when spills are discovered during inspection. Spills on a concrete surface, as opposed to ground, will be more amenable to cleanup. Since individual spills occur randomly, spills that are not noticed at the time of the occurrence will tend to grow in number linearly with time in the interval between inspecton periods.

Many handling operations occur in outdoor areas. The rate at which material volatilizes before it can be cleaned up will then depend on temperature and wind conditions. Similarly, for indoor spills, volatilization rates will depend primarily on temperature and ventilation conditions. Finally, human factors such as management attitudes and worker training and indoctrination must be important in determining the accidental spill rate.

# 10. Particulate Emissions Estimation Techniques for Waste Piles

INTRODUCTION

Particulate air emissions from waste piles occur at several points in the storage cycle:

- transfer of material to and from the pile;

- wind erosion;

- maintenance and traffic activities on the pile.

Particulate emissions from waste piles are influenced by the following factors:

- moisture content;

- rainfall;

- duration of storage;

- compaction of pile;

- amount and size of aggregate fines.

A method of estimating particulate emissions from waste piles is available in EPA publication AP-42. Cuscino, et al. (1981) presents emission factor equations empirically developed by Midwest Research Institute (MRI). Both methods describe emissions of particles smaller than 30 µm in diameter based on a particle density of 2.5 g/cm$^3$.

AP-42 EMISSION FACTOR EQUATION FOR STORAGE PILES

The quantity of suspended particulate emissions from waste piles may be estimated by the following equation:

$$E = \frac{0.33}{\left(\frac{PE}{100}\right)^2}$$

where:  E  = emission rate, pounds per ton of material placed in storage;

   PE  = Thornwaite's precipitation - evaporation index from
         Figure 12.

This equation only considers geographic variations in precipitation/
evaporation.  Emission factor equations developed by MRI based on test data
allow for variances in facility operations.

MRI EMISSION FACTOR EQUATIONS FOR STORAGE PILES

   The emission factor equations empirically developed by MRI for
uncontrolled storage piles are shown in Table 41.  These equations were
developed from limited testing of storage piles at iron and steel facilities
(i.e. coal, iron pellets, coke, slag, etc.).  However, the use of these
predictive equations may provide for greater accuracy than the single value
emission factor equation (from AP-42).  Input parameters for the MRI equations
account for source variability.

   The diversity of materials handled and differences in handling operations
suggest that additional source testing may increase the precision of the
predictive equations.  The estimation accuracy for the storage pile
maintenance and storage pile wind erosion equations are unknown due to limited
test data.

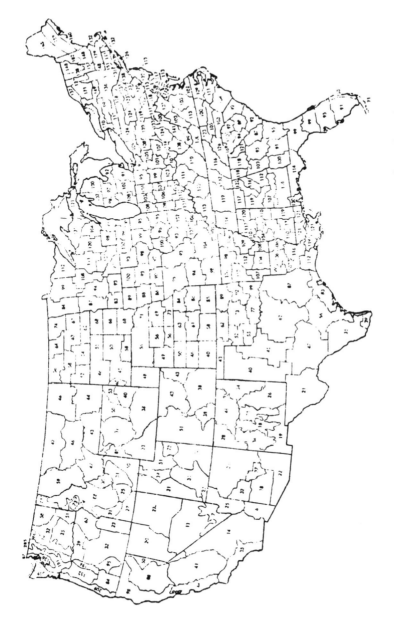

**Figure 12.** Map of Thornthwaite's Precipitation-Evaporation Index values for state climatic divisions. (From AP-42).

TABLE 41.    STORAGE PILE PARTICULATE EMISSION FACTOR EQUATIONS
DEVELOPED BY MRI

---

Batch load-in or load-out:

$$0.0018 \; \frac{\left(\frac{s}{5}\right)\left(\frac{u}{5}\right)\left(\frac{h}{5}\right)}{\left(\frac{M}{2}\right)^2 \left(\frac{Y}{6}\right)^{0.33}} = \text{pounds per ton of material transfered}$$

Continuous load-in:

$$0.0018 \; \frac{\left(\frac{s}{5}\right)\left(\frac{u}{5}\right)\left(\frac{h}{10}\right)}{\left(\frac{M}{2}\right)^2} = \text{pounds per ton of material loaded}$$

Active storage pile maintenance and traffic:

$$0.10K \left(\frac{s}{1.5}\right) \left(\frac{d}{235}\right) = \text{pounds per ton of material put through storage}$$

Active storage pile wind erosion:

$$0.05 \left(\frac{s}{1.5}\right) \left(\frac{d}{235}\right) \left(\frac{f}{15}\right) \left(\frac{D}{90}\right) = \text{pounds per ton of material put through storage}$$

**Parameters**

s = silt content of aggregate (%)

u = mean wind speed at 4m above ground (mph)

h = drop height (ft)

M = unbound moisture content of aggregate (%)

Y = dumping device capacity (cu. yd.)

K = activity factor

d = number of dry day per year

D = duration of material storage (days)

f = percentage of time wind speed exceeds 12 mph measured at one foot above the ground.

# Appendix A: References

American Petroleum Institute, Evaporation Loss Committee. Evaporation Loss from External Floating Roof Tanks. Bull. 2517 (revised), Washington, DC, February 1980.

American Petroleum Institute, Evaporation Loss Committee. Evaporation Loss from Fixed Roof Tanks. Bull. 2518, Washington, DC, June 1962.

American Petroleum Institute, Evaporation Loss Committee. Evaporation Loss from Internal Floating Roof Tanks. Bull. 2519 (revised), Washington, DD, June 1983.

Ames, et al. Suggested Control Measure to Reduce Organic Compound Emissions Associated with Volatile Organic Waste Disposal. Air Resources Board, State of California, Industrial Strategy Development Branch. August 20, 1982.

Arnold, J. H., Unsteady-State Vaporization and Absorption. Transaction of American Institute of Chemical Engineers 40, 361-379. 1944.

Ball, B. C., W. Harris and J. R. Burford. A Laboratory Method to Measure Gas Diffusion and Flow in Soil and Other Porous Materials. Journal of Soil Science, 32. 1981.

Brown, J.A. and M. Weintraub. Bioxidation of Paint Process Wastewater. Journal WPCF, Vol. 54, No. 7, pp. 1127-1130, July 1982.

Calderbank, P. H. "Mass Transfer," Chapter 6 of Mixing, edited by V. W. Uhl and J. B. Gray, Academic Press, New York, 1967.

Chicago Bridge and Iron Company, Plainfield, IL. Hydrocarbon Emission Loss Measurements on a 20 foot Diameter Pilot Test Tank with an Ultraflote and a CBI Weathermaster Internal Floating Roof. R-0113/R-0191, June 1978.

Cohen, Yoram, William Cocchio and Donald Mackay. Laboratory Study of Liquid Phase Controlled Volatilization Rates in Presence of Wind Waves. Environmental Science and Technology, Vol. 12, No. 5. May 1978.

Conway, Richard A. (ed), Environmental Risk Analysis for Chemicals. Van Nostrand Reinhold Company, New York, 1982.

Compilation of Air Pollutant Emission Factors (AP-42), U.S. Environmental Protection Agency. Research Triangle Park, N.C. April 1981.

Cox, Robert D., Jana L. Steinmetz and David L. Lewis. Evaluation of VOC
Emissions from Wastewater Systems (Secondary Emissions), Draft Report.
Radian Corporation. July 1982.

Currie, J. A. Gaseous diffusion in porous media. Part 2 - Dry Granular
Materials. British Journal of Applied Physics, Vol. 11. August 1960.

Currie, J. A. Gaseous diffusion in porous media. Part 3 - Wet Granular
Materials. British Journal of Applied Physics, Vol. 12. June 1961.

Cuscino, Thomas A. Jr., Chatten Cowherd, Jr., and Russel Bonn. Fugitive
Emission Control of Open Dust Sources. Proceedings: Symposium on Iron and
Steel Pollution Abatement Technology for 1980. EPA-600/9-81-071. March 1981.

Danielson, John A. Air Pollution Engineering Manual, AP-40. U.S.
Environmental Protection Agency, Research Triangle Park, NC, May 1973.

Dilling, Wendell L., Nancy B. Tefertiller, and George J. Kallos. Evaporation
Rates and Reactivities of Methylene Chloride, Chloroform,
1,1,1-Trichloroethane, Trichloroethylene, Tetrachloroethylene, and other
Chlorinated Compounds in Dilute Aqueous Solutions. Environmental Science and
Technology, Vol. 9, No. 9. September 1975.

Dilling, Wendell L. Interphase Transfer Processes II. Evaporation Rates of
Chloro Methanes, Ethanes, Ethylenes, Propanes, and Propylenes from Dilute
Aqueous Solutions. Comparisons with Theoretical Predictions. Environmental
Science and Technology, Vol. 11, No. 4. April 1977.

Ehlers, Wilfried, J. Letey, W. F. Spencer and W. J. Farmer. Lindane Diffusion
in Soils: I. Theoretical Consideration and Mechanism of Movement, Soil
Science Society American Proceedings, 33. 1969.

Erikson, D.G. Organic Chemical Manufacturing, Volume 3: Storage, Fugitive
and Secondary Sources. EPA-450/3-80-025. Research Triangle Park, NC,
December 1980.

Farmer, Walter J., M. S. Yang and J. Letey. Land Disposal of Hazardous
Wastes: Controlling Vapor Movement in Soils. Fourth Annual Research
Symposium. EPA-600/9-78-016. August 1978.

Farmer, Walter J., Ming-Shyong Yang, and John Letey. Land Disposal of
Hexachlorobenzene Wastes: Controlling Vapor Movement in Soils.
EPA-600/2-80-119. August 1980.

Freeman, R. A., Stripping of Hazardous Chemicals from Surface Aerated Waste
Treatment Basins. APCA/WPCF Specialty Conference on Control of Specific
(Toxic) Pollutants, Gainesville, Florida, February 13-16 (1979).

Freeman, R. A., J. M. Schroy, J. R. Klieve, and S. R. Archer, Experimental
Studies on the Rate of Air Stripping of Hazardous Chemicals from Waste
Treatment Systems," 73rd APCA Meeting, Montreal, Canada, June 22-27, 1980a.
Paper No. 80-16.7.

Freeman, R. A., Laboratory Study of Biological System Kinetics, unpublished by Catalytic, Inc. under contract with EPA, 1980b.

Freeman, R. A., Comparison of Secondary Emissions from Aerated Treatment Systems, presented at the 1982 Winter AIChE National Meeting, Orlando, Florida, March 3, 1982. Paper No. 5c.

Gerber, V. Y., V. M. Sharafutdinov and A. Ya. Gerceridov, Kinetics of Biochemical Oxidation of Refinery Wastewaters, Chem. Technol. Fuels Oils (USSR) 12: 826 (1976).

German Society for Petroleum Science and Carbon Chemistry (DGMK) and the Federal Ministry of the Interior (BMI). Measurement and Determination of Hydrocarbon Emissions in the Course of Storage and Transfer in Above- Ground Fixed Cover Tanks With and Without Floating Covers. BMI-DGMK Joint Projects 4590-10 and 4590-11; translated for EPA by Literature Research Company, Annandale, VA.

Grau, P., M. Dohanyos, and J. Chudoba, Kinetics of Multicomponent Substrate Removal by Activated Sludge, Water Research, Pergamon Press, Vol. 9, 637, 1975.

Hartley, G. S. Evaporation of Pesticides, published in Advances in Chemistry Series 86, American Chemical Society, Washington, D.C., 1969. pp. 115-134.

Hillenbrand, E., JRB Associates, McLean, VA, private communication based on the Draft Report, "Failure Incidents Analysis: Evaluation of Storage Failure Points," submitted to EPA Office of Solid Waste. March 1982.

Hwang, Seong T. Treatability and Pathways of Priority Pollutants in the Biological Wastewater Treatment. Paper presented at the American Institute of Chemical Engineers Symposium, Chicago, November 1980.

Hwang, S. T. Toxic Emissions from Land Disposal Facilities. Environmental Progress, Vol. I, No. 1. February 1982.

ICF, Inc., Clement Associates, Inc. and SCS Engineers, Inc., "RCRA Risk/Cost Policy Model Project," Phase 2 Report Submitted to EPA, Office of Solid Waste, June 15, 1982.

Kincannon, Donald F. and Enos L. Stover, Stripping Characteristics of Priority Pollutants During Biological Treatment. Paper presented at the Secondary Emissions Session of 74th Annual AICHE Meeting, New Orleans. November 1981.

Kincannon, Donald F., et al. Predicting Treatability of Multiple Organic Priority Pollutants in Wastewater from Single Pollutant Treatability Studies. Paper presented at the 37th Purdue Industrial Waste Conference, Purdue University, West Lafayette, Indiana. May 1982.

Lai, Sung-Ho, James M. Tiedje, and A. Earl Erickson. In situ Measurement of Gas Diffusion Coefficient in Soils. Soil Science Society Proceedings. 40. 1976.

Liss, P. S. and P. G. Slater. Flux of Gases Across the Air-Sea Interface. Nature, Vol. 247. January 1974.

Lyman, Warren J., William F. Rechl and David H. Rosenblatt, Handbook of Chemical Property Estimation Methods. McGraw-Hill, New York, 1982.

Mackay, Donald and Aaron W. Wolkoff. Rate of Evaporation of Low-Solubility Contaminants from Water Bodies to Atmosphere. Environmental Science and Technology, Vol. 7, No. 7. July 1973.

Mackay, Donald and Ronald S. Matsugu. Evaporization Rates of Liquid Hydrocarbon Spills on Land and Water. Canadian Journal of Chemical Engineering, Vol. 51. August 1973.

Mackay, Donald and Paul J. Leinonen. Rate of Evaporation of Low-Solubility Contaminants from Water Bodies to Atmosphere. Environmental Science and Technology, Vol. 9, No. 13. December 1975.

Mackay, Donald and T. K.Yuen. Volatilization Rates of Organic Contaminants from Rivers. Water Pollution Journal of Canada, Vol. 15, No. 2. Pergammon. 1980.

Mackay, Donald. Environmental and Laboratory Rates of Volatilization of Toxic Chemicals from Water. Hazard Assessment of Chemicals, Current Developments, Vol. 1. 1981.

Masser, Charles C. Compilation of Air Pollution Emission Factors, Supplement 12. U.S. Environmental Protection Agency. Research Triangle Park, NC, April 1981.

McCord, Andrew T. A. Study of the Emission Rate of Volatile Compounds from Lagoons, presented at the National Conference of Management of Uncontrolled Hazardous Waste Sites. Washington, D.C. October 28-30, 1981.

McDonald, Randy, EPA Office of Emission Standards Engineering Division, Research Triangle Park, NC, private communication with Ronald Bell, GCA/Technology Division, August 6, 1982.

Millington, R. J. and J. P. Quirk, Permeability of Porous Solids. Trans. Faraday Soc. 57:1200-1207, 1961.

Millington, R. J. and R. C. Shearer. Diffusion in Aggregated Porous Media, Soil Science. III. 1971.

Moryakov, U.S., et al. Losses of Crude Oil and Products in Operation of Refinery Storage Tanks. Translated from Khimiya i Tekhnologiya Toplivi i Masel, No. 4, pp. 8-10, April 1979.

Murphy, B. L., et al. Outer Continental Shelf Oil Spill Probability Assessment, Volume 2, Data Analysis Report. Submitted to U.S. Department of Interior, Bureau of Land Management, November 1981.

Neely, W. Brock. Predicting the Flux of Organics Across the Air/Water Interface. Proceedings of 1976 National Conference on Control of Hazardous Material Spills. New Orleans, LA. April 25-28, 1976.

Nelson, R. W., G. W. Gee, and D. W. Mayer. Control of Radon Emissions from Uranium, Mill Tailings by Multilayered Earth Barriers. Transcripts of American Nuclear Society, 38. 1981.

Owens, M., R. W. Edwards, and J. W. Gibbs. Some Reaeration Studies in Streams. International Journal of Air and Water Pollution, Vol. 8. Pergamon Press. 1964.

Patterson, J. W. and P. S. Kodukala. Biodegradation of Hazardous Organics Pollutants. Chemical Engineering Progress. April 1981.

Rolston, D. E. and B. D. Brown. Measurement of Soil Gaseous Diffusion Coefficients by a Transient-State Method with a Time-dependent Surface Condition. Soil Science Society American Journal, 40. 1976.

Shen, Thomas T. Emission Estimation of Hazardous Organic Compounds from Waste Disposal Sites. Presented at the 73rd Annual Meeting of the Air Pollution Control Association. Montreal, Quebec. June 22-27, 1980.

Shen, Thomas T. Estimating Hazardous Air Emissions from Disposal Sites. Pollution Engineering. August 1981.

Shen, Thomas T. Estimation of Organic Compound Emissions from Waste Lagoons. Journal of the Air Pollution Control Association, Vol. 32, No. 1. January 1982.

Smith, James H., David C. Bomberger Jr. and Daniel L. Haynes. Prediction of the Volatilization Rates of High-Volatility Chemicals from Natural Water Bodies, Environmental Science and Technology, Vol. 14, No. 11. November 1980.

Smith, James H., David C. Bomberger Jr. and Daniel L. Haynes. Volatilization Rates of Intermediate and Low Volatility Chemicals from Water. Chemosphere, Vol. 10, No. 3. 1981.

Springer, Charles, University of Arkansas, Fayetteville, Arkansas, private communication with William Farino, GCA/Technology Division. January 17, 1983.

Steenhuis, T. S., G. D. Bubenzer and J. C. Converse. Ammonia Volatilization of Winter Spread Manure. Transactions of the ASAE, 1979. pp. 152-161.

Sutton, O. G., Micrometerology, McGraw-Hill, New York, 1953.

Taylor, Sterling A. Oxygen Diffusion in Porous Media as a Measure of Soil Aeration, Soil Science Society Proceedings, 14, 55-61. 1949.

Thibodeaux, L. J. and D. G. Parker, "Desorption Limits of Selected Industrial Gases and Liquids from Aerated Basins," paper No. 30d, 76th National Meeting, AIChE, Tulsa, OK. March 1974.

Thibodeaux, L. J., "Air Stripping of Organics from Wastewater: A Compendium," Proceedings of the Second National Conference on Complete Water Use, Chicago, Illinois. May 4-8, 1978.

Thibodeaux, L. J., Chemodynamics: Environmental Movement of Chemicals in Air, Water, and Soil. John Wiley & Sons, New York, 1979.

Thibodeaux, L. J. Estimating the Air Emissions of Chemical from Hazardous Waste Landfills. Journal of Hazardous Materials 4 (1981a). 235-244.

Thibodeaux, L. J., Charles Springer, and Lee M. Riley. Models of Mechanisms for the Vapor Phase Emission of Hazardous Chemicals from Landfills, presented at Symposium of Toxic Substances Management, Atlanta, GA. March 1981b.

Thibodeaux, L. J., D. G. Parker, H. H. Heck and R. Dickerson. Quantifying Organic Emission Rates from Surface Impoundments with Micrometeorological and Concentration Profile Measurements, for presentation at the annual meeting of American Institute of Chemical Engineers. New Orleans, LA. November 8-12, 1981c.

Thibodeaux, L. J., David G. Parker and Howell H. Heck. Measurement of Volatile Chemical Emissions from Wastewater Basins. USEPA-IERL. December 1981d.

Thibodeaux, L. J. and S. T. Hwang. Landfarming of Petroleum Wastes - Modeling the Air Emission Problem. Environmental Progress, Vol. I, No. 1. February 1982a.

Thibodeaux, L. J., C. Springer, T. Hedden, and P. Lunney. Chemical Volatilization Mechanisms from Surface Impoundments in the Absence of Wind. Land Disposal of Hazardous Waste, EPA-600/9-82-002. March 1982b.

Thibodeaux, L. J., C. Springer, P. Lunney, S. C. James and T. T. Shen, Air Emission Monitoring and Hazardous Waste Sites, Proceedings of 3rd National Conference on Management of Uncontrolled Hazardous Waste Sites, November 29 to December 1, 1982c.

Towell, G. C., G. P. Strand and G. H. Ackerman, "Mixing and Mass Transfer in Large Diameter Bubble Columns," AICHE-I, Chem. Eng. Symposium Series No. 10, London: Inst. Chem. Eng., page 10-97, 1965.

TRW Environmental, Inc., Background Documentation for Storage of Organic Liquids. EPA Contract No. 68-02-3174, May 1981.

U.S. Environmental Protection Agency. Emission Test Report-Breathing Loss emissions from Fixed-Roof Petrochemical Storage Tanks. EMB 78-OCM-5, Research Triangle Park, NC, February 1979.

Western Oil and Gas Association. Hydrocarbon Emissions from Fixed-Roof Petroleum Tanks. Prepared by Engineering-Science, Inc., Los Angeles, CA, July 1977.

Wukash, R. F., C. P. Grady and E. J. Kirsch. Prediction of the Fate of Organic Compounds in Biological Wastewater Treatment Systems. AIChe Symposium, Water-1980, pp. 137-143, 1981.

# Appendix B: Derivation of Currie Wet Soil Correlation for Effective Diffusivity

CURRIE (1960) DRY SOIL CORRELATION

$$\frac{D}{D_o} = \gamma \, \varepsilon^\mu$$

where  $D$ = diffusion coefficient in dry soil;

$D_o$ = air diffusion coefficient;

$\varepsilon$ = soil porosity;

$\gamma, \mu$ = constants for a specific type of soil.

CURRIE (1961) WET SOIL CORRELATION

$$\frac{D}{D_v} = \left(\frac{\varepsilon_a}{\varepsilon_v}\right)^\sigma$$

where  $D$ = diffusion coefficient in wet soil;

$D_v$ = diffusion coefficient in air-filled soil;

$\varepsilon_a$ = air-filled soil porosity;

$\varepsilon_v$ = inter-crumb pore spaces;

$\sigma$ = soil specific constant.

Derive $D_v$:

for dry soils, $\varepsilon_a = \varepsilon_T$

where  $\varepsilon_T$ = total porosity

$$D = D_o \, \gamma \, \varepsilon^\mu = D_o \, \gamma \, \varepsilon_T^{\;\mu}$$

$$D = D_v \left(\frac{\varepsilon_a}{\varepsilon_v}\right)^\sigma = D_v \left(\frac{\varepsilon_T}{\varepsilon_v}\right)^\sigma$$

$$D_v = D_o \, \Upsilon \, \varepsilon_T{}^\mu \left(\frac{\varepsilon_v}{\varepsilon_T}\right)^\sigma$$

substituting $D_v$ into wet soil correlation;

$$D = D_v \left(\frac{\varepsilon_a}{\varepsilon_v}\right)^\sigma = D_o \, \Upsilon \, \varepsilon_T{}^\mu \left(\frac{\varepsilon_v}{\varepsilon_T}\right)^\sigma \left(\frac{\varepsilon_a}{\varepsilon_v}\right)^\sigma$$

$$D = D_o \, \Upsilon \, (\varepsilon_T)^{\mu-\sigma} \, (\varepsilon_a)^\sigma$$

check for dry soil, $\varepsilon_a = \varepsilon_T$

$$D = D_o \, \Upsilon \, \varepsilon_T{}^\mu$$

# Part C

# Properties and Categorization of Wastes

The information in Part C is from *Physical-Chemical Properties and Categorization of RCRA Wastes According to Volatility,* prepared by Douglas Dixon and Edwin Rissmann of Versar, Incorporated for the U.S. Environmental Protection Agency, February 1985.

# 1. Introduction

The U.S. Environmental Protection Agency (EPA) Office of Air Quality Planning and Standards (OAQPS) is currently investigating the control of volatile chemical air emissions from hazardous waste treatment, storage, and disposal facilities (TSDFs). Waste handling practices at these facilities include storage and treatment tanks, containers, waste piles, surface impoundments, landfills, and land treatment. As part of this investigation OAQPS will assess the risks of volatile chemical air emissions from TSDFs to human health and the environment and the costs and benefits of controlling these releases. Initial technical investigations include: (1) TSDF site assessments to collect engineering data on volatile air emission control practices, (2) monitoring of volatile air emissions from TSDFs, (3) identification and development of volatile air emissions models for estimating airborne releases of chemical substances from TSDFs, and (4) the gathering of health effects data for volatile chemical wastes. The chemical substances and waste streams which are being investigated are those considered hazardous under the Resource Conservation and Recovery Act (RCRA) of 1976 and listed in 40 CFR Section 261 Subpart D.

Currently, almost 500 chemical substances and chemical waste streams are "listed" as hazardous. Investigation of the potential air releases and resulting health risks of all listed RCRA wastes would be an extremely time consuming operation. Furthermore, the RCRA wastes span many orders of magnitude in volatility (many of the wastes are essentially non-volatile), making such an investigation even more cumbersome. EPA-OAQPS, therefore, must identify those RCRA wastes which are highly volatile and for which further investigation is required. EPA-OAQPS also needs to have the physical-chemical properties related to volatility of the RCRA wastes identified in order to support future

318

modeling efforts.  This document reports on the results of gathering the
physical-chemical properties of the RCRA wastes and the identification of
those wastes handled at the various types of TSDFs which can be
considered highly volatile.

1.2     Purpose and Scope

The purpose of this document is to present (1) the physical-chemical
properties of the RCRA wastes related to volatility and (2) a waste
categorization scheme based on the volatility of RCRA wastes from TSDFs.
The physical-chemical properties were gathered and/or estimated to
support the volatility categorization scheme.  They are also presented to
support future EPA-OAQPS air emission modeling investigations by
providing the basic necessary modeling input parameters.
Physical-chemical properties of RCRA wastes presented include:

- Molecular weight
- Boiling point
- Vapor pressure
- Solubility
- Log P (i.e., octanol/water partition coefficient)
- Henry's constant
- Relative soil volatility

Diffusion coefficients in air and water and water phase mass transfer
coefficients are also presented for RCRA wastes which were identified as
being highly volatile.

The waste categorization scheme is presented to support EPA-OAQPS
efforts in identifying those RCRA wastes that are highly volatile and for
which future monitoring, modeling, and health effects information
gathering will be undertaken.  The volatility categorization scheme is
divided into three parts as follows:

- Volatility of pure substances
- Volatility from aqueous solutions
- Volatility from soil

This approach was necessary because a substance's volatility depends on
the conditions under which it is handled at a TSDF.  While a pure

chemical waste may be highly volatile, it's volatility may be
significantly reduced when present in an aqueous condition or in soil.
The three volatility categorizations are presented to represent the
various types of waste handling practices performed at TSDFs:  storage
and treatment of pure substances in containers and storage tanks,
treatment and/or storage of wastes in lagoons or surface impoundments,
and treatment, storage, and disposal of wastes in soil systems (e.g,
landfills, land treatment).  Each categorization scheme is divided into
four levels of chemical waste volatility as follows:

1. Highly volatile
2. Moderately volatile
3. Slightly volatile
4. Non-volatile

The RCRA wastes investigated, as previously mentioned, are those
listed in 40 CFR Section 261 - Subpart D - List of Hazardous Wastes.
This list specifically includes wastes that are coded as Ps and Us and
individual metals that are coded as Ds.  Generic waste streams that are
coded as Ks, Fs, and Ds are listed but no physical-chemical property data
is presented.  Available information on the chemical constituents of
these waste streams, however, is presented.  For physical-chemical
properties and volatility categorization of a chemical constituent in a
generic waste stream, the investigator must refer to the chemical
specific information presented for its appropriate, P, U, or D waste
code.  Chemical specific data may be used to approximate a chemicals'
volatility behavior in a complex waste solution.  Actual volatility of a
chemical waste in complex solutions or mixtures is dependent on
physical-chemical interactions of solution constituents.  Calculation of
the volatility of the generic waste stream constituents is a complex and
time consuming operation requiring the use of activity coefficients which
are generally unavailable for most RCRA wastes.  Chemical-specific
volatility data (i.e., volatility of pure substance and volatility in
water and soil systems) must currently be used, therefore, to approximate
a constituent's behavior in complex generic waste streams.

This report is divided into six sections.  Following this
introduction, details on the methodology used for gathering and/or
estimating physical-chemical properties and for categorizing the RCRA
wastes are presented (i.e., Section 2).  Data gathering and
categorization results are discussed in Section 3.  Section 4 summarizes
the results and presents recommendations for additional investigation,
and Section 5 lists all references used in the task effort.  Finally, all
gathered data, physical-chemical property estimates, and waste
categorization results are presented in the Appendix to this report.

# 2. Methodology

The following subsections describe the procedures used for gathering and/or estimating the physical-chemical properties of the RCRA wastes (Section 2.1) and the methodology used for categorizing the RCRA wastes according to volatility (Section 2.2). An overview of the required physical-chemical parameters including their relationship to volatility is also provided in Section 2.1.

## 2.1    Physical-Chemical Properties

The primary physical-chemical properties of RCRA wastes required to support a waste volatility categorization scheme are the waste's vapor pressure, solubility in water, Henry's law constant, relative soil volatility, and molecular weight. Following is a review of each of the properties and their relationship to a substance's volatility:

Vapor Pressure - This is defined as the pressure exerted by a gas when in equilibrium with its non-gaseous phase. Vapor pressure provides an indication of the escaping tendency of molecules from pure liquids or solids. A high vapor pressure implies low attractive forces between molecules in the liquid or solid and a high number of molecules being emitted into the vapor phase. These substances are considered volatile. Liquids with strong attractive forces, therefore, have low vapor pressures and are considered nonvolatile. Consequently, vapor pressure is an excellent indicator of emissions from pure substances. However, when liquids are mixed, vapor pressure does not provide an accurate means of estimating emissions of a single compound because of the additional interaction between the various types of molecules.

Solubility - This is defined as the total mass of a substance that will dissolve in a solvent (usually water) at a given temperature and pressure. The solubility of a substance affects the rate at which molecules of the substance will escape from a liquid via vaporization. Substances with high vapor pressures and low solubilities will readily vaporize from solution. Conversely, substances with high water solubilities and low vapor pressures will tend to remain in solution or be nonvolatile. The solubility of a substance is a required input to the calculation of Henry's law constant which is described below.

322

Henry's Law Constant - This is defined as the ratio of the partial
pressure of the solute gas (i.e., the impurity or pollutant) divided
by the mole fraction of the gas in solution (Sienko and Plane 1966).
By taking into account partial pressure and solubility, Henry's law
constant (H) gives an indication of the tendency of molecules to
escape from a solution.

Henry's law constants provide some insight into the volatilization
rate controlling processes. This constant is the best indicator of a
compound's volatility. Some generalizations on the relationship
between Henry's law constant and the emissions rate of compounds in
an aqueous solution are given below (Lyman et al 1982):

- $H < 10^{-7}$ atm-m$^3$/mol. The substance is less volatile than
  water, and its concentration will increase as water evaporates;
  it is essentially nonvolatile.

- $10^{-7} < H < 10^{-5}$ atm-m$^3$/mol. The substance slowly
  volatilizes; the rate is controlled by slow molecular diffusion
  through air.

- $10^{-5} < H < 10^{-3}$ atm-m$^3$/mol. Volatilization starts to
  become a significant transfer mechanism; this range includes
  most polycyclic aromatic hydrocarbons and halogenated aromatics.

- $H > 10^{-3}$ atm-m$^3$/mol. Substances may be released in
  significant quantities; resistance from the water film is the
  rate controlling process.

These criteria will be the basis of the aqueous volatility
categorization of the RCRA wastes (see Section 3.3 and Appendix C).

Molecular Weight - The molecular weight is indicative of the size of
the molecule of a chemical waste. Molecular size affects the rate at
which a chemical will diffuse through a medium (i.e., air, water, and
soil). The molecular weight of a chemical waste is a required input
to the calculation of the waste's ability to volatilize from soil as
described below.

Relative Soil Volatility - The actual rate of volatilization of a
substance from soil (e.g., landfills) depends upon the amount of the
substance in the soil as well as soil type, soil water content, vapor
pressure, molecular weight, and diffusivity (described later) of the
substance, air flow rate over the surface, humidity, and
temperature. Assuming all factors being equal except molecular
weight and vapor pressure, a substance's relative volatility from
soil is controlled by the soil water content. Essentially two
different volatilization processes occur depending on whether the

soil is dry or wet. In a dry soil column, the chemical waste is assumed to move up the soil column via the air pockets within the soil. In a wet soil column, the chemical waste is transported from the soil body to the surface by capillary action, sometimes referred to as the wick effect. In other words, the soil column acts as a wick, and water contaminated with a chemical waste moves up the capillaries of the wick to replenish the contaminated water at the top of the column that is lost by evaporation. These types of transport mechanisms and methods for calculating relative wet and dry soil volatility are further discussed in Section 2.2.

Diffusion Coefficient - The diffusion coefficient is a measure of the molecular diffusion or net transport of a molecule in a liquid or gas medium and is a result of intermolecular collisions rather than turbulence or bulk transport. The process is promoted by gradients, such as pressure, temperature, and concentration. The rate of diffusion is a function of the properties of the chemical of interest and the medium through which it is being transported. Diffusion coefficients are not a required parameter for the categorization scheme used in this report; however, they are required values for actual calculation of a chemical's mass transfer from a TSDF site. Diffusion coefficients will be required in future EPA-OAQPS air emissions modeling efforts and are, therefore, provided in this report for those compounds considered highly volatile from soil and aqueous solutions (see Section 3.5 for further discussion).

Values for the above properties, except diffusion coefficients (as discussed), were required for the categorization of the RCRA wastes according to volatility. Additional properties were also required such as boiling point and octanol/water partition coefficients (i.e., Log P) when experimental values were not available in the literature; these required estimation via the use of these properties. Data gathering and estimation methods for the above properties are discussed in the next subsection.

## 2.2   Data Gathering/Estimation Methods

Experimental values for vapor pressure, solubility, and Henry's constant for the approximately 500 "listed" RCRA wastes were gathered from the readily available scientific literature. Literature sources used to gather experimental property data are listed in the reference section of this document. All the experimental data were normalized to ambient temperature (i.e., at 25°C) before being recorded. Where experimental data were lacking, property values were estimated via the use of the EPA-Office of Toxic Substance's Graphical Exposure Modeling System (GEMS). GEMS is a computerized data base which includes physical-chemical property estimation methods. The principal component of GEMS accessed for this effort was the CHEMEST system. CHEMEST is a computerized version of the estimation methods presented in the Handbook of Chemical Property Estimation Methods (Lyman et al. 1982). The specific approach for gathering and/or estimating values for physical-chemical properties is discussed below according to the specific property:

Vapor Pressure - Vapor pressures at 25°C of the RCRA wastes were obtained from the experimental literature. Vapor pressures gathered at temperatures other than 25°C were estimated at 25°C using the Clausius-Clapeyron equation:

$$\ln \frac{P_2}{P_1} = \frac{\Delta H_v (T_2 - T_1)}{RT_1T_2}$$

where

$P_1$, $P_2$ = vapor pressures
$T_1$, $T_2$ = temperatures (experimental and 25°C)
$R$ = proportionality constant
$\Delta H_v$ = heat of vaporization (as gathered from the experimental literature).

If no experimental data were available, vapor pressures were estimated from a waste's boiling point using the CHEMEST system: either the Antoine equation (for liquids and gases in the vapor pressure range of $10^{-3}$ to 760 mm of Hg) or the Modified Watson

Correlation (for liquids and solids in the vapor pressure range of
$10^{-7}$ to 760 mm of Hg) was used.  Boiling points for estimating
vapor pressures were gathered from the experimental literature or
estimated via CHEMEST using the Miessner Method (Lyman et al. 1982).
The relative error in estimated vapor pressures using the methods of
Lyman et al. is reported as 0 to 3 percent.  The approximate error in
boiling points estimated via the Miessner Method is 0 to 5 percent.

Solubility - Experimental solubility values were also gathered from
the scientific literature.  Where experimental values were lacking,
the solubility of the RCRA wastes were estimated from Log $K_{ow}$
(octanol-water partition coefficients) via standard regression
equations in CHEMEST.  Log $K_{ow}$'s were obtained from Hansch and Leo
(1979) or estimated based on molecular structure using the CLOGP
program in GEMS.  CLOGP is a computerized version of a fragment
additivity method for Log $K_{ow}$ presented in Lyman et al. (1982).
The approximate error in estimated solubility values is reported as
between 0 and 10 percent.  Estimated log $K_{ow}$ values have an
approximate error of ± 0.12 log $K_{ow}$ units.

Henry's Law Constant - Values for Henry's law constant (H) were also
gathered from the experimental literature.  In general, however,
values were extremely limited and the majority of the RCRA wastes
required estimation of H values.  The procedure used for estimating
values of H is as follows (Lyman et al. 1982);

$$H = \frac{P_{vp}}{S}$$

where

$P_{vp}$ = vapor pressure at 25°C in atm.
$S$ = solubility in water at 25°C in mol/m$^3$.

For compounds with extremely high solubilities or those compounds
described as "miscible", a solubility value of $10^6$ mg/l was assumed
when estimating H.  In practical theory, the solubility of an
infinitely miscible substance is the weight of one liter of the
substance.  As most of the miscible RCRA wastes have densities close
to that of water (i.e., 0.8 to 1.2 g/ml), the use of $10^6$ mg/l is a
reasonable value for expressing miscibility and estimating H.
Obviously, the use of this value will result in the introduction of
significant error in estimated H values.,  This indeterminate error,
however, is not considered to significantly affect the relative
categorization of the RCRA wastes.

The methods discussed above are state-of-the-art procedures for estimating solubility, vapor pressure, and Henry's constant. Calculation of relative soil volatility required the development of a procedure specific to available data. Calculation of actual soil volatilization rates depends on concentrations of the waste in soil, data which is unknown and specific to each TSDF. Relative volatility, however, from both wet and dry soils, may be approximated by modifying standard equations for calculating actual wet and dry soil chemical volatilization rates. Several methods have been developed to estimate the volatility of a compound from nonaqueous environments. They vary widely in application and difficulty of calculation; however, most methods are based on a diffusion coefficient (the measure of a compound's ability to migrate through a vapor, liquid, or solid). Furthermore, as previously discussed, these methods can be divided into two major groups: (1) dry soil column methods and (2) wet soil column methods with water flux transport.

The methods selected for modification to calculate relative soil volatility were extracted from Lyman et al. (1982). They are straightforward and are also recommended above other methods evaluated by Lyman et al. during development of the physical-chemical properties estimation handbook. The methods and their modification to derive a relative dry and wet soil volatility are as follows:

Method 1 (dry soil column) (Hamaker Method as presented in W. L. Lyman et al. Handbook of Chemical Property Estimation Methods 1982):

$$Q_t = 2c_0\sqrt{Dt/\pi} \tag{1}$$

where

$Q_t$ = total loss of chemical per unit area over some time t ($M/L^2$)
$c_0$ = initial concentration of chemical in soil ($M/L^3$)
$D$ = diffusion coefficient of vapor through the soil ($L^2/T$)
$\pi$ = 3.14159.

In this equation, $c_0$ is unknown; however, it can be assumed that the soil air space is saturated with the vapor of the chemical waste. By assuming the vapor behaves as an ideal gas, one can apply the ideal gas law:

$$PV = nRT \tag{2}$$

where

    P = pressure (mm Hg)
    V = volume (1)
    n = number of moles
    R = universal gas constant (mm Hg-1/mole-°K)
    T = temperature (°K).

The expression for equilibrium vapor concentration with $P_{vp}$, vapor pressure at 25°C, as the only variable follows:

$$\text{equilibrium vapor concentration (EVC)} = \frac{N}{V} = \frac{P_{vp}}{RT} . \tag{3}$$

In this equation, R and T are constants; therefore,

$$\frac{N}{V} \text{ (or } c_0) = KP_{vp} . \tag{4}$$

Substituting equation (4) in equation (1) yields

$$Q_t = 2KP_{vp} \sqrt{Dt/\tau} . \tag{5}$$

In this equation, 2, K, and $t/\tau$ are constants and are assumed to be a common factor for all RCRA wastes. They can, therefore, be eliminated from further calculation. Equation (5) then reduces to:

$$Q_t \text{ (relative)} = K_1 P_{vp} \sqrt{D} \tag{6}$$

Furthermore, according to Dalton's Law, diffusion coefficients are inversely proportional to the reciprocal of the square roots of molecular weights under fixed conditions. This relationship can be written:

$$D_y = D_x \frac{\sqrt{MW_x}}{\sqrt{MW_y}} \qquad (7)$$

where

$D_y$ = diffusion coefficient of RCRA waste
$D_x$ = measured diffusion coefficient of compound x
$MW_x$ = molecular weight of compound x
$MW_y$ = molecular weight of compound y (i.e., the RCRA waste of interest).

In this equation, $D_x$ and $MW_x$ are also constants, therefore:

$$D_y = \frac{K_2}{\sqrt{MW_y}} \qquad (8)$$

or

$$D_y = K_2 MW^{-1/2} . \qquad (9)$$

Substituting the above into equation (6) yields

$$Q_t \atop (relative) = K_1 P_{vp} \sqrt{K_2 MW^{-1/2}} . \qquad (10)$$

or

$$Q_t \atop (relative) = K_3 P_{vp} MW^{-1/4} . \qquad (11)$$

Finally, since $K_3$ is constant for all RCRA wastes, equation (11) reduces to:

$$Q_t \atop (relative) = P_{vp} MW^{-1/4} \qquad (12)$$

This equation permits the calculation of the _relative_ volatility of RCRA wastes from dry soil.

Method 2 (wet soil column) (Hamaker Method as presented in W.J. Lyman et al. Handbook of Chemical Property Estimation Methods, 1982):

$$Q_t = \frac{P_{vp}}{P_{H_2O}} \cdot \frac{D_v}{D_{H_2O}} \quad (f_w)_v + c(f_w)_L \qquad (13)$$

where

$f_w$ = loss of water per unit area ($M/L^2$)
$P_{vp}$ = vapor pressure of chemical
$P_{H_2O}$ = vapor pressure of water
$D_v$ = diffusion coefficient of chemical in air ($L^2/T$)
$D_{H_2O}$ = diffusion coefficient of water vapor in air ($L^2/T$)
sub V = loss of vapor
sub L = loss of liquid
$c$ = concentration of chemical in soil solution ($M/M$).

In this equation, the following parameters are assumed constant for all RCRA wastes; $P_{H_2O}$, $D_{H_2O}$, $f_w$, and $c$. Therefore, equation (13) can be written as follows:

$$\underset{(relative)}{Q_t} = \frac{P_{vp}}{K_1} \cdot \frac{D_v}{K_2} \cdot K_3 + K_4 (K_5) \qquad (14)$$

or

$$\underset{(relative)}{Q_t} = KP_{vp}D_v + K' \qquad (15)$$

According to Dalton's Law as discussed in the dry soil column method; (see equations (7), (8), and (9));

$$D_v = \frac{K_v}{\sqrt{MW_v}} \cdot$$

Therefore, via substitution, equation (14) can be written as:

$$\underset{(relative)}{Q_t} = K\, P_{vp} \frac{K_v}{\sqrt{MW_v}} + K' \cdot \qquad (16)$$

or

$$\underset{(relative)}{Q_t} = K\, \frac{P_{vp}}{\sqrt{MW_v}} + K' \cdot \qquad (17)$$

Finally, since the constants are assumed to be the same for all RCRA wastes, equation (17) reduces to:

$$Q_t = \frac{P_{vp}}{\sqrt{MW_v}} \quad \text{or} \quad P_{vp}MW_v^{-1/2} \quad . \tag{18}$$

This equation permits the calculation of the relative volatility of RCRA wastes from wet soil.

Other approaches using soil volatilization estimation methods may also have been developed; however, only minor refinements in relative soil volatility estimates would be obtained. It is believed that the approaches discussed essentially result in the identification of those RCRA wastes which will be highly volatile from TSDF soil systems (e.g., landfills and land treatment).

Diffusion coefficients (in air and water) - Diffusion coefficients of RCRA wastes in air and water were gathered or estimated for only those chemicals considered highly volatile in water and soil systems (see results in Section 3). As with the other physical-chemical properties, experimental values for diffusion coefficients were gathered from the scientific literature. Where experimental data were lacking, coefficients were calculated via the methods presented in Lyman et al. (1982). For diffusion coefficients in air, the Fuller, Schettler, and Giddings Method was used. For diffusion coefficients in water the Hayduk and Laudie method was used. Details on both methods are described completely in Lyman et al. (1982).

# 3. Data Gathering and Categorization Results

## 3.1    Physical-Chemical Properties and Relative Soil Volatility

Experimental and estimated values for molecular weight, vapor pressure, solubility, Henry's constant, and relative wet and dry soil volatility of the "listed" RCRA wastes are presented in Appendix A. Boiling points and octanol/water partition coefficients are also listed in Appendix A, however, only when required to estimate vapor pressure and solubility, respectively. Literature references for experimental and/or estimated values are listed in the footnote to Appendix A. The comments section of the table for each chemical provides reasons where no data are recorded.

No data are presented for generic waste streams coded K, F, and D (except individual toxic metals and pesticides). Available information on the constituents of the generic waste streams is, however, presented under comments. Physical-chemical properties of a specific waste stream constituent can be obtained by referring to the data under the chemical-specific P or U waste stream code.

## 3.2    Categorization of RCRA Wastes According to Vapor Pressure

The ambient vapor pressure, as discussed in Section 2.1, provides a good indication of a pure substances volatility. Ambient vapor pressure, therefore, may be used to indicate relative waste volatility as a result of spills and leaks of pure chemical wastes from storage tanks and containers at TSDFs.

All RCRA wastes were ranked according to ambient vapor pressure and then categorized into four volatility groups as follows:

- Highly volatile wastes - those with ambient vapor pressures above 10 torr.

- Moderately volatile wastes - those with ambient vapor pressures in the $10^{-3}$ to 10 torr range.

- Slightly volatile wastes - those with ambient vapor pressures in the $10^{-5}$ to $10^{-3}$ torr range.

332

- Nonvolatile wastes - those with ambient vapor pressures below $10^{-5}$ torr.

Results of the categorization are presented in Appendix B.

## 3.3    Categorization of RCRA Wastes According to Aqueous Volatility

Values of Henry's constant were used to categorize the RCRA wastes according to aqueous volatility. Henry's constant, as discussed in Section 2.1, indicates a chemical's propensity to volatilize from aqueous solution. Henry's constant, therefore, may be used to indicate relative volatility of wastes under aqueous conditions, such as volatilization from impoundments, at TSDFs.

All RCRA wastes were ranked according to values of Henry's constant and then categorized into four volatility groups as follows (Lyman et al. 1982):

- Highly volatile wastes - values of Henry's constant above $10^{-3}$

- Moderately volatile wastes - values of Henry's constant below $10^{-3}$ to $10^{-5}$

- Slightly volatile wastes - values of Henry's constant below $10^{-5}$ to $10^{-7}$

- Nonvolatile wastes - values of Henry's constant below $10^{-7}$

Results of the categorization are presented in Appendix C.

## 3.4    Categorization of RCRA Wastes According to Relative Soil Volatility

Derivation of relative soil volatility values is discussed in Section 2.2. Relative soil volatility may be used to indicate a chemical wastes propensity to volatilize from TSDF landfills and land treatment systems. All RCRA wastes were ranked according to relative soil volatility values and then categorized into four classes as follows:

- Highly volatile - Relative soil volatility greater than 1

- Moderately volatile - Relative soil volatility from 1 to $10^{-3}$

- Slightly volatile - Relative soil volatility from $10^{-3}$ to $10^{-6}$

- Nonvolatile - Relative soil volatility below $10^{-6}$

Results of the relative soil volatility categorization are presented in Appendix D.

## 3.5    Diffusion Coefficients of Highly Volatile RCRA Wastes

Diffusion coefficients in air and water for a subset of those RCRA wastes identified as highly volatile from water and soil were gathered from the scientific literature or estimated by the methods discussed in Section 2.2. Diffusion coefficients are presented to support future EPA-OAQPS air emission modeling efforts for TSDFs. Data were gathered for only a subset of highly volatile RCRA wastes due to the financial and manpower limitations of the task effort. Values for diffusion coefficients in air and water for RCRA wastes identified as highly volatile from water are presented in Appendix E.

Also presented in Appendix E are values for the estimated water phase mass transfer coefficients. Overall water phase mass transfer coefficients ($K_L$) were calculated based on the values of Henry's constant for the subset of highly volatile wastes. The values were calculated using the Southworth Equation presented in Lyman et al. (1982). In general, the Southworth equation can be used for calculating the overall liquid-phase mass transfer coefficient of chemicals:

$$K_L = \frac{(H/RT)\ k_g\ k_1}{(H/RT)\ k_g + k_1} \tag{1.1}$$

where

$k_g$ = gas-phase exchange coefficient
$k_1$ = liquid-phase exchange coefficient
$H$  = Henry's law constant at desired temperature
$R$  = gas constant = $8.2 \times 10^{-5}$ atm-m$^3$/mol
$T$  = desired temperature in °K.

Equations for computing the values $k_g$ and $k_l$ are presented in Lyman et al. (1982). Different equations are recommended depending upon the molecular weight and Henry's law constant of the chemical and wind speed assumed to prevail at the site.

The value of the Henry's law constant for all chemicals for which $K_L$ was calculated was greater than $10^{-3}$ atm-m$^3$/mol. For chemicals having a value for Henry's constant of $10^{-3}$ atm-m$^3$/mol or greater, the resistance of the water film dominates by a factor of at least ten and the mass transfer is liquid phase controlled (Lyman et al. 1982). Therefore, the overall liquid-phase mass transfer coefficient, $K_L$, is equal to the liquid-phase exchange coefficient, $k_l$, or

$$K_L = k_l \text{ for } H > 10^{-3} \text{ atm-m}^3/\text{mol}.$$

Consequently, only the liquid-phase exchange coefficients needed to be calculated in order to determine the overall liquid-phase mass transfer coefficient. Two equations were used to compute $K_L$ (or $k_l$). The following equation was used to compute $k_l$ for chemicals having a value for molecular weight of less than 65:

$$k_l = 20 \sqrt{40/MW}$$

where MW = molecular weight. The equation,

$$k_l = 23.51 \left( \frac{V_{curr}^{0.969}}{Z^{0.673}} \right) \sqrt{32/MW} \; e^{0.526 (V_{wind} - 1.9)}$$

where

$V_{curr}$ = current velocity (m/sec)
$V_{wind}$ = wind velocity (m/sec)
$Z$ = depth of water (m)

was used to compute $k_l$ for chemicals having a value for molecular weight of greater than 65. For all calculations a wind speed, $V_{wind}$, of 3 meters/second, (i.e., average wind speed), a current velocity,

$V_{curr}$ of 0.01 m/s (i.e., negligible current), and a depth, Z, of 1 meter, was assumed. These assumptions were considered to be a reasonable simulation of a TSDF waste impoundment or lagoon.

Values for diffusion coefficients in air and water for a subset of RCRA wastes identified as highly volatile from soil are presented in Appendix F. Soil volatilization rates (i.e., mass transfer coefficients) are not, however, presented. Calculation of soil volatilization rates requires knowledge of the concentration of the RCRA waste in soil. The rate, therefore, must be developed on a TSDF site-specific basis when modeling volatile emissions. Methods for calculating soil volatilization rates are presented in Lyman et al. (1982).

# 4. Summary and Recommendations

This document is a comprehensive catalog of physical-chemical properties of hazardous wastes currently regulated under RCRA. It specifically provides waste properties related to potential air emissions of the wastes from TSDFs. As a catalog of physical-chemical properties related to volatile air emissions, this report should prove to be a valuable reference guide for supplying required data in EPA-OAQPS's future TSDF air emission modeling efforts. The waste volatility categorization scheme also provides a useful guide for identifying whether a specific waste will volatilize from a TSDF and potentially present a health risk to the surrounding community. Furthermore, the waste volatility categorization scheme presents a format for the development of future TSDF waste-specific regulatory controls.

The data presented in this report should be used with the knowledge that a significant number of property values were estimated. In particular, Henry's constant (H) required estimation for the majority of the RCRA wastes. For certain wastes, estimation of boiling points, vapor pressures, and solubilities was also required. While the estimation procedures used are considered state-of-the-art, errors are inherent in the estimated values. Values presented for relative soil volatilities are unitless values; they should only be used for comparing relative volatilities of RCRA wastes from wet and dry soil. They were derived specifically for this task from documented soil volatilization rate formulas. It should also be noted that other sources of experimental property values are available; experimental values were not obtained as a result of an exhaustive literature search. The experimental reference sources used were those readily available or those that could be identified and accessed within the labor and financial resources available to the task effort.

While the data presented in this report are essentially complete and fulfill the requirements of the task objectives, several areas of future work are recommended. Recommendations for future work include:

337

- Estimation of additional gas and liquid phase diffusion coefficients and water phase mass transfer coefficients for highly volatile and moderately volatile RCRA wastes. This report only presents these values for a subset of the highly volatile RCRA wastes because of the time and budgetary restraints of the task effort.

- Identify, gather, and estimate additional physical-chemical properties (e.g., saturation vapor concentration, molar density of chemical vapor, atomic diffusion volume) for the RCRA wastes required as input parameters for TSDF airborne emissions models. This effort should follow final EPA-OAQPS selection of applicable air dispersion models.

- Investigate volatility of waste constituents from complex waste streams (e.g., generic waste streams coded K, D, and F). Intermolecular forces of waste stream constituents affect the volatilization behavior of the constituents. This effort would require the gathering and/or estimation of activity coefficients for each of the RCRA wastes. The activity coefficient is a correction factor compensating for non-ideal behavior and is invaluable in calculations involving multi-component phase equilibria.

- Identify and gather chemical constituent information, including concentration ranges, of generic waste streams. Extremely useful sources of information on generic waste stream constituents are the Hazardous Waste Background Documents prepared by EPA - Office of Solid Waste as required under RCRA.

# 5. References

1.  Aldrich Catalog/Handbook of Fine Chemicals. 1982-1983. Milwaukee, Wisconsin: Aldrich Chemical Company.

2.  Barr RF, Watts H. 1972. Diffusion of some organic and inorganic compounds in air. J. Chem. Eng. Data, 17(1):45-46.

3.  Boublik T, Freid V, Hala E. 1973. The vapour pressures of pure substances. New York, NY: Elsevier Publishing Company.

4.  Briggs GG. 1981. Theoretical and experimental relationship between soil adsorption, octanol-water partitioning coefficients, water solubilities, bioconcentration factors, and the parachor. J. Agric. Food Chem. 29:1050-1059.

5.  Callahan MA, Slimak MW, Gabel NW, et al. 1979. Water-related environmental fate of 129 priority pollutants. Volumes I and II. Washington, DC: U.S. Environmental Protection Agency, Office of Water Planning and Standards. EPA 440/4-79-029a and b.

6.  Dilling WL. 1977. Environmental Science and Technology. 11(4):405-409.

7.  Farm Chemicals Handbook. 1983. Willoughby, Ohio: Meister Publishing Company.

8.  Gile JD, Billett JW. 1981. Transport and fate of organic phosphate insecticides in a laboratory model ecosystem. J. Agric. Food Chem. 129:616-621.

9.  Handbook of Chemistry and Physics. 1971-1983. RC Weast (ed). Cleveland, OH: The Chemical Rubber Company.

10. Handbook of Chemistry. 1949. Lange NA and G Forker, (eds). 7th Edition. Sandusky, Ohio: Hanbook Publishers, Inc.

11. Hansch C, Leo A. 1979. Substituent constants for correlation analysis in chemistry and biology. New York, NY: Wiley Interscience.

12. Hawley GG. 1977. The condensed chemical dictionary, 9th edition. New York, NY: Van Nostrand Publishing Company.

13. Hollifield HC. 1979. Rapid nephalometric estimate of water solubility of highly insoluble organic chemicals of environmental interest. Bull. Environ. Contam. Toxicol. 23:579-586.

14. International Agency for Research in Cancer. 1978. IARC monographs. The evaluation of the carcinogenic risk of chemicals to humans, Vol. 17. Some n-nitroso compounds, IARC. Lyon, France.

15. Kenaga EE, Goring C. 1978. Relationship between water solubility, soil adsorption, octanol-water partitioning and concentration of chemicals in biota. ASTM Philadelphia, PA: ASTM Spec. Tech. Pub. 707:78-109.

16. Kirk-Othmer Encyclopedia of Chemical Technology. 1978. 3rd Edition. New York: Wiley Interscience.

17. Linke WF. 1958. Solubilities of inorganic and metal organic compounds. Washington, DC: American Chemical Society.

18. Lugg GA. 1968. Diffusion coefficients of some organic and other vapors in air. Anal. Chem., 40(7):1072-1077.

19. MacKay D, PJ Leinonen. 1975. Rate of evaporation of low-solubility contaminants from water bodies to atmosphere. Environmental Science and Technology. 9(13):1178-1180.

20. MacKay D, WY Shiu. 1981. A critical review of Henry's law constants for chemicals of environmental interest. J. Phys. Chem. Ref. Data. 10(4).

21. Mackay D, Walkaff A. 1973. Rate of evaporation of low solubility contaminants from water bodies to atmosphere. Envir. Sci. and Technol. 7(1). 611-613.

22. McAulifte C. 1966. Solubility in water of hydrocarbons. Journal of Physical Chemistry 70(4). 1267-77.

23. Mellor JW. 1946. Comprehensive treatise on inorganic chemistry. London: Longmans Green & Company.

24. Shen T. 1982. J. Air Pollut. Control Fed. 32(1):79-82.

25. Sienko MJ, Plane RA. 1966. Chemistry: principles and properties. New York, NY: McGraw-Hill Book Company.

26. Stefan H, Stefan T. 1965. Solubilities of inorganic and organic products, Volume 1 - Binary Systems. New York, NY: Pergamon Press.

27. The Merck Index. 1976. M. Windholz (ed). Rahway, NJ: Merck and Company, Inc.

28. The Pesticide Manual. 1979. CR Worthing (ed). Croydon, England: British Crop. Protection Council.

29. Timmermans J. 1960. The physio-chemical constants of binary systems in concentrated solutions. New York, NY: Interscience Publishing Company.

30. USEPA. 1980. Ambient water quality criteria for chloroalkly ethers. Washington, DC: Office of Water Regulations and Standards, U.S. Environmental Protection Agency. EPA 440/5-80-030.

31. Umweltbundesant. 1981. OECD Hazard Assessment Project - Collection of minimum premarketing sets of data including environmental residue data of existing chemicals, Berlin, Germany: Umweltbundesant.

32. Versar Inc. 1982. Exposure Assessment for DEHP. Washington, DC: U.S. Environmental Protection Agency, Office of Toxic Substances. EPA Contract No. 68-01-6271.

33. Verschueren K. 1977. Handbook of Environmental Data on Organic Chemicals. New York, NY: Van Nostrand Reinhold Company.

34. Wilson G, Deal C. 1962. Activity coefficients and molecular structure. Ind. Eng. Chem. Fundam. 1:20-23.

# Appendix A: Physical-Chemical Properties and Relative Soil Volatility of RCRA Wastes

| Waste Code | Waste Name | Molecular Weight (g/mol) | Boiling Point* (°C) | Vapor Pressure @ 25°C (mm Hg) | Log P** | Solubility in H₂O @ 25°C (mg/l) | Henry's Constant (atm-m³/mol) | Relative Dry Soil Volatility $P_{vp}/Mw^{1/4}$ | Relative Wet Soil Volatility $P_{vp}/Mw^{1/2}$ | Reference/Comments |
|---|---|---|---|---|---|---|---|---|---|---|
| U001 | Acetaldehyde | 44 | 20.8 | 922.57 | --- | Insoluble | --- | 360 | 139 | 1 |
| U002 | Acetone | 58 | 56.2 | 200.1 | -0.24 | $2.3 \times 10^5$ | $6.8 \times 10^{-6}$ | 720 | 26.3 | 1, 6 |
| U003 | Acetonitrile | 41 | 81.6 | 100 | -0.34 | $2.2 \times 10^6$ | $2.4 \times 10^{-6}$ | 40 | 15.6 | 1, 6 |
| U004 | Acetophenone | 120 | 202 | 0.49 | | 5,500 | $1.41 \times 10^{-5}$ | $1.5 \times 10^{-1}$ | $4.5 \times 10^{-2}$ | 1, 4 |
| U005 | 2-Acetylaminofluorene | 223 | 345 | $2.1 \times 10^{-7}$ | 3.255 | 140 | $4.4 \times 10^{-10}$ | $5.4 \times 10^{-8}$ | $1.4 \times 10^{-8}$ | 2 |
| U006 | Acetyl chloride | 79 | 52 | 317 | | reacts with water | --- | 108 | 35.7 | 1, 2, 5 (Reacts with water) |
| U007 | Acrylamide | 71 | 87 @ 2mm | 0.0328 | 1.55 | 205,000 | $1.49 \times 10^{-8}$ | $1.1 \times 10^{-2}$ | $3.9 \times 10^{-3}$ | 3, 6 |
| U008 | Acrylic acid | 72 | 141.6 | 4.24 | | miscible | $4.0 \times 10^{-7}$ | 1.4 | $5.0 \times 10^{-1}$ | 1 |
| U009 | Acrylonitrile | 53 | 77.4 | 91.2 | -0.92 | 73,500 | $9.2 \times 10^{-5}$ | 360 | 13.4 | 4, 6 |
| U010 | Mitomycin C | 334 | 377 | $4.1 \times 10^{-8}$ | -0.38 | 72 | $2.5 \times 10^{-10}$ | $9.4 \times 10^{-9}$ | $2.2 \times 10^{-9}$ | 2 |
| U011 | Amitrole | 84 | $T_m$ 159 | NA | See Note | See Note | --- | --- | --- | 9, Log P cannot be estimated to derive solubility |
| U012 | Aniline | 93 | 184 | 0.6537 | 0.90 | 34,000 | $3.0 \times 10^{-6}$ | $2.7 \times 10^{-1}$ | $8.9 \times 10^{-2}$ | 1, 3, 6 |
| U013 | Asbestos | See Note | | | | | | | | Mineral of various compositions and properties. |
| U014 | Auramine | 304 | NA | NA | NA | NA | --- | --- | --- | 7, Solubility and vapor pressure cannot be estimated for organo-metallics |

*Presented only if required for other calculations.  **Vapor pressure @ 25°C, estimated via extrapolation from experimental value.

Appendix A.  Physical-Chemical Properties and Relative Soil Volatility of RCRA Wastes

| Waste Code | Waste Name | Molecular Weight (g/mol) | Boiling Point (°C) | Vapor Pressure @ 25°C (mm Hg) | Log P* | Solubility in $H_2O$ @ 25°C (mg/l) | Henry's Constant (atm·m³/mol) | Relative Dry Soil Volatility $P_{vp}M^{1/4}$ | Relative Wet Soil Volatility $P_{vp}M^{-1/2}$ | Reference/Comments |
|---|---|---|---|---|---|---|---|---|---|---|
| U015 | Azaserine | 173 | 290 | $0.125 \times 10^{-3}$ | See Note | See Note | --- | $3.3 \times 10^{-5}$ | $9.5 \times 10^{-6}$ | 2. Log P cannot be estimated to derive solubility |
| U016 | 3,4 Benzacridine | 229 | 419 | $8 \times 10^{-10}$ | 4.565 | 0.035 | $7 \times 10^{-9}$ | $2.1 \times 10^{-10}$ | $5.3 \times 10^{-11}$ | 1 |
| U017 | Benzalchloride | 161 | 205 | 0.327 | 2.815 | 410 | $1.7 \times 10^{-4}$ | $9.2 \times 10^{-2}$ | $2.6 \times 10^{-2}$ | 1 |
| U018 | 1,2-Benzanthracene | 228 | 435 | $10^{-10}$ | 5.61 | $8.8 \times 10^{-3}$ | $3.4 \times 10^{-9}$ | $2.6 \times 10^{-11}$ | $6.6 \times 10^{-2}$ | 1 |
| U019 | Benzene | 78 | 80.1 | 101** | 2.13 | 1,780 | $5.5 \times 10^{-3}$ | 34.3 | 11.4 | ***vp = 76 @ 20°C; 60 @ 15°C 1, 6, 23, 30 (H = experimental) |
| U020 | Benzenesulfonyl chloride | 177 | 252 | 0.04** | | Reacts with water | --- | $1.1 \times 10^{-2}$ | $3.0 \times 10^{-3}$ | ***vp = 10 @ 120°C 1 (Reacts with water) |
| U021 | Benzidine | 184 | 402 | $1 \times 10^{-5}$ | 1.34 | 1,550 | $1.91 \times 10^{-8}$ | $2.7 \times 10^{-6}$ | $7.4 \times 10^{-7}$ | 3, 6 |
| U022 | Benzo(a)pyrene | 252 | 310-312 @ 10mm | $1.9 \times 10^{-10}$ | 5.835 | 0.047 | $1.38 \times 10^{-9}$ to $4.16 \times 10^{-10}$ | $4.8 \times 10^{-11}$ | $1.2 \times 10^{-11}$ | 2, 30 |
| U023 | Benzotrichloride | 196 | 221 | 0.157 | 2.92 | 360 | $1.12 \times 10^{-4}$ | $4.2 \times 10^{-2}$ | $1.1 \times 10^{-2}$ | 2, 6 |
| U024 | Bis(2-chloroethoxy)methane | 123 | 181 | 0.815 | 0.469 | 510,000 | $2.77 \times 10^{-7}$ | $2.6 \times 10^{-1}$ | $7.9 \times 10^{-2}$ | 1 |
| U025 | Dichloroethyl ether | 143 | 178 | 1.4 | | 10,200 | $2.58 \times 10^{-5}$ | $4.1 \times 10^{-1}$ | $1.2 \times 10^{-1}$ | 3 |
| U026 | Chlornaphazine | 268 | 210 @ 5 mm | $4.5 \times 10^{-6}$ | 3.69 | 57 | $2.8 \times 10^{-8}$ | $1.1 \times 10^{-6}$ | $2.7 \times 10^{-7}$ | 2 |
| U027 | Bis(2-chloroisopropyl) ether | 171 | 187 | 0.85 | 2.58 | 19 | $1.03 \times 10^{-4}$ | $2.4 \times 10^{-1}$ | $6.5 \times 10^{-2}$ | 1 |
| U028 | Bis(2-ethylhexyl)phthalate | 390 | 384 | $1.38 \times 10^{-5}$ | 9.8 | $2.7 \times 10^{-5}$ | 26.6 | $3.2 \times 10^{-6}$ | $7.0 \times 10^{-7}$ | 2 |

*Presented only if required for other calculations.

**Vapor pressure @ 25°C   estimated via extrapolation from experimental value.

Appendix A.  Physical-Chemical Properties and Relative Soil Volatility of RCRA Wastes

| Waste Code | Waste Name | Molecular Weight (g/mol) | Boiling Point* (°C) | Vapor Pressure @ 25°C (mm Hg) | log P* | Solubility in $H_2O$ @ 25°C (mg/l) | Henry's Constant (atm m³/mol) | Relative Dry Soil Volatility $P_{vp}M^{1/4}$ | Relative Wet Soil Volatility $P_{vp}M^{1/2}$ | Reference/Comments |
|---|---|---|---|---|---|---|---|---|---|---|
| 1029 | Bromomethane | 95 | 3.56 | 5300** | 5.08 | 900 | $5.26 \times 10^{-3}$ | 1,700 | 544 | exp = 761 @ 3.6°C; 1, 3, 6, 30 |
| 1030 | 4 Bromophenyl phenyl ether | 249 | 310 | 0.041 | 5.08 | 0.50 | $2.14 \times 10^{-3}$ | $1.1 \times 10^{-4}$ | $2.6 \times 10^{-3}$ | 1 |
| 1031 | n Butanol | 74 | 117.25 | 6.5 | 0.88 | 91,000 | $7 \times 10^{-6}$ | 2.2 | $7.6 \times 10^{-1}$ | 1, 2, 3, 6 |
| 1032 | Calcium chromate | 156 | dec >200 | NA | — | 163,500 @ 20 | — | — | — | 2, 1. Vapor pressure cannot be estimated for inorganic compounds |
| 1033 | Carbonyl fluoride | 66 | -83 | NA | — | reacts | — | — | — | 1, Substance is permanent gas at room temperature |
| 1034 | Trichloroacetaldehyde | 147 | 97.75 | 35 | 1.7 | 100,000 | $6.7 \times 10^{-5}$ | 10.2 | 2.9 | 1, 3, 10 |
| 1035 | Chlorambucil | 304 | 363 | $1.39 \times 10^{-7}$ | 1.7 | 26,000 | $2.1 \times 10^{-12}$ | $3.3 \times 10^{-8}$ | $8.0 \times 10^{-9}$ | 2, 6 |
| 1036 | Chlordane, tech. | 410 | 175 @ 2 torr | $1 \times 10^{-5}$ | 5.94 | 0.0541 | $4.8 \times 10^{-5}$ | $2.2 \times 10^{-6}$ | $4.9 \times 10^{-7}$ | 6, 8, 31 |
| 1037 | Chlorobenzene | 113 | 132 | 11.8 | 2.18 | 500 | $3.93 \times 10^{-3}$ | 3.6 | 1.11 | 3, 6, 23 (H = experimental), 30 |
| 1038 | Ethyl 4,4'-dichlorobenzilate | 325 | 146.148 @ 0.4 mm | $2.2 \times 10^{-6}$ | 4.812 | 1.6 | $5.89 \times 10^{-7}$ | $5.3 \times 10^{-7}$ | $1.2 \times 10^{-7}$ | 2 |
| 1039 | 4 Chloro m cresol | 142 | 235; bp 66.8 | $5.8 \times 10^{-3}$ | 3.10 | 3,846 | $2.83 \times 10^{-7}$ | $1.1 \times 10^{-3}$ | $4.9 \times 10^{-4}$ | 1, 3 |
| 1040 | 1 Chloro 2,3 epoxypropane | 92.5 | 116.5 | 18.8 | 1.96 | 60,000 | $3.8 \times 10^{-5}$ | 6.02 | 1.96 | 1, 3 |
| 1041 | 2 Chloroethyl vinyl ether | 101 | 109 | 30.5 | 1.28 | 6,000 | $7.35 \times 10^{-4}$ | 9.5 | 2.9 | 2 |
| 1043 | Ethene, chloro. | 62.5 | -13.9 | 2,660 | | 1,100 | 0 1998/20°C 0.023-0.65 @90°C | 958 | 336 | 3 |
| 1044 | Chloroform | 119 | 61 | 172.1 | 1.97 | 9,300 | $3.39 \times 10^{-3}$ | 51.6 | 15.8 | 1, 3, 6, 23 |

*Presented only if required for other calculations.    **Vapor pressure @ 25°C    estimated via extrapolation from experimental value

Appendix A  Physical-Chemical Properties and Relative Soil Volatility of RCRA Wastes

| Waste Code | Waste Name | Molecular Weight (g/mol) | Boiling Point (°C) | Vapor Pressure @ 25°C (mm Hg) | Log P** | Solubility in H₂O @ 25°C (mg/l) | Henry's Constant (atm·m³/mol) | Relative Dry Soil Volatility $P_{vp}/M^{1/4}$ | Relative Wet Soil Volatility $P_{vp}/M^{1/2}$ | Reference/Comments |
|---|---|---|---|---|---|---|---|---|---|---|
| U045 | Chloromethane | 50 | -24.2 | 5 (20) / 6.7 (30) | 0.91 | 4,000 | 0.38 | 1.8 | $7.1 \times 10^{-1}$ | 1, 3, 6, 28 |
| U046 | Chloromethyl methyl ether | 81 | 59.5 | 214 | -0.11 | $2.5 \times 10^{6}$ | $9.12 \times 10^{-6}$ | 10.6 | 23.8 | 1, 3 |
| U047 | β-Chloronaphthalene | 162 | 236 / $I_m$ 61 | $5.15 \times 10^{-3}$ | 4.005 | 12.85 | $3.15 \times 10^{-5}$ | $1.4 \times 10^{-3}$ | $4.1 \times 10^{-4}$ | 1, 30 (H = experimental) |
| U048 | o-Chlorophenol | 128.5 | 174.9 | 0.787 | | 28,500 | $4.1 \times 10^{-6}$ | 23.6 | $6.9 \times 10^{-2}$ | 1, 3 |
| U049 | 4-Chloro-o-toluidine, HC | 142 | 241 | 0.105 | 2.285 | 1,900 | $1.02 \times 10^{-5}$ | $3.1 \times 10^{-2}$ | $8.8 \times 10^{-3}$ | 7 |
| U050 | Chrysene | 228 | 488 | $10^{-13}$ | | 0.011 | $10^{-12}$ | $2.6 \times 10^{-14}$ | $6.6 \times 10^{-15}$ | 1, 13 |
| U051 | Creosote | See Note | | | | | | | | Mixture of substances (cresylic acids) |
| U052 | Cresols and cresylic acid | 108 | o: 191 / m: 202 / p: 201.9 | o: 0.432 / m: 0.18 / p: 0.16 | 1.95 (all isomers) | o: 31,000 / m: 23,500 / p: 24,000 | $2 \times 10^{-6}$ / $1 \times 10^{-6}$ / $9.5 \times 10^{-7}$ | $1.3 \times 10^{-1}$ / $5.6 \times 10^{-2}$ / $5.0 \times 10^{-2}$ | $4.2 \times 10^{-2}$ / $1.7 \times 10^{-2}$ / $1.5 \times 10^{-2}$ | 6, 3 |
| U053 | Crotonaldehyde | 70 | 99-104 | 19 | | 155,000 | $1.13 \times 10^{-5}$ | 6.6 | 2.3 | 3 |
| U055 | Cumene | 120 | 153 | 4.507 | 3.66 | 50 | $1.46 \times 10^{-2}$ | 1.4 | $4.1 \times 10^{-1}$ | 1, 6, 29 |
| U056 | Cyclohexane | 84 | 81 | 6.82 | 3.44 | 55 | $1.78 \times 10^{-1}$ | 2.3 | $7.4 \times 10^{-1}$ | 1, 3, 6, 30 |
| U057 | Cyclohexanone | 98 | 156 | 4.57 | 0.81 | 23,000 | $2.56 \times 10^{-5}$ | 1.5 | $4.6 \times 10^{-1}$ | 1, 6 |
| U058 | Cyclophosphamide | 261 | 329 | $2.8 \times 10^{-6}$ | 0.63 | 4,000 | $2.37 \times 10^{-10}$ | $7.0 \times 10^{-7}$ | $1.7 \times 10^{-7}$ | 2 |
| U059 | Daunomycin | 528 | 533 | $9.4 \times 10^{-15}$ | 1.83 | 30,000 | $2.18 \times 10^{-19}$ | $2.0 \times 10^{-15}$ | $4.1 \times 10^{-16}$ | 2, 6 |

*Presented only if required for other calculations.

**Vapor pressure @ 25°C estimated via extrapolation from experimental value

Appendix A.  Physical-Chemical Properties and Relative Soil Volatility of RCRA Wastes

| Waste Code | Waste Name | Molecular Weight (g/mol) | Boiling Point (°C) | Vapor Pressure @ 25°C (mm Hg) | log P* | Solubility in H₂O @ 25°C (mg/l) | Henry's Constant (atm m³/mol) | Relative Dry Soil Volatility $P_{vp}M^{-1/4}$ | Relative Wet Soil Volatility $P_{vp}M^{-1/2}$ | Reference/Comments |
|---|---|---|---|---|---|---|---|---|---|---|
| U060 | DDD | 320 | 365 | $1.5 \times 10^{-7}$ | | $5 \times 10^{-6}$ | 0.0126 | $3.6 \times 10^{-8}$ | $8.4 \times 10^{-9}$ | 2, 9 |
| U061 | DDT | 354 | 260, $b_m$ 109 | $1.5 \times 10^{-7}$ | 3.98 to 6.19 | $1.28 \times 10^{-3}$ | $5.2 \times 10^{-5}$ | $3.5 \times 10^{-8}$ | $8.0 \times 10^{-9}$ | 6, 8, 20, 30 |
| U062 | Diallate | 270 | 150 @ 9 mm | $7.88 \times 10^{-3}$ | | 14 | $1.99 \times 10^{-4}$ | $2.0 \times 10^{-3}$ | $4.8 \times 10^{-4}$ | 2, 8 |
| U063 | Dibenz[a,h]anthracene | 278 | 441 | $5.2 \times 10^{-11}$ ** | | 0.044 | $10^{-10}$ | $1.3 \times 10^{-11}$ | $3.12 \times 10^{-12}$ | mp = 4 @ 275°C 1, 8 |
| U064 | 1,2:7,8-Dibenzopyrene | 302 | 483 | $10^{-12}$ | 7.22 | $1.6 \times 10^{-5}$ | $2 \times 10^{-8}$ | $2.4 \times 10^{-13}$ | $5.8 \times 10^{-14}$ | 1, 3 |
| U066 | 1,2-Dibromo-3-chloropropane | 236 | 196 | 0.513 | | 1,000 | $1.593 \times 10^{-4}$ | $1.3 \times 10^{-1}$ | $3.3 \times 10^{-2}$ | 1, 3 |
| U067 | Ethylenedibromide | 186 | 131-132, $b_m$ 9.3 | 11 | | 4,310 | $6.25 \times 10^{-4}$ | 3.0 | $8.1 \times 10^{-1}$ | 1, 5 |
| U068 | Methylene bromide | 174 | 97 | 45.8 | | 12,000 | $3.16 \times 10^{-4}$ | 12.8 | 3.47 | 30 |
| U069 | Di-butyl phthalate | 278 | 340 | $10^{-3}$ | | 400 | $1.09 \times 10^{-6}$ | $2.4 \times 10^{-4}$ | $6.0 \times 10^{-5}$ | 1 |
| U070 | o-Dichlorobenzene | 147 | 180.5 | 1.45 | 3.38 | 145 | $1.94 \times 10^{-3}$ | $4.2 \times 10^{-1}$ | $1.2 \times 10^{-1}$ | 3, 6, 23, 30 (H = experimental) |
| U071 | m-Dichlorobenzene | 147 | 173 | 2.1 | 3.38 | 123 | $2.63 \times 10^{-3}$ | $6.1 \times 10^{-1}$ | $1.1 \times 10^{-1}$ | 3, 23, 30 (H = experimental) |
| U072 | p-Dichlorobenzene | 147 | 174 | 0.67 | 3.39 | 69 | $2.31 \times 10^{-3}$ | $1.9 \times 10^{-1}$ | $5.5 \times 10^{-2}$ | 1, 6, 21, 23, 30 (H = experimental) |
| U073 | 3,3'-Dichlorobenzidine | 253 | 334, $b_m$ 133 | $1.12 \times 10^{-7}$ | 2.79 | 38.3 | $10^{-9}$ | $2.8 \times 10^{-8}$ | $1.0 \times 10^{-9}$ | 1, 3 |
| U074 | 1,4-Dichloro-2-butene | 125 | 154 | 4.00 | 1.73 | 9,101 | $6.78 \times 10^{-5}$ | 1.2 | $3.6 \times 10^{-1}$ | 1 |
| U075 | Dichlorodifluoromethane | 121 | -29.8 | 4,830 | | 280 | 0.40 to 0.43 | 1,450 | 439 | 1, 3, 30 |

*Presented only if required for other calculations.    **Vapor pressure @ 25°C estimated via extrapolation from experimental value

Appendix A.  Physical Chemical Properties and Relative Soil Volatility of RCRA Wastes.

| Waste Code | Waste Name | Molecular Weight (g/mol) | Boiling Point (°C) | Vapor Pressure @ 25°C (mm Hg) | log P* | Solubility in H₂O @ 25°C (mg/l) | Henry's Constant (atm·m³/mol) | Relative Dry Soil Volatility $P_{vp}M^{1/4}$ | Relative Wet Soil Volatility $P_{vp}M^{1/2}$ | Reference/Comments |
|---|---|---|---|---|---|---|---|---|---|---|
| U076 | 1,1-Dichloroethane | 99 | 57 | 193.4 | 1.79 | 5,500 | $5.45 \times 10^{-3}$ | 61.9 | 19.4 | 1, 6, 23 (H - experimental) |
| U077 | 1,2-Dichloroethane | 99 | 84 | 75.663 | 1.48 | 8,690 | $1.10 \times 10^{-3}$ | 24.2 | 1.6 | 1, 3, 6, 23 (H = experimental) |
| U078 | 1,1-Dichloroethylene | 97 | 31.9 | 630.1 | | 3,200 | $1.5 \times 10^{-2}$ | 202 | 44.0 | 1, 15, 23 (H = experimental), 30 (H = 0.154 @ 20°C) |
| U079 | 1,2-Dichloroethylene | 97 | 48-60 | cis = 217 trans = 352 | 1.25 | ~ 700 | $6.60 \times 10^{-3}$ $5.32 \times 10^{-3}$ | 69.2 112.3 | 22.0 35.1 | 2, 5, 23 (H = $6.6 \times 10^{-3}$), 30 (H = $5.32 \times 10^{-3}$) |
| U080 | Methylene chloride | 85 | 41 | 427.8 | 1.25 | 16,700 | $3.19 \times 10^{-3}$ | 141.1 | 46.4 | 1, 3, 6, 23 (H - experimental) |
| U081 | 2,4-Dichlorophenol | 163 | 210 | 0.118** | 3.08 to 3.30 | 4,500 | $5.62 \times 10^{-6}$ | $3.3 \times 10^{-2}$ | $9.2 \times 10^{-3}$ | exp. - 110 @ 147°C, 1, 3, 6 |
| U082 | 2,6-Dichlorophenol | 163 | 219-220 $T_m$ 68-69 | 0.0165 | 2.895 | 112 | $2 \times 10^{-5}$ | $4.6 \times 10^{-3}$ | $1.3 \times 10^{-3}$ | 1,5 |
| U083 | 1,2-Dichloropropane | 113 | 96 | 50 | 2.00 | 2,700 | $2.8 \times 10^{-3}$ | 15.5 | 4.1 | 1, 3, 5, 23 (H = experimental) |
| U084 | 1,3-Dichloropropene | 113 | 120.4 | 28 | 2.00 | 1,000 | $1.7 \times 10^{-3}$ | 8.68 | 2.6 | 1, 3, 6, 30 |
| U085 | 1,2:3,4-Diepoxybutane | 86 | 138 | 7.52 | -1.26 | $8.3 \times 10^{7}$ | $1.02 \times 10^{-8}$ | 2.48 | $8.1 \times 10^{-1}$ | 2 |
| U086 | N,N-Diethylhydrazine | 88 | 99 | See Note | See Note | See Note | --- | --- | --- | 1. log P cannot be estimated to derive solubility. Vapor pressure cannot be estimated |
| U087 | O,O-Diethyl S-methyl-dithiophosphate | | See Note | See Note | See Note | See Note | --- | --- | --- | 2. Properties cannot be estimated for thiophosphates |
| U088 | Diethylphthalate | 222 | 298 | $8.1 \times 10^{-3}$ | 3.58 | 50 | $4.15 \times 10^{-5}$ | $2.1 \times 10^{-3}$ | $5.4 \times 10^{-4}$ | 3 |
| U089 | Diethylstilbestrol | 268 | 418 | $2 \times 10^{-12}$ | 5.37 | .015 | $5.07 \times 10^{-11}$ | $5.3 \times 10^{-13}$ | $1.3 \times 10^{-13}$ | 7 |

*Presented only if required for other calculations.    **Vapor pressure @ 25°C; estimated via extrapolation from experimental value

Appendix A.  Physical-Chemical Properties and Relative Soil Volatility of RCRA Wastes

| Waste Code | Waste Name | Molecular Weight (g/mol) | Boiling Point (°C) | Vapor Pressure @ 25°C (mm Hg) | log P** | Solubility in H₂O @ 25°C (mg/l) | Henry's Constant (atm·m³/mol) | Relative Dry Soil Volatility $P_{vp}/W^{1/4}$ | Relative Wet Soil Volatility $P_{vp}/W^{1/2}$ | Reference/Comments |
|---|---|---|---|---|---|---|---|---|---|---|
| U090 | Dihydrosafrole | 164 | 198 | 0.269 | 2.985 | 240 | $2.3 \times 10^{-4}$ | $7.5 \times 10^{-2}$ | $2.1 \times 10^{-2}$ | 2 |
| U091 | 3,3'-Dimethoxybenzidine | 244 | 331 | $1.9 \times 10^{-7}$ | 1.23 | 1,800 | $10^{-11}$ | $4.75 \times 10^{-8}$ | $1.2 \times 10^{-8}$ | 3 |
| U092 | Dimethylamine | 46 | 7.4 | 2.1 | -0.3 | $2.3 \times 10^{6}$ | $5.69 \times 10^{-8}$ | $7.9 \times 10^{-1}$ | $3.0 \times 10^{-1}$ | 3, 4 |
| U093 | Dimethylaminoazobenzene | 225 | sublimes | NA | 4.58 | 160 | --- | --- | --- | 3, 6, Vapor pressure cannot be estimated |
| U094 | 1,12-Dimethylbenz[a]anthracene | 256 | 411 | $3.9 \times 10^{-12}$ | 6.94 | $1.3 \times 10^{-3}$ | $1.03 \times 10^{-9}$ | $9.8 \times 10^{-13}$ | $2.4 \times 10^{-13}$ | 1 |
| U095 | 3,3-Dimethylbenzidine | 212 | 327 | $2.9 \times 10^{-7}$ | 2.69 | 46 | $1.75 \times 10^{-9}$ | $7.5 \times 10^{-8}$ | $2.0 \times 10^{-8}$ | 2 |
| U096 | a,a-Dimethylbenzylhydroperoxide | 152 | 253 | $1.1 \times 10^{-2}$ | See Note | See Note | --- | $3.1 \times 10^{-3}$ | $8.9 \times 10^{-4}$ | *vp = $2.7 \times 10^{-2}$ @ 60°C; 4. log P cannot be estimated to derive solubility |
| U097 | Dimethylcarbamoyl chloride | 107 | 167-168 | 2.49 | See Note | See Note | --- | $7.1 \times 10^{-1}$ | $2.5 \times 10^{-1}$ | 1, 4 (hydrolyzes) |
| U098 | 1,1-Dimethylhydrazine | 60 | 63 | 151 | See Note | See Note | --- | 56.5 | 20.3 | 1, 5. log P cannot be estimated to derive solubility |
| U099 | 1,2-Dimethylhydrazine | 60 | 81 | 67.99 | See Note | See Note | --- | 24.5 | 8.8 | 1, 5. log P cannot be estimated to derive solubility |
| U101 | 2,4-Dimethylphenol | 122 | 210 | 0.118** | 2.30 | 1,600 | $1.18 \times 10^{-5}$ | $3.5 \times 10^{-2}$ | $1.07 \times 10^{-2}$ | *vp = 10 @ 89.3°C, 1 @ 51.8°C; 1 |
| U102 | Dimethylphthalate | 194 | 283.8 | $3.59 \times 10^{-3}$ | | 4,300 | $2.1 \times 10^{-7}$ | $9.1 \times 10^{-4}$ | $2.6 \times 10^{-4}$ | 1, 3 |
| U103 | Dimethyl sulfate | 126 | 188.5 | 0.510** | | 28,000 | $3.37 \times 10^{-6}$ | 0.131 | $5.1 \times 10^{-2}$ | *vp<1 @ 20°C; 1, 2 |
| U105 | 2,4-Dinitrotoluene | 182 | 300 1 atm 70 | $8.54 \times 10^{-5}$ | | 270 | $1.6 \times 10^{-8}$ | $2.3 \times 10^{-5}$ | $6.3 \times 10^{-6}$ | 1, 3 |

*Presented only if required for other calculations.      **Vapor pressure @ 25°C      ***estimated via extrapolation from experimental value

Appendix A. Physical Chemical Properties and Relative Soil Volatility of RCRA Wastes

| Waste Code | Waste Name | Molecular Weight (g/mol) | Boiling Point (°C) | Vapor Pressure @ 25°C (mm Hg) | log P* | Solubility in H$_2$O @ 25°C (mg/l) | Henry's Constant (atm·m$^3$/mol) | Relative Dry Soil Volatility $P_{vp}/M_W^{1/2}$ | Relative Wet Soil Volatility $P_{vp}/M_W^{1/2}$ | Reference/Comments |
|---|---|---|---|---|---|---|---|---|---|---|
| U106 | 2,6-Dinitrotoluene | 182 | 157; $T_m$ 66 | 1.71 | 2.525 | 550 | 7.42x10$^{-4}$ | 4.6x10$^{-1}$ | 1.3x10$^{-1}$ | 1 |
| U107 | Di n octyl phthalate | 391 | 384 | 6.8x10$^{-8}$ | 5.22 | 0.4 | 3x10$^{-7}$ | 1.5x10$^{-8}$ | 3.5x10$^{-9}$ | 11 |
| U108 | 1,4-Diethylene dioxide | 88 | 101 | 31 | -0.42 | 6x10$^7$ | 7.14x10$^{-7}$ | 12.2 | 3.9 | 3, 6 |
| U109 | 1,2-Diphenylhydrazine | 184 | 178; $T_m$ 131 | 5.23x10$^{-5}$ | 2.95 | 180,000 | 10$^{-11}$ | 1.5x10$^{-5}$ | 3.9x10$^{-6}$ | 1 |
| U110 | Dipropylamine | 68 | 110 | 30 | 1.46 to 1.73 | 12,000 | 3.32x10$^{-4}$ | 12.0 | 4.3 | 1, 5, 6 |
| U111 | Di n propylnitrosamine | 130 | 206 | 0.419** | | 10,000 | 7.2x10$^{-6}$ | 1.3x10$^{-1}$ | 3.7x10$^{-2}$ | mvp = 13 @ 89°C; 1, 3, 6 |
| U112 | Ethyl acetate | 88 | 77 | 82.2 | 0.73 | 79,000 | 1.2x10$^{-4}$ | 27.1 | 8.8 | 1, 3, 6 |
| U113 | Ethyl acrylate | 114 | 118 | 36.141 | | 20,000 | 2.71x10$^{-4}$ | 10.8 | 3.4 | 1, 3 |
| U114 | Ethylenebis(dithiocarbamic acid) | 212 | 319 | 3.67x10$^{-4}$ | <-3 | >10$^6$ | <10$^{-10}$ | 9.5x10$^{-5}$ | 2.5x10$^{-5}$ | 23 |
| U115 | Ethylene oxide | 44 | 13.5 | 1,294 | -0.30 | 2.1x10$^6$ | 3.63x10$^{-5}$ | 505 | 195 | 1, 5, 6 |
| U116 | Ethylene thiourea | 102 | $T_m$ ~200 | NA | -0.66 | 10,418 | NA | --- | --- | 1, 6, Properties cannot be estimated due to thiourea fragment |
| U117 | Ethyl ether | 74 | 34.51 | 540 | 0.77; 0.83; 0.89 | 60,500 | 8.69x10$^{-4}$ | 184 | 62.8 | 1, 3, 6 |
| U118 | Ethylmethacrylate | 114 | 117 | 29 | 1.485 | 19,000 | 1.49x10$^{-4}$ | 5.9 | 1.78 | 1 |

*Presented only if required for other calculations

**vapor pressure @ 25°C; estimated via extrapolation from experimental value

Appendix A.  Physical Chemical Properties and Relative Soil Volatility of RCRA Wastes

| Waste Code | Waste Name | Molecular Weight (g/mol) | Boiling Point (°C) | Vapor Pressure @ 25°C (mm Hg) | log P* | Solubility in H₂O @ 25°C (mg/l) | Henry's Constant (atm m³/mol) | Relative Dry Soil Volatility $P_{vp}M^{1/4}$ | Relative Wet Soil Volatility $P_{vp}M^{1/2}$ | Reference/Comments |
|---|---|---|---|---|---|---|---|---|---|---|
| U119 | Ethyl methanesulfonate | 124 | 85.06 @ 10 mm | 0.328 | 0.09 | $1.1\times10^6$ | $3.14\times10^{-8}$ | $9.8\times10^{-2}$ | $3.0\times10^{-2}$ | 1 |
| U121 | Trichloromonofluoromethane | 137 | 23.7 | 768 | | 1,100 | $5.83\times10^{-2}$ | 223 | 65.6 | 1, 3, 5, 23 |
| U122 | Formaldehyde | 30 | -21 | 4,433 | | 600,000 | $2.92\times10^{-4}$ | 1,906 | 809 | 1, 4 |
| U123 | Formic acid | 46 | 100.7 | 33.4 | -0.54 | $4.6\times10^6$ | $4.4\times10^{-7}$ | $1.5\times10^{-5}$ | 4.9 | 1, 6 |
| U124 | Furan | 68 | 31.36 | 633.99 | | 10,000 | $5.7\times10^{-3}$ | 722 | 76.9 | 1, 3 |
| U125 | Furfural | 96 | 161.7 | 2.39 | | 83,000 | $3.6\times10^{-6}$ | $7.7\times10^{-1}$ | $2.6\times10^{-1}$ | 1, 3 |
| U126 | Glycidylaldehyde | 72 | 91 | 42.6 | -0.529 | $7\times10^6$ | $5.8\times10^{-7}$ | 14.5 | 5.02 | 5 |
| U127 | Hexachlorobenzene | 285 | 322 $t_m$ 230 | $1.25\times10^{-5}$ | 4.13 | 0.035 | $6\times10^{-5}$ @20°C $1.7\times10^{-3}$ @25°C | $2.7\times10^{-4}$ | $6.5\times10^{-7}$ | 1, 6, 16, 23, 30 ($4.9\times10^{-5}$) |
| U128 | Hexachlorobutadiene | 261 | 215 | 0.269 | 4.14 | 10 | $9.16\times10^{-3}$ @20°C $10.3\times10^{-3}$ @25°C | $6.7\times10^{-2}$ | $1.7\times10^{-2}$ | 1, 23 |
| U129 | α-Hexachlorocyclohexane | 291 | 323 $t_m$ 113 | $3.3\times10^{-6}$ | 1.19 | 10 | $3.16\times10^{-7}$ | $7.9\times10^{-7}$ | $1.9\times10^{-7}$ | 1, 3, 29 (H = $4.93\times10^{-7}$), 30 |
| U130 | Hexachlorocyclopentadiene | 273 | 239 | 0.08 | 4.3 | 6.4 | 0.0164 | $2.0\times10^{-2}$ | $4.8\times10^{-3}$ | 1, 3, 23 |
| U131 | Hexachloroethane | 237 | $t_{sub}$ 186 | 0.6 | 3 | 50 | $9.85\times10^{-3}$ | $1.5\times10^{-1}$ | $3.9\times10^{-2}$ | 1, 3, 5, 23, 28, 30 (H = $1.3\times10^{-2}$) |
| U132 | Hexachlorophene | 407 | 412 $t_m$ 166-167 | $10^{-12}$ | 1.54 | 410,000 | $10^{18}$ | $2.2\times10^{-13}$ | $5.0\times10^{-14}$ | 1, 3, 6 |
| U133 | Hydrazine | 32 | 113.5 | 14.38 | | miscible | $6.0\times10^{-7}$ | 6.1 | 2.5 | 1, 3 |
| U134 | Hydrofluoric acid | 20 | 19.7 | 900 | | miscible | $2.0\times10^{-5}$ | 378 | 179 | 2 |

*Presented only if required for other calculations.
**Vapor pressure @ 25°C
***estimated via extrapolation from experimental value

Appendix A   Physical-Chemical Properties and Relative Soil Volatility of RCRA Wastes

| Waste Code | Waste Name | Molecular Weight (g/mol) | Boiling Point (°C) | Vapor Pressure @ 25°C (mm Hg) | Log P** | Solubility in H₂O @ 25°C (mg/l) | Henry's Constant (atm·m³/mol) | Relative Dry Soil Volatility $P_{vp}/M^{1/4}$ | Relative Wet Soil Volatility $P_{vp}/M^{1/2}$ | Reference/Comments |
|---|---|---|---|---|---|---|---|---|---|---|
| U135 | Hydrogen sulfide | 34 | -85.5 | 20 | | 4,000 | | 8.2 | 3.4 | 1, 2 |
| U136 | Hydroxydimethylarsineoxide | 138 | $T_m$=196 | NA | | 661,000 | | | | 1, Vapor pressure cannot be estimated for organometallics |
| U137 | Indeno[1,2,3-cd]pyrene | 276 | $T_m$=162.5 -164 | ~$10^{-10}$ | 7.66 | $5\times10^{-5}$ | $7.2\times10^{-7}$ | $2.5\times10^{-11}$ | $6.0\times10^{-12}$ | 31 |
| U138 | Methyl iodide | 142 | 42.4 | 400 | | 14,000 | $5\times10^{-3}$ | 116 | 33.6 | 1, 3 |
| U139 | Iron dextran | See Note | See Note | See Note | | See Note | | | | 2, (This substance is actually a mixture of five dissimilar constituents) |
| U140 | Isobutanol | 74 | 99.5 | 10 | 0.61 | 95,000 | $1.03\times10^{-5}$ | 3.4 | 1.2 | 1, 3, 6 |
| U141 | Isosafrole | 162 | 253 | 0.026 | 2.435 | 1,600 | $4.00\times10^{-6}$ | $1.3\times10^{-3}$ | $2.0\times10^{-3}$ | 1 |
| U142 | Kepone | | | $2.6\times10^{-8}$ | | | $5.6\times10^{-5}$ | | | |
| U143 | Lasiocarpine | See Note | | | | | | | | Mixture of chemical substances |
| U144 | Lead acetate | 325 | $T_m$=280 | NA | | 625,000 | | | | 1, Vapor pressure cannot be estimated for inorganic salt |
| U145 | Lead phosphate | 365 | $T_m$ = 800 | NA | | 0.14 | | | | 1, Vapor pressure cannot be estimated for inorganic salt |
| U146 | Lead subacetate | 808 | $T_m$=75 | NA | | 656,000 | | | | 1, Vapor pressure cannot be estimated for inorganic salt |
| U147 | Maleic anhydride | 98 | 197-199 $T_m$=60 | $5\times10^{-5}$(20) $2\times10^{-4}$(30) | | 163,000 | $~10^{-10}$ | $1.6\times10^{-5}$ | $5.1\times10^{-6}$ | 1, 3 |
| U148 | Maleic hydrazide | 112 | $T_{dec}$=300 $T_m$ = 292 | See Note | | 6,000 | | | | 1, 9, Compound decomposes |

*Presented only if required for other calculations.    **Vapor pressure @ 25°C estimated via extrapolation from experimental value

Appendix A.  Physical-Chemical Properties and Relative Soil Volatility of RCRA Wastes

| Waste Code | Waste Name | Molecular Weight (g/mol) | Boiling Point (°C) | Vapor Pressure @ 25°C (mm Hg) | log P* | Solubility in H₂O @ 25°C (mg/l) | Henry's Constant (atm·m³/mol) | Relative Dry Soil Volatility $P_{vp} \cdot M_w^{1/4}$ | Relative Wet Soil Volatility $P_{vp} \cdot M_w^{1/2}$ | Reference/Comments |
|---|---|---|---|---|---|---|---|---|---|---|
| U149 | Malononitrile | 66 | 219 | 0.148 | | 130,000 | $10^{-7}$ | $5.2 \times 10^{-2}$ | $1.82 \times 10^{-2}$ | 1, 10 |
| U150 | Melphalan | 305 | 376 $I_m - 183$ | $4 \times 10^{-8}$ | 1.283 | 400 | $10^{-11}$ | $9.6 \times 10^{-9}$ | $2.3 \times 10^{-9}$ | 5 |
| U151 | Mercury | 200 | 357 | $1.3 \times 10^{-3}$ | | 0.03 | $1.14 \times 10^{-2}$ | $3.51 \times 10^{-4}$ | $9.2 \times 10^{-5}$ | 1, 20, 29 |
| U152 | 1,4-Dioxane | 88 | 101 | 37 | -0.42 | $6 \times 10^{5}$ | $7 \times 10^{-7}$ | 12.2 | 4.0 | 1, 5, 6 |
| U153 | Methanethiol | 48 | 6.2 | 1,520 | | 23,300 | $4 \times 10^{-3}$ | 578 | 219 | 1, 3, 5 |
| U154 | Methyl alcohol | 32 | 65 | 113.9 | -0.64 | $4.4 \times 10^{6}$ | $1.1 \times 10^{-6}$ | 47.8 | 20.1 | 1, 6 |
| U155 | Methapyrilene | 266.39 | 173.5 | 1.89 | 2.004 | 8,500 | $7.6 \times 10^{-5}$ | $4.7 \times 10^{-1}$ | $1.2 \times 10^{-1}$ | 1 |
| U156 | Methyl chlorocarbonate | 94.5 | 71 | 113 | See Note | Reacts with water | --- | 36.2 | 11.6 | 1 (Reacts with water) |
| U157 | 3-Methylcholanthrene | 268 | $I_m$ 180 280 @ 80mm | $3.79 \times 10^{-6}$ | 6.965 | $1.7 \times 10^{-4}$ | $7.7 \times 10^{-3}$ | $9.5 \times 10^{-7}$ | $2.3 \times 10^{-7}$ | 1 |
| U158 | 4,4'-Methylenebis(2-chloroaniline) | 269 | $I_m$ 99-101 | $6 \times 10^{-6}$ | | 15 | $1.6 \times 10^{-7}$ | $1.50 \times 10^{-6}$ | $3.7 \times 10^{-7}$ | 3, 4 |
| U159 | 2-Butanone | 72 | 79.6 | 89.6** | 0.29 | 353,000 | $2.6 \times 10^{-5}$ | 30.5 | 10.6 | **vp. = 77.5 @ 20; 199.28 @ 42.8 and 241.7 @ 48.1  1, 4, 6 |
| U160 | Methylethylketone peroxide | 90 | 19 | See Note | | | | --- | --- | 4, (Substance decomposes in gas phase) |
| U161 | Methyl-2 pentanone | 100 | 116.85 | 19** | | 19,000 | $1.32 \times 10^{-4}$ | 6.08 | 1.9 | **vp. = 6 @ 20°C  1, 3 |
| U162 | Methyl methacrylate | 100 | 101 | 35.5** | | 15,000 | $3.11 \times 10^{-4}$ | 11.4 | 3.6 | **vp. = 32 @ 24°C, 40 @ 26°C  1, 3, 4 |

*Presented only if required for other calculations          **vapor pressure @ 25°C          ***estimated via extrapolation from experimental value

Appendix A.  Physical Chemical Properties and Relative Soil Volatility of RCRA Wastes

| Waste Code | Waste Name | Molecular Weight (g/mol) | Boiling Point* (°C) | Vapor Pressure @ 25°C (mm Hg) | Log P* | Solubility in H2O @ 25°C (mg/l) | Henry's Constant (atm·m³/mol) | Relative Dry Soil Volatility $P_{vp} \cdot M^{1/4}$ | Relative Wet Soil Volatility $P_{vp} \cdot M^{1/2}$ | Reference/Comments |
|---|---|---|---|---|---|---|---|---|---|---|
| U163 | N Methyl-N'-nitro-N-nitrosoguanidine | 147 | 89 | 69.1 | | IM | | 20 | 5.7 | 5, Solubility cannot be estimated for guanidine compounds |
| U164 | Methylthiouracil | 142 | $T_{dec}$ 300 | $4.8 \times 10^{-5}$ | 0.22 | 51 | $1.8 \times 10^{-7}$ | $1.4 \times 10^{-5}$ | $4.0 \times 10^{-6}$ | 2, 6 |
| U165 | Naphthalene | 81 | 218 | $0.232^{**}$ | 3.06 to 330 | 30 | $4.8 \times 10^{-4}$ | $7.7 \times 10^{-2}$ | $2.6 \times 10^{-2}$ | **vp = 1 @ 53°C / 4, 6, 30 (H - experimental) |
| U166 | 1,4 Naphthalenedione | 158 | 213 | 0.201 | 1.78 / 1.71 | 11,695 | $3.6 \times 10^{-6}$ | $5.6 \times 10^{-2}$ | $1.6 \times 10^{-2}$ | 4, 6 |
| U167 | 1 Naphthalamine | 143 | 246 | 0.0559 | 2.24 | 2221 | $4.7 \times 10^{-6}$ | $1.6 \times 10^{-2}$ | $4.7 \times 10^{-3}$ | 1, 5, 6 |
| U168 | 2 Naphthalamine | 143 | 246 | 0.0559 | 2.28 | 1,958 | $5.4 \times 10^{-6}$ | $1.6 \times 10^{-2}$ | $4.7 \times 10^{-3}$ | 1, 5, 6 |
| U169 | Nitrobenzene | 123 | 210.8 | 0.209 | | 1,900 | $2.4 \times 10^{-5}$ | $6.3 \times 10^{-1}$ | $1.9 \times 10^{-2}$ | 1, 3, 23 (H - experimental) |
| U170 | 4 Nitrophenol | 139 | 279, $T_m$ 115 | $3.88 \times 10^{-5}$ | | 16,000 | $10^{-10}$ | $1.15 \times 10^{-5}$ | $3.3 \times 10^{-4}$ | 1, 3 |
| U171 | 2 Nitropropane | 89 | 120 | 11.5 | 2.5 | 11 | 0.121 | 5.8 | 1.6 | 1, 3 |
| U172 | N Nitroso di n butylamine | 158 | 157 | 4.11 | 2.5 | 1,100 | $7.9 \times 10^{-4}$ | 1.15 | $3.3 \times 10^{-1}$ | 4 |
| U173 | N Nitrosodiethanolamine | 124 | 128 | 14.2 | -1.539 | $3 \times 10^{8}$ | $8 \times 10^{-9}$ | 4.26 | 1.28 | 5 |
| U174 | N Nitrosodiethylamine | 102 | 115-117 | 1.73 | 0.48 | 468,320 | $6 \times 10^{-7}$ | $5.4 \times 10^{-1}$ | $1.7 \times 10^{-1}$ | 3, 6 |
| U176 | N Nitroso N ethylurea | 117 | 125 | 16.3 | 0.48 | 470,000 | $5.4 \times 10^{-6}$ | 4.9 | 1.5 | 5 |
| U177 | N Nitroso N methylurea | 103 | 107 | 33.5 | 0.03 | $2 \times 10^{6}$ | $2.2 \times 10^{-6}$ | 10.4 | 3.3 | 5, 6 |

*Presented only if required for other calculations

**Vapor pressure @ 25°C estimated via extrapolation from experimental value

Appendix A.  Physical-Chemical Properties and Relative Soil Volatility of RCRA Wastes

| Waste Code | Waste Name | Molecular Weight (g/mol) | Boiling Point (°C) | Vapor Pressure @ 25°C (mm Hg) | log P* | Solubility in H2O @ 25°C (mg/l) | Henry's Constant (atm·m³/mol) | Relative Dry Soil Volatility $P_{vp}M^{1/4}$ | Relative Wet Soil Volatility $P_{vp}M^{1/2}$ | Reference/Comments |
|---|---|---|---|---|---|---|---|---|---|---|
| U178 | N Nitroso N methylurethane | 132 | 109 | 31.2** | | 13,000 | $4.1 \times 10^{-4}$ | 9.4 | 2.7 | ***vp = 13 @ 65°C; 1, 10, 20 |
| U179 | N Nitrosopiperidine | 114 | 217 | 0.244 | 0.63 | 284,318 | $9.78 \times 10^{-5}$ | $7.6 \times 10^{-2}$ | $2.3 \times 10^{-2}$ | 6 |
| U180 | N Nitrosopyrrolidine | 100 | 52 | 284 | -0.19 | $3 \times 10^6$ | $1.13 \times 10^{-5}$ | $7.6 \times 10^{-2}$ | 28.4 | 5 |
| U181 | 5 Nitro-o toluidine | 152 | 178 $I_m$ 108 | 0.280 | 2.085 | 340 | $1.67 \times 10^{-4}$ | $7.8 \times 10^{-2}$ | $2.3 \times 10^{-2}$ | 1, 5 |
| U182 | Paraldehyde | 132 | 128 | 25.3 | | 120,000 | $3.66 \times 10^{-5}$ | 7.6 | 2.2 | 1, 2, 4 (Substance is a formaldehyde polymer and decomposes in gas phase) |
| U183 | Pentachlorobenzene | 250 | 277 $I_m$ 86 | $4.12 \times 10^{-4}$ | 5.105 | 0.10 | $1.3 \times 10^{-3}$ | $1.0 \times 10^{-4}$ | $2.7 \times 10^{-5}$ | 1 |
| U184 | Pentachloroethane | 202 | 162 | 3.5 | | 500 | $2.17 \times 10^{-3}$ | $7.6 \times 10^{-3}$ | $2.5 \times 10^{-1}$ | 1, 3, 5, 28, 30 |
| U185 | Pentachloronitrobenzene | 295 | 240 $I_m$ 144 | $2.38 \times 10^{-3}$ | 5.45 | 0.032 | 0.0288 | $5.79 \times 10^{-4}$ | $1.6 \times 10^{-4}$ | 1, 5 |
| U186 | 1,3 Pentadiene | 68 | 42 | 414 | 2.30 | 870 | 0.0424 | 145 | 50.2 | 1 |
| U187 | Phenacetin | 179 | 266 $I_m$ 130 | $3.16 \times 10^{-3}$ | 1.57 (no MCl) | 530 | $1.6 \times 10^{-6}$ | $8.5 \times 10^{-4}$ | $2.6 \times 10^{-4}$ | 1, , 5, 6 |
| U188 | Phenol | 94 | 182 | 0.62 | 1.46 | 82,000 | $1.3 \times 10^{-6}$ | $2.0 \times 10^{-1}$ | $6.4 \times 10^{-2}$ | 1, 6, 23 |
| U189 | Phosphorus sulfide | 222 | 514 | See Note | See Note | $1.11 \times 10^{-4}$ | See Note | --- | --- | 1, 12 (Substance slowly reacts with water) |
| U190 | Phthalic anhydride | 148 | 295.1 $I_m$ 132 | $2 \times 10^{-4}$ @ 20° | -0.62 | 152,990 | $-10^{-10}$ | $5.4 \times 10^{-5}$ | $1.2 \times 10^{-5}$ | ***vp = 0.001 @ 30°C; 1, 2, 3, 4, 6 (Substance hydrolyzes in water) |
| U191 | 2 Picoline | 93 | 128.8 | 10 | 1.11 | 51,000 | $2.6 \times 10^{-5}$ | 3.2 | 1.0 | 1, 6 |

*Presented only if required for other calculations    **Vapor pressure @ 25°C    estimated via extrapolation from experimental value

Appendix A.  Physical-Chemical Properties and Relative Soil Volatility of RCRA Wastes

| Waste Code | Waste Name | Molecular Weight (g/mol) | Boiling Point (°C) | Vapor Pressure @ 25°C (mm Hg) | log P* | Solubility in H2O @ 25°C (mg/l) | Henry's Constant (atm-m³/mol) | Relative Dry Soil Volatility $P_{vp}/N_d^{1/4}$ | Relative Wet Soil Volatility $P_{vp}/N_d^{1/2}$ | Reference/Comments |
|---|---|---|---|---|---|---|---|---|---|---|
| U192 | Pronamide | 256 | 321 | $4.02 \times 10^{-4}$ | | 15 | $9 \times 10^{-6}$ | $1.0 \times 10^{-4}$ | $2.5 \times 10^{-5}$ | 9, 23 |
| U193 | 1,3 Propane sultone | 122 | 180 @ 30mm, $T_m$ 30-33 | $6.37 \times 10^{-4}$ | -0.4 | $2.3 \times 10^6$ | $10^{-11}$ | $1.9 \times 10^{-4}$ | $5.8 \times 10^{-5}$ | 2 |
| U194 | 1 Propanamine | 59 | 49 | 280** | | miscible | $2.0 \times 10^{-5}$ | 101 | 36 | ** VP = 400 @ 31.5°C |
| U196 | Pyridine | 79 | 115.5 | 20 | -1.69 | $3 \times 10^8$ | $1 \times 10^{-9}$ | 6.8 | 2.3 | 1, 3, 6 |
| U197 | p Benzoquinone | 108 | $T_m$ 115.7 | 0.09 | 0.20 | 23,655 | $5 \times 10^{-7}$ | $2.8 \times 10^{-2}$ | $8.7 \times 10^{-3}$ | 1, 3, 6 |
| U200 | Reserpine | 609 | $T_m$ 265 | $>1.02 \times 10^{-3}$ | 0.14 | 190 | 4.28 | $2.0 \times 10^{-4}$ | $>4.1 \times 10^{-5}$ | 1, 3 |
| U201 | Resorcinol | 110 | 280, $T_m$ 111 | $1.2 \times 10^{-5}$ | | $2.20 \times 10^6$ | $10^{-13}$ | $3.7 \times 10^{-6}$ | $1.1 \times 10^{-5}$ | 1, 3 |
| U202 | Saccharin and salts | 183 | 243, $T_{dec}$ 229 | $2.69 \times 10^{-3}$ | | 3,446 | $1.9 \times 10^{-7}$ | $7.3 \times 10^{-4}$ | $2.0 \times 10^{-4}$ | 1, 4 |
| U203 | Safrole | 162 | 234.5 | 0.0109 | 2.435 | 1,400 | $1.08 \times 10^{-5}$ | $2.0 \times 10^{-2}$ | $5.6 \times 10^{-3}$ | 1 |
| U204 | Selenium dioxide | 111 | $T_m$ 340 | 0.16** | | 38,400 | $6.0 \times 10^{-7}$ | $4.9 \times 10^{-2}$ | $1.5 \times 10^{-2}$ | ***VP = 1 @ 157°C, 1 |
| U205 | Sulfur selenide | | See Note | See Note | | See Note | | | | 12 (Sulfur and selenium form semi-continuous solid solutions containing SeS, $SeS_2$ & $SSe_2$ as actual constituents) |
| U206 | Streptozotocin | 265 | 205, $T_m$ 115 | $1.21 \times 10^{-4}$ | -1.45 | $8 \times 10^4$ | $10^{-11}$ | $3.2 \times 10^{-5}$ | $7.8 \times 10^{-6}$ | 2, 6 |
| U207 | 1,2,4,5 tetrachlorobenzene | 216 | 243-246, $T_m$ 140 | $2.14 \times 10^{-3}$ | | 6 | $1.0 \times 10^{-4}$ | $5.6 \times 10^{-4}$ | $1.5 \times 10^{-4}$ | 1, 16 |
| U208 | 1,1,1,2 tetrachloroethane | 168 | 130.5 | 10 | | 560 | $2.76 \times 10^{-3}$ | 2.8 | $7.7 \times 10^{-1}$ | 1, 4, 30 |
| U209 | 1,1,2,2 tetrachloroethane | 168 | 146.2 | 4.2 | | 3,000 | $4.1 \times 10^{-4}$ | 1.18 | $3.2 \times 10^{-1}$ | 1, 2, 5, 30 |

*Presented only if required for other calculations    ***Vapor pressure @ 25°C estimated via extrapolation from experimental value.

Appendix A.  Physical-Chemical Properties and Relative Soil Volatility of RCRA Wastes

| Waste Code | Waste Name | Molecular Weight (g/mol) | Boiling Point (°C) | Vapor Pressure @ 25°C (mm Hg) | Log P* | Solubility in H2O @ 25°C (mg/l) | Henry's Constant (atm·m³/mol) | Relative Dry Soil Volatility $P_{vp} \cdot M^{-1/2}$ | Relative Wet Soil Volatility $P_{vp} \cdot M^{-1/2}$ | Reference/Comments |
|---|---|---|---|---|---|---|---|---|---|---|
| U210 | Tetrachloroethylene | 166 | 121 | ** | 2.60 | 100 | $2.87 \times 10^{-2}$ | — | — | *vp - 10 @ 14°C; 1,2,6,23,30 (H - 0.012 @ 20°C) |
| U211 | Tetrachloromethane | 154 | 76.54 | 115.3 | 2.70 | 500 | $2.13 \times 10^{-2}$ | 32.3 | 9.3 | 1,5,6,12,23,30 (H - experimental) |
| U212 | 2,3,4,6-Tetrachlorophenol | 232 | 150 $T_m = 70$ | 1.55 | 4.10 | 10.6 | $4.46 \times 10^{-2}$ | $4.0 \times 10^{-1}$ | $1.0 \times 10^{-1}$ | 1,6 |
| U213 | Tetrahydrofuran | 72 | 67 | 149 | 0.73 | 130,988 | $1.06 \times 10^{-4}$ | 50.7 | 17.6 | 1,5,6 |
| U214 | Thallium(I)acetate | 263 | $T_m - 110$ | NA | — | soluble | — | — | — | 1, Properties cannot be estimated for inorganic compounds |
| U215 | Thallium(I)carbonate | 468 | $T_m - 272$ | Insignificant (< $10^{-6}$) | — | 52,000 | — | — | — | *vp - 10 @ 517°C; 1, Properties cannot be estimated for inorganic compounds |
| U216 | Thallium(I)chloride | 240 | 720 $T_m - 430$ | Insignificant (< $10^{-5}$) | — | 2,900 | — | 2.5 | — | 1, Properties cannot be estimated for inorganic compounds |
| U217 | Thallium(I)nitrate | 266 | 430 $T_m - 206$ | See Note | — | 95,500 | — | — | — | 1,12 (Substance decomposes to $Tl_2O_3 + NO_x$, higher temp measurements may include contributions from decomposition products) |
| U218 | Thioacetamide | 75 | $T_m - 114$ | NA | — | 163,000 | — | — | — | 1,4, Vapor pressure cannot be estimated because of the thioacetamide group |
| U219 | Thiourea | 76 | $T_m - 182$ | NA | — | 91,800 | — | — | — | 1,3, Properties cannot be estimated due to thio fragment |
| U220 | Toluene | 92 | 110.6 | 28.8 | 2.69 | 515 | $6.64 \times 10^{-3}$ | 8.6 | 2.8 | 1,3,6,23,30 (H - experimental) |

*Presented only if required for other calculations          **Vapor pressure @ 25°C estimated via extrapolation from experimental value

Appendix A  Physical-Chemical Properties and Relative Soil Volatility of RCRA Wastes

| Waste Code | Waste Name | Molecular Weight (g/mol) | Boiling Point (°C) | Vapor Pressure @ 25°C (mm Hg) | log P* | Solubility in $H_2O$ @ 25°C (mg/l) | Henry's Constant (atm·m³/mol) | Relative Dry Soil Volatility $P_{sp}/M^{1/4}$ | Relative Wet Soil Volatility $P_{sp}/M^{1/2}$ | Reference/Comments |
|---|---|---|---|---|---|---|---|---|---|---|
| U221 | Toluenediamine | 122 | 255 $T_m=61$ | $1.7 \times 10^{-3}$ | 0.345 | 120,000 | $2.3 \times 10^{-9}$ | $5.3 \times 10^{-4}$ | $1.6 \times 10^{-3}$ | 1 |
| U222 | o-Toluidine hydrochloride | 143 | 242.2 $T_m=215$ | $1.19 \times 10^{-3}$ | 1.57 (no HCl) | 30 | $7.55 \times 10^{-6}$ | $3.5 \times 10^{-4}$ | $1.0 \times 10^{-4}$ | 1 |
| U223 | Toluenediisocyanate | 174 | 251 | 0.021** | --- | reacts with water | --- | $6.1 \times 10^{-3}$ | $1.65 \times 10^{-3}$ | **vp = 11 @ 126°C; 2, 4 (Reacts with water) |
| U224 | Toxaphene | 258 | $T_m=65-90$ | NA | --- | 0.40 | $4.89 \times 10^{-3}$ | --- | --- | 1, 16, 5, 23 (M = experimental) Vapor pressure cannot be estimated |
| U225 | Bromoform | 253 | 149.5 | 5.6 | 2.69 | 3,190 | $5.32 \times 10^{-4}$ | 1.40 | $3.5 \times 10^{-1}$ | 1, 3, 5, 23, 30 (H = $6.12 \times 10^{-4}$) |
| U226 | 1,1,1-Trichloroethane | 133 | 74.1 | 117** | 2.49 | 950 | $4.92 \times 10^{-3}$ | 33.9 | 10.1 | **vp = 99.1 @ 20°C; 1, 4, 6, 23, 30 (H = $3.42 \times 10^{-2}$ @ 20°C) |
| U227 | 1,1,2-Trichloroethane | 187 | 113.67 | 22.4** | 3.12 | 4,500 | $1.18 \times 10^{-3}$ | 6.05 | 1.6 | **vp = 15.06 @ 50°C; 1, 3, 5, 30 |
| U228 | Trichloroethylene | 131 | 87 | 71.6 | 2.29 | 1,100 | $8.92 \times 10^{-3}$ | 21.5 | 6.3 | 1, 2, 6, 23, 30 (H = experimental) |
| U230 | 2,4,5-Trichlorophenol | 197 | $T_{sub}=252$ $T_m=69$ | 0.0496 | 3.72 | <2,000 | $6 \times 10^{-6}$ | $1.3 \times 10^{-2}$ | $3.5 \times 10^{-3}$ | 2, 3, 6 |
| U231 | 2,4,6-Trichlorophenol | 197 | 246 | 0.0149** | 3.69 | 800 | $4.82 \times 10^{-6}$ | $4.0 \times 10^{-3}$ | $1.06 \times 10^{-3}$ | **vp = .1 @ 76.5°C; 1, 3, 6 |
| U232 | 2,4,5-T | 255 | 294 $T_m=150, 153$ | $2.44 \times 10^{-5}$ | --- | 238 | $3.44 \times 10^{-8}$ | $6.1 \times 10^{-6}$ | $1.5 \times 10^{-6}$ | 8 |
| U233 | Silvex | 269.5 | 310 $T_m=181.6$ | $7.12 \times 10^{-6}$ | --- | 140 | $1.8 \times 10^{-8}$ | $1.8 \times 10^{-6}$ | $4.3 \times 10^{-7}$ | 2, 4 |

*Presented only if required for other calculations

**Vapor pressure @ 25°C

***estimated via extrapolation from experimental value

Appendix A.  Physical-Chemical Properties and Relative Soil Volatility of RCRA Wastes

| Waste Code | Waste Name | Molecular Weight (g/mol) | Boiling Point (°C) | Vapor Pressure @ 25°C (mm Hg) | log P* | Solubility In H2O @ 25°C (mg/l) | Henry's Constant (atm m³/mol) | Relative Dry Soil Volatility $P_{dry}M^{-1/4}$ | Relative Wet Soil Volatility $P_{wet}M^{-1/2}$ | Reference/Comments |
|---|---|---|---|---|---|---|---|---|---|---|
| U234 | Benzene, 1,3,5-trinitro- | 213 | 315 $T_m$=122 | 2.86** | 1.18 | 350 | $2.3 \times 10^{-3}$ | 0.75 | 0.20 | ***vp = 2 @ 11.5°C 1,2,6 |
| U235 | Tris (2,3 dibromopropyl)phosphate | 697 | $T_{dec}$= 255 | $<1 \times 10^{-3}$ | 2.83 to 3.21 | 6.3 | $1.46 \times 10^{-4}$ | $1.9 \times 10^{-4}$ | $<3.8 \times 10^{-5}$ | 15 |
| U236 | Trypan blue | 961 | decomposes | | See Note | See Note | --- | --- | --- | 1, log P cannot be estimated to derive solubility. |
| U237 | Uracil, 5[bis(2-chloromethyl) amino] | 224 | 224 | 0.168 | -1.505 | $6 \times 10^8$ | $10^{-10}$ | $4.6 \times 10^{-2}$ | $1.1 \times 10^{-2}$ | 1 |
| U238 | Ethyl carbamate (urethane) | 89 | 185 $T_m$=48 | 0.3515 | -0.15 | $2 \times 10^6$ | $2 \times 10^{-8}$ | $1.2 \times 10^{-1}$ | $3.8 \times 10^{-2}$ | 1,2,6 |
| U239 | Xylene | 106 | 137-140 | o = 2.11 p = 3.15 m = 3.20 | 5.0 6.5 6.0 | 175 | $5.27 \times 10^{-3}$ $2.51 \times 10^{-3}$ $2.55 \times 10^{-3}$ | 2.05 $9.8 \times 10^{-1}$ $9.9 \times 10^{-1}$ | $2.1 \times 10^{-1}$ $3.1 \times 10^{-1}$ $3.1 \times 10^{-1}$ | 3, 30 (N = experimental) |
| U240 | 2,4-D Salts & Esters | | See Note | See Note | See Note | See Note | --- | --- | --- | Mixture of compounds |
| U242 | Pentachlorophenol | 266 | 109-110 | $1.1 \times 10^{-4}$ | 4.30 | 80 | $4.8 \times 10^{-7}$ | $1.1 \times 10^{-4}$ | $6.7 \times 10^{-6}$ | 1, 3 |
| U243 | Hexachloropropene | 249 | 210 | 0.344 | 4.38 | 4.5 | $2.5 \times 10^{-2}$ | $8.6 \times 10^{-2}$ | $2.9 \times 10^{-2}$ | 2, 7 |
| U244 | Bis(dimethylthiocarbamoyl)disulfide | NA | NA | NA | NA | NA | NA | --- | --- | Properties cannot be estimated for disulfides |
| U246 | Bromine cyanide | 106 | 61.5 | 100 | See Note | Reacts with water | --- | 31.1 | 9.7 | 1, 4 (hydrolyzes in water to HOCN + HBr) |

*Presented only if required for other calculations.

***Vapor pressure @ 25°C   ***estimated via extrapolation from experimental value.

Appendix A. Physical-Chemical Properties and Relative Soil Volatility of RCRA Wastes

| Waste Code | Waste Name | Molecular Weight (g/mol) | Boiling Point[a] (°C) | Vapor Pressure @ 25°C (mm Hg) | log P[a] | Solubility in H₂O @ 25°C (mg/l) | Henry's Constant (atm m³/mol) | Relative Dry Soil Volatility $P_q/M^{-1/4}$ | Relative Wet Soil Volatility $P_q/M^{-1/2}$ | Reference/Comments |
|---|---|---|---|---|---|---|---|---|---|---|
| P001 | Warfarin | 308 | 432 | $10^{-11}$, $t_m = 161$ | 0.05 to 2.52 | 860 | $10^{-15}$ | $2.6 \times 10^{-12}$ | $5.7 \times 10^{-13}$ | 6, 8 |
| P002 | 1-Acetyl-2-thiourea | 118 | 206 | 0.14 | NA | NA | NA | $4.2 \times 10^{-2}$ | $1.3 \times 10^{-2}$ | 1, (log P and solubility cannot be estimated due to presence of thiourea fragment) |
| P003 | Acrolein | 56 | 53 | 258 | NA | 200,000 | $6.79 \times 10^{-5}$ | 95.5 | 34.5 | 1, 3 |
| P004 | Aldrin | 365 | $t_m = 104$ | $6 \times 10^{-6}$ | 3.01 | 0.2 | $4.96 \times 10^{-4}$ | $1.6 \times 10^{-6}$ | $3.1 \times 10^{-7}$ | 6, 20, 23 (H=4.96×10⁻⁴), 28, 30 (H=2.8×10⁻⁵), 29 (H=1.44×10⁻⁵) |
| P005 | Allyl alcohol | 58 | 97 | 20.10 | 0.17 | 62,000 | $3.47 \times 10^{-6}$ | 10.1 | 3.7 | 1, 5, 6 |
| P006 | Aluminum phosphide | 58 | See Note | See Note | See Note | Reacts | See Note | --- | --- | 12, (Substance has high melting point (m.p. 1250°C), vapor pressure negligible to m.p. Reacts with water). |
| P007 | 5-(Aminomethyl)-3-isoxazol | 114 | 256 $t_{dec} = 75$ | $1.4 \times 10^{-3}$ | -1.815 | $6.4 \times 10^{8}$ | $10^{-13}$ | $4.3 \times 10^{-4}$ | $1.3 \times 10^{-4}$ | 2 |
| P008 | 4-Pyridinamine | 94 | 108 | 25.8 | 0.26, 0.28 | 730,000 | $4.4 \times 10^{-6}$ | 8.3 | 2.1 | 1, 5, 6 |
| P009 | Ammonium picrate | 246 | 209 | 0.230 | See Note | 10,000 | $7.4 \times 10^{-6}$ | $5.8 \times 10^{-2}$ | $1.5 \times 10^{-2}$ | 4 |
| P010 | Arsenic acid | 142 | decomposes | See Note | | 167,000 | | --- | --- | 12, (Arsenic acid decomposes to As₂O₅ + H₂O in gas phase.) |
| P011 | Arsenic V oxide | 230 | $t_{dec} = 300$ | See Note | | $1.5 \times 10^{6}$ | | --- | --- | 1, 3, (As₂O₅ decomposes to As₂O₃ + O₂ in gas phase.) |
| P012 | Arsenic II oxide | 198 | 457.2 | $8.3 \times 10^{-8}$ ** | | 17,000 | $\sim 1.1 \times 10^{-12}$ | $2.2 \times 10^{-8}$ | $5.9 \times 10^{-9}$ | ***vp. 2×10⁻¹ @ 60°C, 1, 4, 12 |

*Presented only if required for other calculations

**Vapor pressure @ 25°C    ***estimated via extrapolation from experimental value

Appendix A  Physical-Chemical Properties and Relative Soil Volatility of RCRA Wastes

| Waste Code | Waste Name | Molecular Weight (g/mol) | Boiling Point* (°C) | Vapor Pressure @ 25°C (mm Hg) | log P* | Solubility in $H_2O$ @ 25°C (mg/l) | Henry's Constant (atm·m³/mol) | Relative Dry Soil Volatility $P_{vp}M^{1/4}$ | Relative Wet Soil Volatility $P_{vp}M^{1/2}$ | Reference/Comments |
|---|---|---|---|---|---|---|---|---|---|---|
| P013 | Barium cyanide | 189 | NA | NA | NA | 800,000 | ... | ... | ... | 1, Boiling point and vapor pressure cannot be estimated for inorganic substances |
| P014 | Benzenethiol | 110 | 168.3 | -1.5 | 2.52 | 706 | $3.1 \times 10^{-4}$ | $4.7 \times 10^{-3}$ | $1.4 \times 10^{-1}$ | 1, 3, 6 |
| P015 | Beryllium dust | 9 | 2,500 $T_m$=1,287 | $2.7 \times 10^{-1}$** | ... | Insoluble (<1 mg/l) | ... | $1.6 \times 10^{-3}$ | $9.0 \times 10^{-8}$ | 1, svp = $10^{-5}$ @ 942°C |
| P016 | Bis(chloromethyl)ether | 115 | 104 | 36.3** | -0.38 | 22,000 | $2.5 \times 10^{-4}$ | 11.3 | 3.4 | svp = 30 @ 22°C; 1, 19 |
| P017 | Bromoacetone | 137 | 136.5 $T_m$=36.50 | 5.51** | ... | miscible | $9.9 \times 10^{-7}$ | 1.6 | $4.7 \times 10^{-1}$ | 1, ** vp.8 mm @ 31.5°C |
| P018 | Brucine | 394 | 470 | $10^{-12}$ | 0.22 | 33,620 | $10^{-17}$ | $2.2 \times 10^{-12}$ | $5.0 \times 10^{-14}$ | 1 |
| P020 | Dinoseb(2,4-dinitro-6-sec-butylphenol) | 240 | 223 | 0.0516 | | 100 | $1.82 \times 10^{-4}$ | $1.4 \times 10^{-2}$ | $3.7 \times 10^{-3}$ | 8 |
| P021 | Calcium cyanide | 92 | $T_{dec}$=350 | See Note | See Note reacts | See Note | See Note | ... | ... | 1, 12, (Substance decomposes to carbon + calcium cyanamide. Also reacts with water to yield HCN.) |
| P022 | Carbon disulfide | 76 | 46.25 | 351 | NA | 2,940 | 0.012 | 121 | 41.0 | 1, 3, 5, (log P cannot be estimated due to structure) |
| P023 | Chloroacetaldehyde | 78.5 | 85 | 311 | -0.5 | $7 \times 10^5$ | $4.1 \times 10^{-6}$ | 108 | 35.8 | 1, 3 |
| P024 | p-Chloroaniline | 127.5 | 232 $T_m$=12.5 | 0.015 | 1.83 | 1,855 | $3 \times 10^{-6}$ | $4.5 \times 10^{-3}$ | $1.3 \times 10^{-3}$ | 1, 3, 6 |
| P026 | 1-(o-Chlorophenyl) thiourea | 181 | 275 | $4.0 \times 10^{-4}$ | 1.19 | 3,900 | $2.51 \times 10^{-8}$ | $1.1 \times 10^{-4}$ | $3.0 \times 10^{-5}$ | ? |
| P027 | 3-Chloropropionitrile | 90 | 114-176 | 1.33 | NA | 45,000 | $3.5 \times 10^{-6}$ | $4.3 \times 10^{-1}$ | $1.4 \times 10^{-1}$ | 2 |
| P028 | Benzylchloride | 126 | 179 | 1.25 | 2.30 | 1,619 | $2.36 \times 10^{-4}$ | $3.8 \times 10^{-1}$ | $1.1 \times 10^{-1}$ | 1, 6 |

*Presented only if required for other calculations    **vapor pressure @ 25°C    ***estimated via extrapolation from experimental value

Appendix A. Physical-Chemical Properties and Relative Soil Volatility of RCRA Wastes

| Waste Code | Waste Name | Molecular Weight (g/mol) | Boiling Point (°C) | Vapor Pressure @ 25°C (mm Hg) | log P* | Solubility in $H_2O$ @ 25°C (mg/l) | Henry's Constant (atm·m³/mol) | Relative Dry Soil Volatility $P_{vp}/P_w^{1/4}$ | Relative Wet Soil Volatility $P_{vp}/P_w^{1/2}$ | Reference/Comments |
|---|---|---|---|---|---|---|---|---|---|---|
| P029 | Copper cyanides ($CuCN$, $CuCN_2$) | 89, 115 | See Note | See Note | See Note | 2.6, 1.8 | --- | --- | --- | 25, 12, (Substance decomposes to $Cu + C_2N_2$; In equilibrium with these products.) |
| P030 | Cyanides | | | | | | | | | See specific compound |
| P031 | Cyanogen | 52 | -21 | 3,900** | | 21.5 | 9.91 | 1,473 | 552 | bvp = 5 atm @ 21.4 / 1, 2 |
| P033 | Cyanogen chloride | 61 | 12.66 | 1,190 | | 30,000 | $3.2 \times 10^{-3}$ reacts slowly | 428 | 152.4 | 1, 3, 4, (hydrolyzes to HOCN + HCl) |
| P034 | 2-Cyclohexyl-4,6-dinitrophenol | 266 | 291.5 | $2.87 \times 10^{-3}$** | 4.81 | 1.2 | $8.37 \times 10^{-4}$ | $7.2 \times 10^{-4}$ | $1.7 \times 10^{-4}$ | bvp = 1mm @ 132.86 / 1 |
| P035 | 2,4-D | 221 | 278, $b_m$140.5 | $1.59 \times 10^{-4}$** | | 620 | $7.5 \times 10^{-8}$ | $4.1 \times 10^{-5}$ | $1.1 \times 10^{-5}$ | bvp = 0.4 @ 160°C / 1, 8 |
| P036 | Dichlorophenylarsine | 223 | 255 | $2.7$** | | hydrolyzes | --- | $7.0 \times 10^{-1}$ | --- | bvp = 14 @ 131°C / 1, Compound hydrolyzes in water and wet soil |
| P037 | Dieldrin | 381 | $b_m$175 | $7.78 \times 10^{-7}$ | 6.20 | 0.19 | $1.1 \times 10^{-5}$ | $1.8 \times 10^{-7}$ | $4.0 \times 10^{-8}$ | 1, 7, 8, 15, 23, 30, 31 (H - experimental) |
| P038 | Diethylarsine | 134 | 105 | 30 | See Note | See Note | --- | 8.7 | 2.6 | 1, Solubility of inorganic chemicals cannot be estimated |
| P039 | Disulfoton | 274 | 108 @ 0.01 mm | $1.8 \times 10^{-4}$ | | 25 | $2.59 \times 10^{-6}$ | $4.5 \times 10^{-5}$ | $1.1 \times 10^{-5}$ | 2, 9 |
| P040 | O,O-Diethyl o pyrazinyl Phosphorothioate | 248 | 80 | $3 \times 10^{-3}$ | | 100 | $8.58 \times 10^{-6}$ | $7.5 \times 10^{-4}$ | $1.9 \times 10^{-4}$ | 1, 8 |
| P041 | Diethyl p nitrophenyl phosphate | 275 | 169-170 | $7.8 \times 10^{-5}$ | 1.69 | 10,000 | $2.83 \times 10^{-9}$ | $2.0 \times 10^{-5}$ | $4.1 \times 10^{-6}$ | 2 |
| P042 | 1,2 Benzenediol, 4-[1-hydroxy 2-(methylamino)ethyl] epinephrine | 211 | 306 | $4 \times 10^{-4}$ | NA | NA* | --- | $1.0 \times 10^{-4}$ | $2.8 \times 10^{-5}$ | 2, *Log P cannot be estimated due to epinephrine fragment |

*Presented only if required for other calculations

**vapor pressure @ 25°C    ***vapor pressure @ 25°C estimated via extrapolation from experimental value.

Appendix A.  Physical-Chemical Properties and Relative Soil Volatility of RCRA Wastes

| Waste Code | Waste Name | Molecular Weight (g/mol) | Boiling Point* (°C) | Vapor Pressure @ 25°C (mm Hg) | log P* | Solubility in H₂O @ 25°C (mg/l) | Henry's Constant (atm·m³/mol) | Relative Dry Soil Volatility $P_{vp}M^{1/4}$ | Relative Wet Soil Volatility $P_{vp}M^{1/2}$ | Reference/Comments |
|---|---|---|---|---|---|---|---|---|---|---|
| P043 | Diisopropylfluorophosphate | 184 | 183 | 0.519 | -- | 15,000 | $9.1 \times 10^{-6}$ | $1.6 \times 10^{-1}$ | $4.3 \times 10^{-2}$ | 2 |
| P044 | Dimethoate | 229 | $T_m = 52$ | $8.5 \times 10^{-6}$ | 0.5 | 25,000 | $10^{-10}$ | $2.2 \times 10^{-6}$ | $5.6 \times 10^{-7}$ | 6, 8, 9 |
| P045 | Thiofanox | 218 | 296 | $1.1 \times 10^{-4}$ | 2.11 | 52,000 | $9.33 \times 10^{-10}$ | $4.4 \times 10^{-5}$ | $1.2 \times 10^{-4}$ | 8 |
| P046 | α,α-Dimethylphenethylamine | 149 | 205 @ 750mm | 0.276 | 2.11 | 2,900 | $1.86 \times 10^{-5}$ | $1.9 \times 10^{-2}$ | $2.3 \times 10^{-2}$ | 2 |
| P047 | 4,6-Dinitro-o-cresol (and salts) | 198 | 214 $T_m = 86.5$ | 0.018 | 1.50 | 130 | $1.4 \times 10^{-5}$ | $4.8 \times 10^{-3}$ | $1.3 \times 10^{-3}$ | 1, 9, 23 (H = experimental) |
| P048 | 2,4-Dinitrophenol | 184 | $T_m = 116$ | NA | 1.50 | 19,500 | -- | -- | -- | 1, 6, Boiling point and vapor pressure cannot be estimated due to presence of nitro groups |
| P050 | Endosulfan | 407 | $T_m = 70-100$ | $1 \times 10^{-5}$ | 5.34 | 0.22 | $2.5 \times 10^{-5}$ | $2.2 \times 10^{-4}$ | $5.0 \times 10^{-1}$ | 8, 9 |
| P051 | Endrin | 381 | $T_m = 245$ | $2 \times 10^{-7}$ | 5.3 | 0.024 | $4.2 \times 10^{-6}$ | $4.6 \times 10^{-8}$ | $1.0 \times 10^{-8}$ | 1, 2, 8, 16 |
| P054 | Ethyleneimine | 43 | 56 | 200 | -- | miscible | $1.1 \times 10^{-5}$ | 78 | 30 | bp = 160 @ 20°C & 250 @ 30°C; 1 |
| P056 | Fluorine | 19 | permanent gas | permanent gas | -1.05 | reacts | -- | -- | -- | 1, (Reacts with water) |
| P057 | Fluoroacetimide | 77 | 181 | 0.875 | -1.05 | $3.8 \times 10^{-1}$ | $2.33 \times 10^{-9}$ | $3.0 \times 10^{-1}$ | $1.0 \times 10^{-1}$ | 2, 6, 7 |
| P058 | Fluoroacetic acid, sodium salt | decomposes | NA | NA | -- | NA | -- | -- | -- | 1, (decomposes) |
| P059 | Heptachlor | 373 | $T_m = 96$ | $3 \times 10^{-4}$ | 0.56 | 0.56 | $1.49 \times 10^{-3}$ | $6.9 \times 10^{-5}$ | $1.6 \times 10^{-5}$ | 8, 20, 23 |
| P060 | Hexachlorohexahydro, exo, exo dimethanonaphthalene | 365 | 379 | $1.4 \times 10^{-4}$ | -0.39 | $8.5 \times 10^6$ | $1.1 \times 10^{-7}$ | $3.2 \times 10^{-5}$ | $7.3 \times 10^{-6}$ | 1 |

*Presented only if required for other calculations

**Vapor pressure @ 25°C estimated via extrapolation from experimental value

Appendix A. Physical-Chemical Properties and Relative Soil Volatility of RCRA Wastes

| Waste Code | Waste Name | Molecular Weight (g/mol) | Boiling Point (°C) | Vapor Pressure @ 25°C (mm Hg) | log P* | Solubility In H2O @ 25°C (mg/l) | Henry's Constant (atm·m³/mol) | Relative Dry Soil Volatility $P_{vp}/M^{1/4}$ | Relative Wet Soil Volatility $P_{vp}/M^{1/2}$ | Reference/Comments |
|---|---|---|---|---|---|---|---|---|---|---|
| P062 | Hexaethyltetraphosphate | — | See Note | See Note | | reacts | — | — | — | Boiling point and vapor pressure cannot be estimated |
| P063 | Hydrocyanic acid | 27 | 25.7 | 730 | -1.5 | $5.6\times10^7$ | $4.65\times10^{-7}$ | 321 | 140 | 1, 3, 6 |
| P064 | Methyl isocyanate | 57 | 59.6 | 201 | | Reacts with water | — | 12.3 | 26.6 | 1, 4 (Reacts with water) |
| P065 | Mercury fulminate | 284 | explodes | See Note | | 100** | — | — | — | ≈≈ 12°C  1, 10, 12 (Hg(ONC)2 detonates at fairly low temp. Gas phase contains decomposition products in addition to chemical) |
| P066 | Methomyl | 162 | $T_m$ 78-79 | $5\times10^{-5}$ | | 58,000 | $10^{-10}$ | $1.4\times10^{-5}$ | $3.9\times10^{-6}$ | 9, 8 |
| P067 | 2-Methylaziridine | 57 | 20 | 92.0 | -0.35 | $3.1\times10^6$ | $2.22\times10^{-5}$ | 33.1 | 12.2 | 1 |
| P068 | Methyl hydrazine | 46 | 87.5 | 49.6 | | miscible | $3.0\times10^{-6}$ | 16.4 | 7.3 | 4 |
| P069 | 2-Methyllactonitrile | 85 | 251 | 0.0244 | -0.508 | $7.7\times10^6$ | $10^{-10}$ | $7.4\times10^{-2}$ | $2.6\times10^{-3}$ | 5 |
| P070 | Aldicarb | 190 | | $10^{-4}$ | | 60,000 | $10^{-10}$ | $2.7\times10^{-4}$ | $7.3\times10^{-6}$ | 9 |
| P071 | Methyl parathion | 263 | 236 $T_m$ 38 | 0.0249 | | 50 | $1.97\times10^{-7}$ | $6.5\times10^{-3}$ | $1.6\times10^{-3}$ | 9, 30 |
| P072 | α-Naphthylthiourea | 202 | 331 | $5.1\times10^{-7}$ | | 600 | $2.26\times10^{-10}$ | $1.4\times10^{-7}$ | $3.6\times10^{-8}$ | 3 |
| P073 | Nickel carbonyl | 171 | 43 | 400 | | 180 | -0.5 | 112 | 30.6 | 1 |
| P074 | Nickel cyanide | 111 | decomposes NA | | | $9.1\times10^3$ | | — | — | 25, (decomposes) |

*Presented only if required for other calculations.

**Vapor pressure @ 25°C; estimated via extrapolation from experimental value.

Appendix A.  Physical-Chemical Properties and Relative Soil Volatility of RCRA Wastes

| Waste Code | Waste Name | Molecular Weight (g/mol) | Boiling Point (°C) | Vapor Pressure @ 25°C (mm Hg) | Log P* | Solubility in H₂O @ 25°C (mg/l) | Henry's Constant (atm·m³/mol) | Relative Dry Soil Volatility $P_{vp}M^{1/4}$ | Relative Wet Soil Volatility $P_{vp}M^{1/2}$ | Reference/Comments |
|---|---|---|---|---|---|---|---|---|---|---|
| P075 | Nicotine | 162 | 246 | 0.055** | 1.11 | 14,000 | $1.6 \times 10^{-7}$ | $1.6 \times 10^{-2}$ | $4.4 \times 10^{-3}$ | bvp = 28 @ 133.8°C  1, 5, 6 |
| P076 | Nitric oxide | 30 | -151.7 | permanent gas @ 20°C | --- | 4.6 ml/ 100ml @ 20 | --- | --- | --- | 2 |
| P077 | p-Nitroaniline | 138 | 331.7 $T_m = 149$ | $4 \times 10^{-3}$ | --- | 800 | $10^{-6}$ | $1.2 \times 10^{-3}$ | $3.4 \times 10^{-4}$ | 1, 3 |
| P078 | Nitrogen dioxide | 46 | 21.0 | 800 | --- | reacts | --- | 301 | 118 | 1, 12 (reacts with water to yield NO an nitric acid) |
| P081 | 1,2,3-Propanetriol, trinitrate | 227 | 256 | 0.26** | --- | 1,200 1,800 | $5.18 \times 10^{-5}$ | $6.8 \times 10^{-2}$ | $1.73 \times 10^{-2}$ | bvp = 2 @ 125°C  1, 2, 3 |
| P082 | N-Nitrosodimethylamine (Dimethylnitrosamine) | 74 | 153 | 4.87** | -0.74 | $1.4 \times 10^7$ | $3 \times 10^{-8}$ | 1.6 | $5.7 \times 10^{-1}$ | bvp = 14 @ 50°C  3 |
| P084 | N-Nitrosomethylvinylamine | 86 | --- | 16** | 0.03 | 30,000 | $5.65 \times 10^{-5}$ | 5.3 | 1.7 | bvp = 30 mm @ 47°C  2, 3 |
| P085 | Octamethylpyrophosphoramide | 286 | 154 @ 2mm | $10^{-3}$ | --- | miscible | $3.7 \times 10^{-10}$ | $2.4 \times 10^{-3}$ | $5.9 \times 10^{-5}$ | 2 |
| P087 | Osmium tetraoxide | 254 | $T_m = 40$ | 10 | --- | 57,000 | $5.86 \times 10^{-5}$ | 2.5 | $6.3 \times 10^{-1}$ | 1, 10 |
| P088 | Endothall | 230 | See Note | NA | See Note | 100,000 | --- | --- | --- | Boiling point and log P cannot be estimated  2 |
| P089 | Parathion | 291 | 113 | $3.78 \times 10^{-5}$ | --- | 24 | $1.21 \times 10^{-6}$ | $9.1 \times 10^{-6}$ | $2.2 \times 10^{-6}$ | 1, 8, 30 |
| P092 | Phenylmercuric acetate | 336 | $T_m = 149$ | $9 \times 10^{-6}$ | --- | 4,370 | $10^{-9}$ | $2.2 \times 10^{-6}$ | $4.9 \times 10^{-7}$ | 2, 8 |
| P093 | N-Phenylthiourea | 152 | 255 | 0.0162 | 0.73 | 2,200 | $1.41 \times 10^{-7}$ | $4.5 \times 10^{-3}$ | $1.3 \times 10^{-3}$ | 2, 6 |
| P094 | Phorate | 260 | 15.78 @ .01 mm | $8 \times 10^{-4}$ | --- | 50 | $5.47 \times 10^{-6}$ | $2.0 \times 10^{-4}$ | $5.0 \times 10^{-5}$ | 2 |

*Presented only if required for other calculations.

**Vapor pressure @ 25°C estimated via extrapolation from experimental value

Appendix A.  Physical-Chemical Properties and Relative Soil Volatility of RCRA Wastes

| Waste Code | Waste Name | Molecular Weight (g/mol) | Boiling Point* (°C) | Vapor Pressure @ 25°C (mm Hg) | Log P* | Solubility in H2O @ 25°C (mg/l) | Henry's Constant (atm·m³/mol) | Relative Dry Soil Volatility $P_{vp}P_w^{-1/4}$ | Relative Wet Soil Volatility $P_{vp}P_w^{-1/2}$ | Reference/Comments |
|---|---|---|---|---|---|---|---|---|---|---|
| P095 | Phosgene | 99 | 7.56 | 1,430 | See Note | slightly | --- | 458 | 144 | 1. Log P cannot be estimated to derive solubility |
| P096 | Phosphine | 34 | -87.7 | $1.6\times10^{-4}$ | --- | 2,600 | 0.189 | $5.1\times10^{-5}$ | $2.4\times10^{-5}$ | 1 |
| P097 | Dimethylcarbamyl chloride | 101 | 167-168 @ 775 mm | 2.6 | --- | 20,000 | $1.81\times10^{-3}$ | $8.1\times10^{-1}$ | $2.5\times10^{-1}$ | 2 |
| P098 | Potassium cyanide | 65 | NA | insignificant (< $10^{-4}$) | --- | 500,000 | --- | --- | --- | 4. Boiling points and vapor pressures cannot be estimated for inorganic chemicals |
| P099 | Potassium silver cyanide | 199 | See Note | See Note | --- | 250,000** | --- | --- | --- | neg 20°C; 3, 10. Boiling point cannot be estimated to derive vapor pressure. |
| P101 | Propane nitrile | 55 | 97.35 | 40 | --- | 119,000 | $2.4\times10^{-5}$ | 14.8 | 5.4 | 2 |
| P102 | 2-Propyn-1-ol | 56 | 113.6 | 9.56 | -1.09 | $3.2\times10^{7}$ | $2\times10^{-8}$ | 3.5 | 1.28 | 1 |
| P103 | Selenourea | 123 | dec ~213 | NA | --- | 100,000** | --- | --- | --- | **19°C; 1, 10. (Decomposes) |
| P104 | Silver cyanide | 134 | dec ~320 | See Note | See Note | 23 @ 20°C | See Note | --- | --- | 1, 12 (Substance decomposes to Ag + $C_2N_2$. Equilibrium with metal + $C_2N_2$ exists) |
| P105 | Sodium azide | 65 | ~350 | See Note | See Note | 417,000 | --- | --- | --- | 1, 27 (Decomposes to $N_2$ + sodium nitride in gas phase) |
| P106 | Sodium cyanide | 49 | 1,496 | insignificant (< $10^{-4}$) | --- | 600,000 | --- | --- | --- | 1, Boiling points and vapor pressures cannot be estimated for inorganic chemicals |
| P107 | Strontium sulfide | 119 | ~2,000 | insignificant (< $10^{-4}$) | --- | insoluble reacts at low temp | See Note | --- | --- | 1, 12 (Substance reacts with water or moisture to yield $H_2S$ + $Sr(OH)_2$) |

*Presented only if required for other calculations.

**Vapor pressure @ 25°C estimated via extrapolation from experimental value.

Appendix A. Physical-Chemical Properties and Relative Soil Volatility of RCRA Wastes

| Waste Code | Waste Name | Molecular Weight (g/mol) | Boiling Point (°C) | Vapor Pressure @ 25°C (mm Hg) | log P* | Solubility in $H_2O$ @ 25°C (mg/l) | Henry's Constant (atm·m³/mol) | Relative Dry Soil Volatility $P_{vp}M^{1/4}$ | Relative Wet Soil Volatility $P_{vp}M^{1/2}$ | Reference/Comments |
|---|---|---|---|---|---|---|---|---|---|---|
| P108 | Strychnine and salts | 334 | 463 | $1.2 \times 10^{-11}$** | 1.93 | 156 | $10^{-14}$ | $2.76 \times 10^{-12}$ | $6.6 \times 10^{-13}$ | ***vp = 5 @ 210°C; 1, 2, 6 |
| P109 | Tetraethyldithiopyrophosphate | 222 | 265 | 0.0148 | See Note | See Note | --- | $3.8 \times 10^{-3}$ | $9.9 \times 10^{-4}$ | 1, log P cannot be estimated to derive solubility |
| P110 | Tetraethyl lead | 323 | decomposes | 0.19** | See Note | See Note | --- | $4.5 \times 10^{-2}$ | $1.1 \times 10^{-2}$ | ***vp = 0.15 @ 20°C; 3. log P cannot be estimated to derive solubility |
| P111 | Tetraethylpyrophosphate | 290 | 124 @ 1 mm | $4.7 \times 10^{-4}$ | See Note | See Note | --- | $1.1 \times 10^{-4}$ | $2.8 \times 10^{-5}$ | 2. log P cannot be estimated to derive solubility |
| P112 | Tetranitromethane | 196 | 126 | 13** | 0.606 | 530,000 | $6.33 \times 10^{-6}$ | 3.5 | $9.3 \times 10^{-1}$ | ***vp = 10 @ 23°C; 8.4 @ 20°C; 1, 5 |
| P113 | Thallic oxide | 457 | $T_m$ = 717 | Insignificant (< $10^{-6}$) | | IA | --- | --- | --- | 3. Vapor pressure and solubility cannot be estimated for inorganic chemicals |
| P114 | Thallium selenate | 551 | $T_m$ = 7,600 | Insignificant (< $10^{-6}$) | | 21,300** | --- | --- | --- | **10°C; 1 |
| P115 | Sulfuric acid, thallium(I)salt | 504 | $T_m$ = 632 | Insignificant (< $10^{-6}$) | | 48,700 | --- | --- | --- | 1. Vapor pressure and solubility cannot be estimated for inorganic chemicals |
| P116 | Thiosemicarbazide | 91 | $T_m$ = 183 | IA | | soluble | --- | --- | --- | 1. Vapor pressure and solubility cannot be estimated for inorganic chemicals |
| P118 | Trichloromethyl mercaptan | 151 | 136 | 5.78 | 1.74 | 11,000 | $1.04 \times 10^{-4}$ | 1.7 | $4.7 \times 10^{-1}$ | 1. Vapor pressure and solubility cannot be estimated for inorganic chemicals |
| P119 | Ammonium vanadate | 117 | $T_m$ = 200 | IA | | 4,400 | --- | --- | --- | 1. Vapor pressure and solubility cannot be estimated for inorganic chemicals |

*Presented only if required for other calculations.

**Vapor pressure @ 25°C   estimated via extrapolation from experimental value.

Appendix A   Physical Chemical Properties and Relative Soil Volatility of RCRA Wastes

| Waste Code | Waste Name | Molecular Weight (g/mol) | Boiling Point[a] (°C) | Vapor Pressure @ 25°C (mm Hg) | log P** | Solubility in H2O @ 25°C (mg/l) | Henry's Constant (atm m³/mol) | Relative Dry Soil Volatility $P_{vp}/M^{1/4}$ | Relative Wet Soil Volatility $P_{vp}/M^{1/2}$ | Reference/Comments |
|---|---|---|---|---|---|---|---|---|---|---|
| P120 | Vanadium pentoxide | 182 | $T_m$ 690 $T_{dec}$ 1,750 | insignificant (<10⁻⁶) | – | 8,000 | ... | ... | ... | 1, Vapor pressure and solubility cannot be estimated for inorganic chemicals |
| P121 | Zinc cyanide | 111 | See Note | See Note | – | 50 | ... | ... | ... | 4, 12 (Substance decomposes to Zn + C₂N₂; in equilibrium with those products) |
| P122 | Zinc phosphide | 258 | 1100 | See Note | – | reacts | ... | ... | ... | 1, 12 (Substance reacts with water to yield PH₃ + Zn(OH)₂) |
| P123 | Octachlorocamphene | 412 | 309 | 8.08x10⁻⁶ | 6.91 | 2.1x10⁻³ | 2.01x10⁻³ | 1.8x10⁻⁴ | 4.0x10⁻⁷ | 5 |

**Presented only if required for other calculations.

***Vapor pressure @ 25°C estimated via extrapolation from experimental value

Appendix A. Physical-Chemical Properties and Relative Soil Volatility of RCRA Wastes

| Waste Code | Waste Name | Molecular Weight (g/mol) | Boiling Point* (°C) | Vapor Pressure @ 25°C (mm Hg) | log P* | Solubility in H₂O @ 25°C (mg/l) | Henry's Constant (atm·m³/mol) | Relative Dry Soil Volatility $P_{vp}M^{1/4}$ | Relative Wet Soil Volatility $P_{vp}M^{1/2}$ | Reference/Comments |
|---|---|---|---|---|---|---|---|---|---|---|
| D000 | Any combination of arsenic, barium, cadmium, and chromium waste. | | | | | | | | | |
| D001 | Solid waste that exhibits the characteristic of ignitability, but is not listed as a hazardous waste in Subpart D. | | | | | | | | | No information available on wastestream constituents |
| D002 | Solid waste that exhibits the characteristic of corrosivity, but is not listed as a hazardous waste in Subpart D. | | | | | | | | | No information available on wastestream constituents |
| D003 | Solid waste that exhibits the characteristics of reactivity, but is not listed as a hazardous waste in Subpart D. | | | | | | | | | No information available on wastestream constituents |
| D004 | Arsenic | 75 | 75 | 1 @ 372°C | | Insoluble | | nonvolatile | nonvolatile | 1 |
| D005 | Barium | 137 | 1,600 | NA | | reacts slowly with water | | nonvolatile | nonvolatile | 1, Vapor pressure cannot be estimated, however should be less than $10^{-4}$ at 25°C |
| D006 | Cadmium | 112 | 765 | $10^{-5}$ @ 148°C | | Insoluble | | nonvolatile | nonvolatile | 1 |
| D007 | Chromium | 52 | 2,672 | $10^{-5}$ @ 907°C | | Insoluble | | nonvolatile | nonvolatile | 1 |
| D008 | Lead | 207 | 1,740 | $10^{-5}$ @ 483°C | | Insoluble | | nonvolatile | nonvolatile | 1 |
| D009 | Mercury | 200 | 357 | $1.3 \times 10^{-3}$ | | $3 \times 10^{-2}$ | $1.14 \times 10^{-2}$ | $3.5 \times 10^{-4}$ | $1.2 \times 10^{-4}$ | 1, 29 |
| D010 | Selenium | 79 | 685 | NA | | Insoluble | | nonvolatile | nonvolatile | 1, Vapor pressure cannot be estimated, however should be less than $10^{-4}$ at 25°C |
| D011 | Silver | 108 | 2,060 | NA | | Insoluble | | nonvolatile | nonvolatile | 1, Vapor pressure cannot be estimated, however should be less than $10^{-4}$ at 25°C |

*Required only if required for other calculations.     **Vapor pressure @ 25°C     ***estimated via extrapolation from experimental value

Appendix A. Physical-Chemical Properties and Relative Soil Volatility of RCRA Wastes

| Waste Code | Waste Name | Molecular Weight (g/mol) | Boiling Point (°C) | Log P* | Vapor Pressure @ 25°C (mm Hg) | Solubility in H₂O @ 25°C (mg/l) | Henry's Constant (atm·m³/mol) | Relative Dry Soil Volatility $P_{vp}/V\cdot 1/4$ | Relative Wet Soil Volatility $P_{vp}/V\cdot 1/2$ | Reference/Comments |
|---|---|---|---|---|---|---|---|---|---|---|
| D012 | Endrin (1,2,3,4,10,10-hexachloro-1,7-epoxy-1,4,4a,5,6,7,8,8a-octahydro-1,4-endo, endo-5,8-dimethane naphthalene) | 381 | $T_m$=235° | | $2\times10^{-7}$ | 0.024 0.26 | $3.8\times10^{-7}$ | $4.5\times10^{-8}$ | $1.0\times10^{-8}$ | 31 |
| D013 | Lindane (1,2,3,4,5,6-hexachlorocyclohexane, gamma isomer) | 291 | $T_m$=112.9° | | $9.4\times10^{-6}$ | 7.3 | $4.93\times10^{-7}$ | $2.3\times10^{-6}$ | $5.5\times10^{-7}$ | 29, 31 |
| D014 | Methoxychlor (1,1,1-Trichloro-2,2-bis(p-methoxy phenol)) ethane | 346 | $T_{m.p.}$=89° | NA | | < 1 mg/l | --- | --- | --- | 8, Vapor pressure cannot be estimated |
| D015 | Toxaphene (technical chlorinated camphene, 67-69 percent chlorine) | 343-517 | $T_{dec}$=120° | | 0.2-0.4 | ~3 | $~7.5\times10^{-2}$ | $~3.0\times10^{-1}$ | $~8.9\times10^{-2}$ | 31 |
| D016 | 2,4-D(2,4-Dichlorophenoxyacetic acid) | 221 | $T_m$=140.5° | | $1.59\times10^{-4}$ mm | 620 | $7.5\times10^{-8}$ | $4.1\times10^{-5}$ | $1.1\times10^{-5}$ | vp ~ 0.4 mm @ 160°C |
| D017 | 2,4,5-TP Silvex (2,4,5-Trichloro-phenoxypropionic acid) | 255 | $T_m$=154° | | NA | 238 @ 30° | --- | --- | --- | 8, Vapor pressure cannot be estimated |

*Presented only if required for other calculations.

**vapor pressure @ 25°C estimated via extrapolation from experimental value.

Appendix A.  Physical-Chemical Properties and Relative Soil Volatility of RCRA Wastes

| Waste Code | Waste Name | Molecular Weight (g/mol) | Boiling Point* (°C) | Vapor Pressure @ 25°C (mm Hg) | Log P* | Solubility in H₂O @ 25°C (mg/l) | Henry's Constant (atm-m³/mol) | Relative Dry Soil Volatility $P_{vp}/K_d^{1/4}$ | Relative Wet Soil Volatility $P_{vp}/K_d^{1/2}$ | Reference/Comments |
|---|---|---|---|---|---|---|---|---|---|---|
| F001 | Spent halogenated solvents used in degreasing, tetrachloroethylene, trichloroethylene, methylene chloride, 1,1,1-trichloroethane, carbon tetrachloride, and the chlorinated fluorocarbons; sludges from the recovery of these solvents in degreasing operations. | | | | | SEE SPECIFIC WASTE UNDER P OR U WASTE CODES FOR PROPERTY DATA. | | | | |
| F002 | Spent halogenated solvents, tetrachloroethylene, methylene chloride, trichloroethylene, 1,1,1-trichloroethane, chlorobenzene, 1,1,2-trichloro-1,2,2-trifluoroethane, o-dichlorobenzene, trichlorofluoromethane, the still bottoms from the recovery of these solvents. | | | | | SEE SPECIFIC WASTE UNDER P OR U WASTE CODES FOR PROPERTY DATA. | | | | |
| F003 | Spent non-halogenated solvents, xylene, acetone, ethyl acetate, ethyl benzene, ethyl ether, n-butyl alcohol, cyclohexanone; the still bottoms from the recovery of these solvents. | | | | | SEE SPECIFIC WASTE UNDER P OR U WASTE CODES FOR PROPERTY DATA. | | | | |
| F004 | Spent non-halogenated solvents: cresols and cresylic acid, nitrobenzene; and the still bottoms from the recovery of these solvents. | | | | | SEE SPECIFIC WASTE UNDER P OR U WASTE CODES FOR PROPERTY DATA. | | | | |
| F005 | Spent non-halogenated solvents, methanol, toluene, methyl ethyl ketone, methyl isobutyl ketone, carbon disulfide, isobutanol, pyridine; the still bottoms from the recovery of these solvents. | | | | | SEE SPECIFIC WASTE UNDER P OR U WASTE CODES FOR PROPERTY DATA. | | | | |
| F006 | Wastewater treatment sludges from electroplating operations. | | | | | SEE SPECIFIC WASTE UNDER P OR U WASTE CODES FOR PROPERTY DATA. | | | | Wastestream includes cadmium, hexavalent chromium, nickel, and complexed cyanide. |

*Presented only if required for other calculations.    **Vapor pressure @ 25°C estimated via extrapolation from experimental value

Appendix A.  Physical-Chemical Properties and Relative Soil Volatility of RCRA Wastes

| Waste Code | Waste Name | Molecular Weight (g/mol) | Boiling Point (°C) | Vapor Pressure @ 25°C (mm Hg) | log p* | Solubility in H₂O @ 25°C (mg/l) | Henry's Constant (atm·m³/mol) | Relative Dry Soil Volatility $P_{vp}M_t^{-1/4}$ | Relative Wet Soil Volatility $P_{vp}M_t^{-1/2}$ | Reference/Comments |
|---|---|---|---|---|---|---|---|---|---|---|
| F007 | Spent plating bath solutions from electroplating operations. | | | SEE SPECIFIC WASTE UNDER P OR U WASTE CODES FOR PROPERTY DATA. | | | | | | Wastestream includes cyanide salts, copper, nickel, and other heavy metals depending on plating operation. |
| F008 | Plating bath sludges from the bottom of plating baths from electroplating operations. | | | SEE SPECIFIC WASTE UNDER P OR U WASTE CODES FOR PROPERTY DATA. | | | | | | Wastestream includes cyanide salts and heavy metals depending on the operation. |
| F009 | Spent stripping and cleaning bath solutions from electroplating operations. | | | SEE SPECIFIC WASTE UNDER P OR U WASTE CODES FOR PROPERTY DATA. | | | | | | Wastestream includes cyanide salts and heavy metal salts. |
| F010 | Quenching bath sludge from oil baths from metal heat treating operations where cyanides are used in the process (except for precious metals heat-treating quenching bath sludges). | | | SEE SPECIFIC WASTE UNDER P OR U WASTE CODES FOR PROPERTY DATA. | | | | | | Wastestream includes cyanide salts and heavy metal salts. |
| F011 | Spent cyanide solutions from salt bath pot cleaning from metal heat treating operations (except for precious metals heat treating spent cyanide solutions from salt bath pot cleaning). | | | SEE SPECIFIC WASTE UNDER P OR U WASTE CODES FOR PROPERTY DATA. | | | | | | Wastestream includes cyanide salts and heavy metal salts. |
| F012 | Quenching wastewater treatment sludges from metal heat treating operations. | | | SEE SPECIFIC WASTE UNDER P OR U WASTE CODES FOR PROPERTY DATA. | | | | | | Wastestream includes complexed cyanide. |
| F014 | Cyanidation wastewater treatment tailing pond sediment from mineral metals recovery operations. | | | SEE SPECIFIC WASTE UNDER P OR U WASTE CODES FOR PROPERTY DATA. | | | | | | Wastestream includes complexed cyanide. |
| F015 | Spent cyanide bath solutions from mineral metals recovery operations. | | | SEE SPECIFIC WASTE UNDER P OR U WASTE CODES FOR PROPERTY DATA. | | | | | | Wastestream includes cyanide salts |
| F019 | Wastewater treatment sludges from chemical conversion coating of aluminum | | | SEE SPECIFIC WASTE UNDER P OR U WASTE CODES FOR PROPERTY DATA. | | | | | | Wastestream includes hexavalent chromium, and complexed cyanides |

*Presented only if required for other calculations.    **Vapor pressure @ 25°C   estimated via extrapolation from experimental value.

Appendix A. Physical-Chemical Properties and Relative Soil Volatility of RCRA Wastes

| Waste Code | Waste Name | Molecular Weight (g/mol) | Boiling Point* (°C) | Vapor Pressure @ 25°C (mm Hg) | log P* | Solubility in H₂O @ 25°C (mg/l) | Henry's Constant (atm·m³/mol) | Relative Dry Soil Volatility $P_{vap} \cdot 1/4$ | Relative Wet Soil Volatility $P_{vap} \cdot 1/2$ | Reference/Comments |
|---|---|---|---|---|---|---|---|---|---|---|
| K001 | Bottom sludge from the treatment of wastewaters from wood preserving processes that use creosote and/or pentachlorophenol | | | SEE SPECIFIC WASTE UNDER P OR U WASTE CODES FOR PROPERTY DATA. | | | | | | Wastestream includes pentachlorophenol, phenol, 2-chlorophenol, p-chloro-m-cresol, 2,4-dimethylphenol, 2,4-dinitrophenol, trichlorophenols, tetrachlorophenols, 2,4-dinitrophenol, creosote, chrysene, naphthalene, fluoranthene, benzo(b)fluoranthene, benzo(a)pyrene, indeno(1,2,3-cd)pyrene, benzo(a)anthracene, dibenz(a)anthracene, acenaphthalene. |
| K002 | Wastewater treatment sludge from the production of chrome yellow and orange pigments. | | | SEE SPECIFIC WASTE UNDER P OR U WASTE CODES FOR PROPERTY DATA. | | | | | | Wastestream includes hexavalent chromium and lead. |
| K003 | Wastewater treatment sludge from the production of molybdate orange pigments. | | | SEE SPECIFIC WASTE UNDER P OR U WASTE CODES FOR PROPERTY DATA. | | | | | | Wastestream includes hexavalent chromium and lead. |
| K004 | Wastewater treatment sludge from the production of zinc yellow pigments. | | | SEE SPECIFIC WASTE UNDER P OR U WASTE CODES FOR PROPERTY DATA. | | | | | | Wastestream includes hexavalent chromium. |
| K005 | Wastewater treatment sludge from the production of chrome green pigments. | | | SEE SPECIFIC WASTE UNDER P OR U WASTE CODES FOR PROPERTY DATA. | | | | | | Wastestream includes hexavalent chromium and lead. |
| K006 | Wastewater treatment sludge from the production of chrome oxide green pigments (anhydrous and hydrated). | | | SEE SPECIFIC WASTE UNDER P OR U WASTE CODES FOR PROPERTY DATA. | | | | | | Wastestream includes hexavalent chromium. |
| K007 | Wastewater treatment sludge from the production of iron blue pigments. | | | SEE SPECIFIC WASTE UNDER P OR U WASTE CODES FOR PROPERTY DATA. | | | | | | Wastestream includes complexed cyanide and hexavalent chromium. |
| K008 | Oven residue from the production of chrome oxide green pigments. | | | SEE SPECIFIC WASTE UNDER P OR U WASTE CODES FOR PROPERTY DATA. | | | | | | Wastestream includes hexavalent chromium. |

*Presented only if required for other calculations.       ***Vapor pressure @ 25°C       **estimated via extrapolation from experimental value.

Appendix A.  Physical-Chemical Properties and Relative Soil Volatility of RCRA Wastes

| Waste Code | Waste Name | Molecular Weight (g/mol) | Boiling Point (°C) | Vapor Pressure @ 25°C (mm Hg) | log P* | Solubility in H₂O @ 25°C (mg/l) | Henry's Constant (atm m³/mol) | Relative Dry Soil Volatility $P_v/M^{1/4}$ | Relative Wet Soil Volatility $P_v/M^{1/2}$ | Reference/Comments |
|---|---|---|---|---|---|---|---|---|---|---|
| K009 | Distillation bottoms from the production of acetaldehyde from ethylene. | | | SEE SPECIFIC WASTE UNDER P OR U WASTE CODES FOR PROPERTY DATA. | | | | | | Wastestream includes chloroform, formaldehyde, methylene chloride, methyl chloride, paraldehyde, and formic acid. |
| K010 | Distillation side cuts from the production of acetaldehyde from ethylene. | | | SEE SPECIFIC WASTE UNDER P OR U WASTE CODES FOR PROPERTY DATA. | | | | | | Wastestream includes chloroform, formaldehyde, methylene chloride, paraldehyde, methyl chloride, formic acid, and chloroacetaldehyde. |
| K011 | Bottom stream from the wastewater stripper in the production of acrylonitrile. | | | SEE SPECIFIC WASTE UNDER P OR U WASTE CODES FOR PROPERTY DATA. | | | | | | Wastestream includes acrylonitrile, acetonitrile, and hydrocyanic acid. |
| K013 | Bottom stream from the acetonitrile column in the production of acrylonitrile. | | | SEE SPECIFIC WASTE UNDER P OR U WASTE CODES FOR PROPERTY DATA. | | | | | | Wastestream includes acrylonitrile, acetonitrile, and hydrocyanic acid. |
| K014 | Bottoms from the acetonitrile purification column in the production of acrylonitrile. | | | SEE SPECIFIC WASTE UNDER P OR U WASTE CODES FOR PROPERTY DATA. | | | | | | Wastestream includes acetonitrile and acrylamide. |
| K015 | Still bottoms from the distillation of benzyl chloride. | | | SEE SPECIFIC WASTE UNDER P OR U WASTE CODES FOR PROPERTY DATA. | | | | | | Wastestream includes benzyl chloride, chlorobenzene, toluene, and benzotrichloride. |
| K016 | Heavy ends or distillation residues from the production of carbon tetrachloride. | | | SEE SPECIFIC WASTE UNDER P OR U WASTE CODES FOR PROPERTY DATA. | | | | | | Wastestream includes hexachlorobenzene, hexachloro-butadiene, carbon tetrachloride, hexachloroethane, and perchloroethylene. |
| K017 | Heavy ends from purification in production of epichlorohydrin | | | SEE SPECIFIC WASTE UNDER P OR U WASTE CODES FOR PROPERTY DATA. | | | | | | Wastestream includes epichloro-hydrin, chloroethers, [bis(chloromethyl) ether and bis (2-chloroethyl) ethers], tri-chloropropane, dichloropropanols |
| K018 | Heavy ends from the fractionation column in ethyl chloride production. | | | SEE SPECIFIC WASTE UNDER P OR U WASTE CODES FOR PROPERTY DATA. | | | | | | Wastestream includes 1,2-dichloroethane, trichloroethylene, hexachlorobutadiene, hexachloro-benzene |

*Presented only if required for other calculations.    **Vapor pressure @ 25°C    estimated via extrapolation from experimental value.

Appendix A. Physical-Chemical Properties and Relative Soil Volatility of RCRA Wastes

| Waste Code | Waste Name | Molecular Weight (g/mol) | Boiling Point (°C) | Vapor Pressure @ 25°C (mm Hg) | log P* | Solubility In $H_2O$ @ 25°C (mg/l) | Henry's Constant (atm·m³/mol) | Relative Dry Soil Volatility $P_{vp}W^{-1/2}$ | Relative Wet Soil Volatility $P_{vp}W^{-1/2}$ | Reference/Comments |
|---|---|---|---|---|---|---|---|---|---|---|
| K019 | Heavy ends from the distillation of ethylene dichloride in ethylene dichloride production. | | | SEE SPECIFIC WASTE UNDER P OR U WASTE CODES FOR PROPERTY DATA. | | | | | | Wastestream includes ethylene dichloride; 1,1,1-trichloroethane; 1,1,2-trichloroethane; 1,1,2,2-tetrachloroethane; 1,1,1,2-tetrachloroethane; trichloroethylene, tetrachloroethylene, carbon tetrachloride, chloroform, vinyl chloride, and vinylidene chloride. |
| K020 | Heavy ends from the distillation of vinyl chloride in vinyl chloride monomer production. | | | SEE SPECIFIC WASTE UNDER P OR U WASTE CODES FOR PROPERTY DATA. | | | | | | Wastestream includes ethylene dichloride, 1,1,1-trichloroethane, 1,1,2-trichloroethane, tetrachloroethanes (1,1,2,2-tetrachloroethane and 1,1,1,2-tetrachloroethane), trichloroethylene, tetrachloroethylene, carbon tetrachloride, chloroform, vinyl chloride, vinylidene chloride. |
| K021 | Aqueous spent antimony catalyst wastes from fluoromethanes production | | | SEE SPECIFIC WASTE UNDER P OR U WASTE CODES FOR PROPERTY DATA. | | | | | | Wastestream includes antimony, carbon tetrachloride, and chloroform. |
| K022 | Distillation bottom tars from the production of phenol/acetone from cumene | | | SEE SPECIFIC WASTE UNDER P OR U WASTE CODES FOR PROPERTY DATA. | | | | | | Wastestream includes phenol and tars (polycyclic aromatic hydrocarbon). |
| K023 | Distillation light ends from the production of phthalic anhydride from naphthalene. | | | SEE SPECIFIC WASTE UNDER P OR U WASTE CODES FOR PROPERTY DATA. | | | | | | Wastestream includes phthalic anhydride and maleic anhydride. |
| K024 | Distillation bottoms from the production of phthalic anhydride from naphthalene. | | | SEE SPECIFIC WASTE UNDER P OR U WASTE CODES FOR PROPERTY DATA. | | | | | | Wastestream includes phthalic anhydride and 1,4-naphthoquinone |
| K025 | Distillation bottoms from the production of nitrobenzene by the nitration of benzene | | | SEE SPECIFIC WASTE UNDER P OR U WASTE CODES FOR PROPERTY DATA. | | | | | | Wastestream includes meta-dinitrobenzene and 2,4-dinotrotoluene. |
| K026 | Stripping still tails from the production of methyl ethyl pyridines. | | | SEE SPECIFIC WASTE UNDER P OR U WASTE CODES FOR PROPERTY DATA. | | | | | | Wastestream includes paraldehyde, pyridines, and 2-picoline. |

*Presented only if required for other calculations.    **vapor pressure @ 25°C    estimated via extrapolation from experimental value

Appendix A. Physical-Chemical Properties and Relative Soil Volatility of RCRA Wastes

| Waste Code | Waste Name | Molecular Weight (g/mol) | Boiling Point (°C) | Vapor Pressure @ 25°C (mm Hg) | log P* | Solubility in H₂O @ 25°C (mg/l) | Henry's Constant (atm·m³/mol) | Relative Dry Soil Volatility P_soil(VP) | Relative Wet Soil Volatility P_soil(1/2) | Reference/Comments |
|---|---|---|---|---|---|---|---|---|---|---|
| K027 | Centrifuge residue from toluene diisocyanate production | SEE SPECIFIC WASTE UNDER P OR U WASTE CODES FOR PROPERTY DATA. | | | | | | | | Wastestream includes toluene diisocyanate, and toluene-2,4-diamine |
| K028 | Spent catalyst from the hydrochlorinator reactor in 1,1,1 TCE production | SEE SPECIFIC WASTE UNDER P OR U WASTE CODES FOR PROPERTY DATA. | | | | | | | | Wastestream includes 1,1,1-trichloroethane and vinyl chloride |
| K029 | Waste from production steam stripper in 1,1,1-trichloroethane production. | SEE SPECIFIC WASTE UNDER P OR U WASTE CODES FOR PROPERTY DATA. | | | | | | | | Wastestream includes 1,1,1-trichloroethane, 1,1,1-tri-chloroethane, vinyl chloride, vinylidene chloride, and chloroform |
| K030 | Column bottoms or heavy ends from the combined production of trichloroethylene and perchloroethylene. | SEE SPECIFIC WASTE UNDER P OR U WASTE CODES FOR PROPERTY DATA. | | | | | | | | Wastestream includes hexachloro-benzene, hexachlorobutadiene, hexachloroethane; 1,1,2-tetra-chloroethane; 1,1,2,2-tetra-chloroethane; ethylene dichloride |
| K031 | By products salts generated in the production of MSMA and cacodylic acid. | SEE SPECIFIC WASTE UNDER P OR U WASTE CODES FOR PROPERTY DATA. | | | | | | | | Wastestream contains arsenic |
| K032 | Wastewater treatment sludge from chlordane production | SEE SPECIFIC WASTE UNDER P OR U WASTE CODES FOR PROPERTY DATA. | | | | | | | | Wastestream includes hexachloro-cyclopentadiene |
| K033 | Wastewater and scrub water from the chlorination of cyclopentadiene in the production of chlordane. | SEE SPECIFIC WASTE UNDER P OR U WASTE CODES FOR PROPERTY DATA. | | | | | | | | Wastestream includes hexachloro-cyclopentadiene |
| K034 | Filter solids from filtration of hexachlorocyclopentadiene | SEE SPECIFIC WASTE UNDER P OR U WASTE CODES FOR PROPERTY DATA. | | | | | | | | Wastestream includes hexachloro-cyclopentadiene |
| K035 | Wastewater treatment sludges generated in the production of creosote | SEE SPECIFIC WASTE UNDER P OR U WASTE CODES FOR PROPERTY DATA. | | | | | | | | Wastestream includes creosote, chrysene, naphthalene, fluoran-thene, benzo(b)fluoranthene, benzo(a)pyrene, indeno(1,2,3-cd) pyrene, benzo(a)anthracene, dibenzo(a)anthracene, acenaph-thalene |

*Presented only if required for other calculations.    **vapor pressure @ 25°C    estimated via extrapolation from experimental value

Appendix A.  Physical-Chemical Properties and Relative Soil Volatility of RCRA Wastes

| Waste Code | Waste Name | Molecular Weight (g/mol) | Boiling Point (°C) | Vapor Pressure @ 25°C (mm Hg) | Log P* | Solubility in H₂O @ 25°C (mg/l) | Henry's Constant (atm-m³/mol) | Relative Dry Soil Volatility $P_{vap}M^{1/2}$ | Relative Wet Soil Volatility $P_{vap}M^{1/2}/S_w$ | Reference/Comments |
|---|---|---|---|---|---|---|---|---|---|---|
| K036 | Still bottoms from toluene reclamation distillation in the production of disulfoton. | | | SEE SPECIFIC WASTE UNDER P OR U WASTE CODES FOR PROPERTY DATA. | | | | | | Wastestream includes toluene, phosphorodithioic and phosphoro-thioic acid esters |
| K037 | Wastewater treatment sludges from the production of disulfoton. | | | SEE SPECIFIC WASTE UNDER P OR U WASTE CODES FOR PROPERTY DATA. | | | | | | Wastestream includes toluene, phosphorodithioic and phosphoro-thioic acid esters |
| K038 | Wastewater from the washing and stripping of phorate production. | | | SEE SPECIFIC WASTE UNDER P OR U WASTE CODES FOR PROPERTY DATA. | | | | | | Wastestream includes phorate, formaldehyde, phosphorodithioic and phosphorothioic acid esters |
| K039 | Filter cake from the filtration of diethylphos-phorodithioic acid in the production of phorate. | | | SEE SPECIFIC WASTE UNDER P OR U WASTE CODES FOR PROPERTY DATA. | | | | | | Wastestream includes phosphoro-dithioic and phosphorothioic acid esters |
| K040 | Wastewater treatment sludge from the production of phorate. | | | SEE SPECIFIC WASTE UNDER P OR U WASTE CODES FOR PROPERTY DATA. | | | | | | Wastestream includes phorate, formaldehyde, phosphorodithioic and phosphorothioic acid esters |
| K041 | Wastewater treatment sludge from the production of toxaphene. | | | SEE SPECIFIC WASTE UNDER P OR U WASTE CODES FOR PROPERTY DATA. | | | | | | Wastestream includes toxaphene |
| K042 | Heavy ends or distillation residues from the distillation of tetrachlorobenzene in the production of 2,4,5-T. | | | SEE SPECIFIC WASTE UNDER P OR U WASTE CODES FOR PROPERTY DATA. | | | | | | Wastestream includes hexachloro-benzene and ortho-dichlorobenzene |
| K043 | 2,6-Dichlorophenol waste from the production of 2,4-D | | | SEE SPECIFIC WASTE UNDER P OR U WASTE CODES FOR PROPERTY DATA. | | | | | | Wastestream includes 2,4-dichlorophenol, 2,6-dichlorophenol, and 2,4,6-trichlorophenol |
| K044 | Wastewater treatment sludges from the manufacturing and processing of explosives. | | | SEE SPECIFIC WASTE UNDER P OR U WASTE CODES FOR PROPERTY DATA. | | | | | | No information on constituents. |
| K045 | Spent carbon from wastewater treatment in explosives production | | | SEE SPECIFIC WASTE UNDER P OR U WASTE CODES FOR PROPERTY DATA. | | | | | | No information on constituents. |

*Presented only if required for other calculations.    **vapor pressure @ 25°C  estimated via extrapolation from experimental value.

Appendix A.  Physical-Chemical Properties and Relative Soil Volatility of RCRA Wastes

| Waste Code | Waste Name | Molecular Weight (g/mol) | Boiling Point[a] (°C) | Vapor Pressure @ 25°C (mm Hg) | Log P* | Solubility in H₂O @ 25°C (mg/l) | Henry's Constant (atm·m³/mol) | Relative Dry Soil Volatility $P_{vp}/W·1/2$[b] | Relative Wet Soil Volatility $P_{vp}/W·1/2$[b] | Reference/Comments |
|---|---|---|---|---|---|---|---|---|---|---|
| K046 | Wastewater treatment sludges from the manufacturing, formulation, and loading of lead-based initiating compounds. | SEE SPECIFIC WASTE UNDER P OR U WASTE CODES FOR PROPERTY. | | | | | | | | Wastestream contains lead |
| K047 | Pink/red water from TNT operations. | SEE SPECIFIC WASTE UNDER P OR U WASTE CODES FOR PROPERTY DATA. | | | | | | | | Wastestream includes dinitrotoluene isomers and dinitrobenzoic acid isomers. |
| K048 | Dissolved air flotation (DAF) float from the petroleum refining industry. | SEE SPECIFIC WASTE UNDER P OR U WASTE CODES FOR PROPERTY DATA. | | | | | | | | Wastestream includes hexavalent chromium and lead. |
| K049 | Slop oil emulsion solids from the petroleum refining industry. | SEE SPECIFIC WASTE UNDER P OR U WASTE CODES FOR PROPERTY DATA. | | | | | | | | Wastestream includes hexavalent chromium and lead. |
| K050 | Heat exchanger bundle cleaning sludge from the petroleum refining industry. | SEE SPECIFIC WASTE UNDER P OR U WASTE CODES FOR PROPERTY DATA. | | | | | | | | Wastestream includes hexavalent chromium. |
| K051 | API separator sludge from the petroleum refining industry. | SEE SPECIFIC WASTE UNDER P OR U WASTE CODES FOR PROPERTY DATA. | | | | | | | | Wastestream includes hexavalent chromium and lead. |
| K052 | Tank bottoms (leaded) from the petroleum refining industry. | SEE SPECIFIC WASTE UNDER P OR U WASTE CODES FOR PROPERTY DATA. | | | | | | | | Wastestream includes lead. |
| K060 | Ammonia still lime sludge from coking operations. | SEE SPECIFIC WASTE UNDER P OR U WASTE CODES FOR PROPERTY DATA. | | | | | | | | Wastestream contains cyanide, naphthalene, phenolic compounds, and arsenic. |

[a]Presented only if required for other calculations.    [b]Vapor pressure @ 25°C  estimated via extrapolation from experimental value.

Appendix A. Physical-Chemical Properties and Relative Soil Volatility of RCRA Wastes

| Waste Code | Waste Name | Molecular Weight (g/mol) | Boiling Point[a] (°C) | Vapor Pressure @ 25°C (mm Hg) | log P[a] | Solubility in H₂O @ 25°C (mg/l) | Henry's Constant (atm·m³/mol) | Relative Dry Soil Volatility $P_{vp}M^{-1/4}$ | Relative Wet Soil Volatility $P_{vp}M^{-1/2}$ | Reference/Comments |
|---|---|---|---|---|---|---|---|---|---|---|
| K061 | Emission control dust/sludge from the electric furnace production of steel | SEE SPECIFIC WASTE UNDER P OR U WASTE CODES FOR PROPERTY DATA. | | | | | | | | Wastestream includes hexavalent chromium, lead, cadmium, and possibly nickel. |
| K062 | Spent pickle liquor from steel finishing operations | SEE SPECIFIC WASTE UNDER P OR U WASTE CODES FOR PROPERTY DATA. | | | | | | | | Wastestream includes hexavalent chromium, lead, and dilute spent HCl. |
| K069 | Emission control dust/sludge from secondary lead smelting | SEE SPECIFIC WASTE UNDER P OR U WASTE CODES FOR PROPERTY DATA. | | | | | | | | Wastestream includes hexavalent chromium, lead, and cadmium. |
| K071 | Brine purification muds from the mercury cell process in chlorine production, where separately prepurified brine is not used. | SEE SPECIFIC WASTE UNDER P OR U WASTE CODES FOR PROPERTY DATA. | | | | | | | | Wastestream includes mercury |
| K073 | Chlorinated hydrocarbon waste from purification step of diaphragm cell process using graphite anodes in chlorine production. | SEE SPECIFIC WASTE UNDER P OR U WASTE CODES FOR PROPERTY DATA. | | | | | | | | Wastestream contains chloroform, carbon tetrachloride, hexachloroethane, trichloroethane, tetrachloroethylene, dichloroethylene, 1,1,2,2-tetrachloroethane |
| K083 | Still bottoms from aniline production | SEE SPECIFIC WASTE UNDER P OR U WASTE CODES FOR PROPERTY DATA. | | | | | | | | Wastestream contains aniline, diphenylamine, nitrobenzene, phenylenediamine |
| K084 | Wastewater treatment sludges generated during the production of veterinary pharmaceuticals from arsenic or organo-arsenic compounds | SEE SPECIFIC WASTE UNDER P OR U WASTE CODES FOR PROPERTY DATA. | | | | | | | | Wastestream contains arsenic |
| K085 | Distillation or fractionation column bottoms from the production of chlorobenzenes | SEE SPECIFIC WASTE UNDER P OR U WASTE CODES FOR PROPERTY DATA. | | | | | | | | Wastestream contains benzene, dichlorobenzenes, trichlorobenzenes, tetrachlorobenzenes, pentachlorobenzene, hexachlorobenzene, benzyl chloride |

[a]Presented only if required for other calculations.    [b]Vapor pressure @ 25°C estimated via extrapolation from experimental value.

Appendix A.  Physical-Chemical Properties and Relative Soil Volatility of RCRA Wastes

| Waste Code | Waste Name | Molecular Weight (g/mol) | Boiling Point (°C) | Vapor Pressure @ 25°C (mm Hg) | Log $P^{ow}$ | Solubility in H₂O @ 25°C (mg/l) | Henry's Constant (atm·m³/mol) | Relative Dry Soil Volatility $P_s g^{/w_s - 1/4}$ | Relative Wet Soil Volatility $P_s g^{/w_s - 1/2}$ | Reference/Comments |
|---|---|---|---|---|---|---|---|---|---|---|
| K091 | Decanter tank tar sludge from coking operations. | | | SEE SPECIFIC WASTE UNDER P OR U WASTE CODES FOR PROPERTY DATA. | | | | | | Wastestream includes phenol, naphthalene, and polynuclear aromatics |
| K093 | Distillation light ends from production of phthalic anhydride from o-xylene | | | SEE SPECIFIC WASTE UNDER P OR U WASTE CODES FOR PROPERTY DATA. | | | | | | Wastestream contains phthalic anhydride and maleic anhydride |
| K094 | Distillation bottoms from production of phthalic anhydride from o-xylene | | | SEE SPECIFIC WASTE UNDER P OR U WASTE CODES FOR PROPERTY DATA. | | | | | | Wastestream contains phthalic anhydride |
| K095 | Distillation bottoms from the production of 1,1,1-trichloroethane | | | SEE SPECIFIC WASTE UNDER P OR U WASTE CODES FOR PROPERTY DATA. | | | | | | Wastestream contains 1,1,2-trichloroethane, 1,1,2,2-tetrachloroethane, 1,1,2,2-tetrachloroethane |
| K096 | Heavy ends from heavy ends column from the production of 1,1,1-trichloroethane | | | SEE SPECIFIC WASTE UNDER P OR U WASTE CODES FOR PROPERTY DATA. | | | | | | Wastestream contains 1,2-dichloroethane, 1,1,1-dichloroethane, 1,1,2-trichloroethane |
| K097 | Vacuum stripper discharge from the chlordane chlorinator in the production of chlordane | | | SEE SPECIFIC WASTE UNDER P OR U WASTE CODES FOR PROPERTY DATA. | | | | | | Wastestream contains chlordane and heptachlor |
| K098 | Untreated process wastewater from production of toxaphene | | | SEE SPECIFIC WASTE UNDER P OR U WASTE CODES FOR PROPERTY DATA. | | | | | | Wastestream contains toxaphene |
| K099 | Untreated wastewater from production of 2,4-D | | | SEE SPECIFIC WASTE UNDER P OR U WASTE CODES FOR PROPERTY DATA. | | | | | | Wastestream contains 2,4-dichlorophenol and 2,4,5-trichlorophenol |
| K100 | Waste leaching solution from acid leaching of emission control dust/sludge from secondary lead smelting | | | SEE SPECIFIC WASTE UNDER P OR U WASTE CODES FOR PROPERTY DATA. | | | | | | Wastestream contains hexavalent chromium, lead, and cadmium |
| K101 | Distillation tar residues from the distillation of aniline-based compounds in the production of veterinary pharmaceuticals from arsenic or organo arsenic compounds | | | SEE SPECIFIC WASTE UNDER P OR U WASTE CODES FOR PROPERTY DATA. | | | | | | Wastestream contains arsenic |

*Presented only if required for other calculations.

**vapor pressure @ 25°C   estimated via extrapolation from experimental value.

Appendix A. Physical-Chemical Properties and Relative Soil Volatility of RCRA Wastes

| Waste Code | Waste Name | Molecular Weight (g/mol) | Boiling Point[a] (°C) | Vapor Pressure @ 25°C (mm Hg) | Log P[a] | Solubility in H$_2$O @ 25°C (mg/l) | Henry's Constant (atm·m³/mol) | Relative Dry Soil Volatility $P_{vp}/W_s^{-1/4}$ | Relative Wet Soil Volatility $P_{vp}/W_s^{-1/2}$ | Reference/Comments |
|---|---|---|---|---|---|---|---|---|---|---|
| K102 | Residue from the use of activated carbon for decolorization in the production of veterinary pharmaceuticals from arsenic or organoarsenic compounds | | SEE SPECIFIC WASTE UNDER P OR U WASTE CODES FOR PROPERTY DATA. | | | | | | | Wastestream contains arsenic |
| K103 | Process residues from aniline extraction from the production of aniline | | SEE SPECIFIC WASTE UNDER P OR U WASTE CODES FOR PROPERTY DATA. | | | | | | | Wastestream contains aniline, nitrobenzene, and phenylenediamine |
| K104 | Combined wastewater streams generated from nitrobenzene/aniline production | | SEE SPECIFIC WASTE UNDER P OR U WASTE CODES FOR PROPERTY DATA. | | | | | | | Wastestream contains aniline, benzene, diphenylamine, nitrobenzene, phenylenediamine |
| K105 | Separated aqueous stream from the reactor product washing step in the production of chlorobenzenes | | SEE SPECIFIC WASTE UNDER P OR U WASTE CODES FOR PROPERTY DATA. | | | | | | | Wastestream contains benzene, monochlorobenzene, dichlorobenzenes, and 2,4,6-trichlorophenol |
| K106 | Wastewater treatment sludge from the mercury cell process in chlorine production | | SEE SPECIFIC WASTE UNDER P OR U WASTE CODES FOR PROPERTY DATA. | | | | | | | Wastestream contains mercury |

[a]Presented only if required for other calculations.    [a]Vapor pressure @ 25°C    estimated via extrapolation from experimental value.

## Footnote to Appendix A

## References used for Physical-Chemical Property Data

1. Handbook of Chemistry and Physics. 1971-1983. RC Weast (ed). Cleveland, OH: The Chemical Rubber Company.

2. The Merck Index. 1976. M. Windholz (ed). Rahway, NJ: Merck and Company, Inc.

3. Verschueren K. 1977. Handbook of Environmental Data on Organic Chemicals. New York, NY: Van Nostrand Reinhold Company.

4. Kirk-Othmer. 1978. Encyclopedia of Chemical Technology. 3rd Edition. New York, NY: Wiley Interscience.

5. Boublik T, Freid V, Hala E. 1973. The vapour pressures of pure substances. New York, NY: Elsevier Publishing Company.

6. Hansch C, Leo A. 1979. Substituent constants for correlation analysis in chemistry and biology. New York, NY: Wiley Interscience.

7. Aldrich Catalog/Handbook of Fine Chemicals. 1982-1983. Milwaukee, Wisconsin: Aldrich Chemical Company.

8. The Pesticide Manual. 1979. CR Worthing (ed). Croydon, England: British Crop. Protection Council.

9. Farm Chemicals Handbook. 1983. Willoughby, Ohio: Meister Publishing Company.

10. Handbook of Chemistry. 1949. Lange NA and G Forker, (eds). 7th Edition. Sandusky, Ohio: Hanbook Publishers, Inc.

11. Versar Inc. 1982. Exposure Assessment for DEHP. Washington, DC: U.S. Environmental Protection Agency, Office of Toxic Substances. EPA Contract No. 68-01-6271.

12. Mellor JW. 1946. Comprehensive treatise on inorganic chemistry. London: Longmans Green & Company.

13. Hollifield HC. 1979. Rapid nephalometric estimate of water solubility of highly insoluble organic chemicals of environmental interest. Bull. Environ. Contam. Toxicol. 23:579-586.

14. Wilson G, Deal C. 1962. Activity coefficients and molecular structure. Ind. Eng. Chem. Fundam. 1:20-23.

15. Umweltbundesant. 1981. OECD Hazard Assessment Project - Collection of minimum premarketing sets of data including environmental residue data of existing chemicals, Berlin: Umweltbundesant.

16. Kenaga EE, Goring C. 1978. Relationship between water solubility, soil adsorption, octanol-water partitioning and concentration of chemicals in biota. ASTM Philadelphia, PA: ASTM Spec. Tech. Pub. 707:78-109.

17. Briggs GG. 1981. Theoretical and experimental relationship between soil adsorption, octanol-water partitioning coefficients, water solubilities, bioconcentration factors, and the parachor. J. Agric. Food Chem. 29:1050-1059.

18. Gile JD, Billett JW. 1981. Transport and fate of organic phosphate insecticides in a laboratory model ecosystem. J. Agric. Food Chem. 129:616-621.

19. USEPA. 1980. Ambient Water Quality Criteria for chloroalkly ethers. Washington, DC: Office of Water Regulations and Standards, U.S. Environmental Protection Agency. EPA 440/5-80-030.

20. International Agency for Research in Cancer. 1978. IARC monographs the evaluation of the carcinogenic risk of chemicals to humans, Vol. 17. Some n-nitroso compounds, IARC. Lyon, France.

21. Stefan H, Stefan T. 1965. Solubilities of inorganic and organic products, Volume 1 - Binary Systems. New York, NY: Pergamon Press.

22. McAulifte C. 1966. Solubility in water of hydrocarbons. Journal of Physical Chemistry 70(4). 1267-77.

23. Shen T. 1982. J. Air Pollut. Control Fed. 32(1):79-82

24. Mackay D, Walkaff A. 1973. Rate of evaporation of low solubility contaminants from water bodies to atmosphere. Envir. Sci. and Technol. 7(1). 611-613.

25. Linke WF. 1958. Solubilities of inorganic and metal organic compounds. Washington, DC: American Chemical Society.

26. Timmermans J. 1960. The physio-chemical constants of binary systems in concentrated solutions. New York, NY: Interscience Publishing Company.

27. Hawley GG. 1977. The condensed chemical dictionary, 9th edition. New York, NY: Van Nostrand Publishing Company.

28. Dilling WL. 1977. Environmental Science and Technology. 11(4):405-409.

29. MacKay D, PJ Leinonen. 1975. Rate of evaporation of low-solubility contaminants from water bodies to atmosphere. Environmental Science and Technology. 9(13):1178-1180.

30. MacKay D, WY Shiu. 1981. A critical review of Henry's law constants for chemicals of environmental interest. J. Phys. Chem. Ref. Data. 10(4).

31. Callahan MA, Slimak MW, Gabel NW, et al. 1979. Water-related environmental fate of 129 priority pollutants. Volumes I and II. Washington, DC: U.S. Environmental Protection Agency, Office of Water Planning and Standards. EPA 440/4-79-029a and b.

# Appendix B: RCRA Waste Categorization Based on Vapor Pressure (Volatility of Pure Substances)

| Waste Name | Ambient Vapor Presssure (torr) @ 25°C | Waste Code |
|---|---|---|

Highly Volatile Wastes - Those with ambient vapor pressures above 10 torr

| | | |
|---|---|---|
| Carbonyl fluoride | permanent gas @ 25°C | U033 |
| Fluorine | permanent gas @ 25°C | P056 |
| Phosphine | permanent gas @ 25°C | P096 |
| Nitric oxide | permanent gas @ 25°C | P076 |
| Hydrogen sulfide | 15,200 | U135 |
| Bromomethane | 5,300 | U029 |
| Dichlorodifluoromethane | 4,830 | U075 |
| Formaldehyde | 4,433 | U122 |
| Ethene, chloro | 2,660 | U043 |
| Dimethylamine | 1,596 | U092 |
| Methanethiol | 1,520 | U153 |
| Ethylene oxide | 1,294 | U115 |
| Phosgene | 1,215 | P095 |
| Cyanogen chloride | 1,000 | P033 |
| Methyl ethyl ketone peroxide | 944 (est) | U160 |
| Acetaldehyde | 922.6 (est) | U001 |
| 2-Methylaziridine | 920 | P067 |
| Hydrofluoric acid | 800 | U134 |
| Nitrogen dioxide | 800 | P078 |
| Trichloromonofluoromethane | 768 | U121 |
| Nitrogen (IV) oxide | 760 | P078 |
| Hydrofluoric acid | 760 | U134 |
| Hydrogen cyanide | 730 | P063 |
| Furan | 634 | U124 |
| 1,1-Dichloroethylene | 630.1 | U078 |
| Ethyl ether | 439.8 | U117 |
| Methylene chloride | 427.8 | U080 |
| 1,3-Pentadiene | 414 | U186 |
| Nickel carbonyl | 400 | P076 |
| Methyl iodide | 400 | U138 |
| Carbon disulfide | 357 | P022 |
| Methylene bromide | 340 | U068 |
| Acetyl chloride | 317 | U006 |
| Chloroacetaldehyde | 317 | P023 |
| N-Nitrosopyrrolidine | 284 | U180 |
| 1-Propanamine | 280 | U194 |
| Acrolein | 220 | P003 |
| 1,2-Dichloroethylene | 217 | U079 |
| Chloromethyl methyl ether | 214 | U046 |
| Methyl isocyanate | 201 | P064 |
| Acetone | 200.1 | U002 |
| Ethylenimine | 200 | P054 |
| 1,1-Dichloroethane | 193.4 | U076 |

Appendix B.   (Continued)

| Waste Name | Ambient Vapor Presssure (torr) @ 25°C | Waste Code |
|---|---|---|

Highly Volatile Wastes - (continued)

| Waste Name | Ambient Vapor Presssure (torr) @ 25°C | Waste Code |
|---|---|---|
| Chloroform | 172.1 | U044 |
| 1,1-Dimethylhydrazine | 157 | U098 |
| Tetrahydrofuran | 149 | U213 |
| Tetrachloromethane | 115.3 | U211 |
| Methanol | 113.9 | U154 |
| Methylchlorocarbonate | 113 | U156 |
| Acetonitrile | 100 | U003 |
| Cyanogen bromide | 100 | U246 |
| 1,1,1-Trichloroethane | 99.7 | U226 |
| Acrylonitrile | 97.2 | U009 |
| Ethyl acetate | 82.2 | U112 |
| 2-Butanone | 77.5 | U159 |
| Benzene | 76 | U019 |
| 1,2-Dichloroethane | 75.7 | U077 |
| 1,1,2-Trichloroethane | 75 | U227 |
| Trichloroethylene | 70 | U228 |
| N-Methyl-N'-nitro-N-nitroso-guanidine | 69.1 | U163 |
| 1,2-Dimethylhydrazine | 68 | U099 |
| 1,2-Dichloropropane | 50 | U083 |
| Methylhydrazine | 49.6 | P068 |
| Glycidylaldehyde | 42.6 | U126 |
| Propanenitrile | 40 | P101 |
| Ethyl acrylate | 40 | U113 |
| 1,4-Dioxane | 37 | U152 |
| 1,4-Diethylenedioxide | 37 | U108 |
| Trichloroacetaldehyde | 35 | U034 |
| N-Nitroso-N-methyl urea | 33.5 | U177 |
| Formic acid | 33.4 | U123 |
| Methyl methacrylate | 32 | U162 |
| N-nitroso-N-methyl urethane | 31.2 | U178 |
| Dipropylarsine | 30 | U110 |
| Diethylarsine | 30 | P038 |
| Bis(chloromethyl) ether | 30 | P016 |
| Allyl alcohol | 28.1 | P005 |
| 1,3-Dichloropropane | 28 | U084 |

Appendix B.    (Continued)

| Waste Name | Ambient Vapor Presssure (torr) @ 25°C | Waste Code |
|---|---|---|
| Highly Volatile Wastes - (continued) | | |
| Toluene | 26.8 | U220 |
| 4-Pyridinamine | 25.8 | P008 |
| Paraldehyde | 25.3 | U182 |
| Pyridine | 20 | U196 |
| Ethylmethacrylate | 19 | U118 |
| Crotonaldehyde | 19 | U053 |
| N-nitroso-N-ethylurea | 16.3 | U176 |
| N-Nitrosomethyl vinyl amine | 16 | P084 |
| Hydrazine | 14.38 | U133 |
| N-nitrosodiethanolamine | 14.2 | U173 |
| Tetranitromethane | 13 | P112 |
| 2-Nitropropane | 13 | U171 |
| 1-Chloro-2,3-epoxypropane | 12 | U041 |
| Chlorobenzene | 11.8 | U037 |
| Ethylene dibromide | 11 | U067 |
| Osmium tetroxide | 10 | P087 |
| Isobutanol | 10 | U140 |
| 4-Methyl-2-pentanone | 10 | U161 |
| 2-Picoline | 10 | U191 |
| 1,1,1,2-Tetrachloroethane | 10 | U209 |
| 2-Chloroethyl vinyl ether | 10 | U042 |
| Tetrachloroethylene | 10 | U210 |
| Moderately Volatile Wastes - Those with ambient vapor pressures in the $10^{-3}$ to 10 torr range | | |
| 2-Propyn-1-ol | 9.56 | P012 |
| Bromoacetone | 8.0 | P017 |
| 1,2:3,4-Diepoxybutane | 7.52 | U085 |
| Cyclohexane | 6.82 | U056 |
| o-xylene | 6.60 | U239 |
| n-Butanol | 6.50 | U031 |
| Trichloromethanethiol | 5.78 | P118 |
| Bromoform | 5.60 | U225 |
| Chloromethane | 5.0 | U045 |
| Dimethylnitrosamine | 4.87 | P082 |
| Cyclohexanone | 4.57 | U057 |
| Cumene | 4.50 | U055 |
| Acrylic acid | 4.24 | U008 |

Appendix B.  (Continued)

| Waste Name | Ambient Vapor Presssure (torr) @ 25°C | Waste Code |
|---|---|---|

Moderately Volatile Wastes -  (continued)

| Waste Name | Ambient Vapor Presssure (torr) @ 25°C | Waste Code |
|---|---|---|
| N-Nitroso-di-n-butylamine | 4.11 | U172 |
| 1,4-Dichloro-2-Butene | 4.00 | U074 |
| Pentachloroethane | 3.50 | U184 |
| m-xylene | 3.20 | U239 |
| p-xylene | 3.15 | U239 |
| Furfural | 3.0 | U125 |
| Benzene, 1,3,5-trinitro | 2.86 | U234 |
| Dichlorophenylarsine | 2.70 | P036 |
| Dimethylcarbamoyl chloride | 2.60 | U097 |
| m-Dichlorobenzene | 2.10 | U071 |
| Methapyrilene | 1.89 | U155 |
| N-Nitrosodiethylamine | 1.73 | U174 |
| 2,3,4,6-Tetrachlorophenol | 1.55 | U212 |
| o-Dichlorobenzene | 1.45 | U070 |
| Dichloroethylether | 1.40 | U025 |
| 3-Chloropropionitrile | 1.33 | P027 |
| Benzyl chloride | 1.25 | P028 |
| Benzenethiol | 1.0 | P014 |
| Benzotrichloride | 1.0 (@ 45.8°C) | U023 |
| Fluoroacetamide | 0.88 | P057 |
| Bis(2-chloroethoxy)methane | 0.88 | U024 |
| Bis(2-chloroisopropyl)ether | 0.85 | U027 |
| Aniline | 0.85 | U012 |
| p-Dichlorobenzene | 0.67 | U072 |
| Phenol | 0.62 | U188 |
| Hexachloroethane | 0.60 | U131 |
| Diisopropyl fluorophosphate | 0.58 | P043 |
| 1,2-Dibromo-3-chloropropane | 0.513 | U066 |
| Acetophenone | 0.49 | U004 |
| Cresols | 0.43 | U052 |
| 2,4-Dichlorophenoxyacetic acid | 0.40 | P035 |
| Toxaphene | 0.40 | D015 |
| Ethyl carbamate | 0.36 | U238 |
| Hexachloropropene | 0.34 | U243 |
| Ethylmethanesulfonate | 0.33 | U119 |
| Benzyl chloride | 0.30 | U017 |
| Nitrobenzene | 0.30 | U169 |
| 5-Nitro-o-toluidine | 0.28 | U181 |
| α,α-Dimethylphenethylamine | 0.28 | P046 |
| Selenium dioxide | 0.16 | U204 |
| N-Nitrosopiperidine | 0.244 | U179 |
| Ammonium picrate | 0.23 | P009 |

Appendix B.   (Continued)

| Waste Name | Ambient Vapor Presssure (torr) @ 25°C | Waste Code |
|---|---|---|
| **Moderately Volatile Wastes -**   (continued) | | |
| 1,4-Naphthalenedione | 0.20 | U166 |
| Tetraethyl lead | 0.19 | P110 |
| Hexachlorobutadiene | 0.17 | U128 |
| Uracil,5(bis-2-chloromethyl-amino-) | 0.168 | U237 |
| Malononitrile | 0.148 | U149 |
| 1-Acetyl-2-thiourea | 0.14 | P002 |
| 2,4-Dichlorophenol | 0.118 | U081 |
| 2,4-Dimethylphenol | 0.118 | U101 |
| 4-Chloro-o-toluidine | 0.105 | U049 |
| p-Benzoquinone | $9 \times 10^{-2}$ | U197 |
| Hexachlorocyclopentadiene | $8 \times 10^{-2}$ | U130 |
| Safrole | $7 \times 10^{-2}$ | U203 |
| Dinoseb | $5.8 \times 10^{-2}$ | P020 |
| 1-Naphthylamine | $5.6 \times 10^{-2}$ | U167 |
| 2-Naphthylamine | $5.6 \times 10^{-2}$ | U168 |
| Nicotine | $5.0 \times 10^{-2}$ | P075 |
| Naphthalene | $5.3 \times 10^{-2}$ | U165 |
| 2,4,5-Trichlorophenol | $5.0 \times 10^{-2}$ | U230 |
| p-Chloroaniline | $5.0 \times 10^{-2}$ | P024 |
| Benzenesulfonylchloride | $4.0 \times 10^{-2}$ | U020 |
| Acrylamide | $3.0 \times 10^{-2}$ | U007 |
| Isosafrole | $2.6 \times 10^{-2}$ | U141 |
| Methyl parathion | $2.5 \times 10^{-2}$ | P071 |
| 2-Methyl acetonitrile | $2.4 \times 10^{-2}$ | P069 |
| 4,6-Dinitro-o-cresol | $1.8 \times 10^{-2}$ | P047 |
| 2,6-Dichlorophenol | $1.7 \times 10^{-2}$ | U082 |
| Tetraethyldithiopyrophosphate | $1.5 \times 10^{-2}$ | P0109 |
| Dimethyl phthalate | $1.0 \times 10^{-2}$ | U102 |
| Diallate | $8.8 \times 10^{-3}$ | U062 |
| Diethylphthalate | $8.1 \times 10^{-3}$ | U088 |
| 4-Chloro-m-cresol | $5.8 \times 10^{-3}$ | U039 |
| B-Chloronaphthalene | $5.6 \times 10^{-3}$ | U)47 |
| Phenacetin | $3.2 \times 10^{-3}$ | U187 |
| o,o-Diethyl(o-pyrazinyl) phosphorothioate | $3.0 \times 10^{-3}$ | P040 |
| 2-Cyclohexyl-4,6-dinitrophenol | $2.9 \times 10^{-3}$ | U034 |
| Saccharin | $2.7 \times 10^{-3}$ | U202 |
| Pentachloronitrobenzene | $2.4 \times 10^{-3}$ | U185 |

Appendix B.   (Continued)

| Waste Name | Ambient Vapor Presssure (torr) @ 25°C | Waste Code |
|---|---|---|

**Moderately Volatile Wastes** -   (continued)

| 1,2,4,5-Tetrachlorobenzene | $2.1 \times 10^{-3}$ | U207 |
| Toluenediamine | $1.8 \times 10^{-3}$ | U221 |
| N-Phenyl thiourea | $1.6 \times 10^{-3}$ | P093 |
| p-Nitroaniline | $1.5 \times 10^{-3}$ | P077 |
| 5-Aminomethyl-3-isoxazolol | $1.4 \times 10^{-3}$ | P007 |
| Mercury | $1.3 \times 10^{-3}$ | U151 |
| o-Toluidine hydrochloride | $1.2 \times 10^{-3}$ | U222 |
| Octamethylpyrophosphoramide | $1.0 \times 10^{-3}$ | P085 |

**Slightly Volatile Wastes** -   Those with ambient vapor pressures in the $10^{-5}$ to $10^{-3}$ torr range

| Reserpine | $<10^{-3}$ | U200 |
| Tris(2,3-dibromopropyl) phosphate | $<10^{-3}$ | U235 |
| Phorate | $8.0 \times 10^{-4}$ | P094 |
| 1,3-Propanesultone | $6.4 \times 10^{-4}$ | U193 |
| Tetraethylpyrophosphate | $4.7 \times 10^{-4}$ | P111 |
| Pentachlorobenzene | $4.1 \times 10^{-4}$ | U183 |
| Pronamide | $4.0 \times 10^{-4}$ | U192 |
| 1-(o-Chlorophenyl)thiourea | $4.0 \times 10^{-4}$ | P026 |
| 1,2-Benzenediol,4[1-hydroxy-2(methylaminoethyl)- | $4.0 \times 10^{-4}$ | P042 |
| Ethylenebis(dithiocarbamic acid) | $3.7 \times 10^{-4}$ | U114 |
| Heptachlor | $3.0 \times 10^{-4}$ | P081 |
| Nitroglycerin | $2.6 \times 10^{-4}$ | P081 |
| Phthalic anhydride | $2.0 \times 10^{-4}$ | U190 |
| Disulfoton | $1.8 \times 10^{-4}$ | P039 |
| Thiofanax | $1.7 \times 10^{-4}$ | P045 |
| 2,4-D | $1.6 \times 10^{-4}$ | D016 |
| Aldrin | $1.4 \times 10^{-4}$ | P060 |
| Azaserine | $1.3 \times 10^{-4}$ | U015 |
| Streptozotocin | $1.3 \times 10^{-4}$ | U206 |
| Pentachlorophenol | $1.1 \times 10^{-4}$ | U242 |
| ALdicarb | $1.0 \times 10^{-4}$ | P070 |
| Diethyl-p-nitrophenylphosphate | $9.8 \times 10^{-5}$ | P041 |
| 2,4-Dinitrotoluene | $8.5 \times 10^{-5}$ | U105 |
| 1,2-Diphenyl hydrazine | $5.2 \times 10^{-5}$ | U109 |
| Maleic anhydride | $5.0 \times 10^{-5}$ | U147 |

Appendix B. (Continued)

| Waste Name | Ambient Vapor Presssure (torr) @ 25°C | Waste Code |
|---|---|---|
| **Slightly Volatile Wastes** - (continued) | | |
| Methomyl | $5.0 \times 10^{-5}$ | P066 |
| Methylthiouracil | $4.8 \times 10^{-5}$ | U164 |
| Parathion | $3.8 \times 10^{-5}$ | P089 |
| 2,4,5-T | $2.4 \times 10^{-5}$ | U232 |
| Bis(2-ethylhexyl)phthalate | $1.4 \times 10^{-5}$ | U028 |
| Resorcinol | $1.2 \times 10^{-5}$ | U201 |
| Hexachlorobenzene | $1.1 \times 10^{-5}$ | U127 |
| Chlordane | $1.0 \times 10^{-5}$ | U036 |
| Endosulfan | $1.0 \times 10^{-5}$ | P050 |
| Benzidine | $1.0 \times 10^{-5}$ | U021 |
| Lindane | $9.4 \times 10^{-6}$ | D013 |
| **Nonvolatile Wastes** - Those with ambient vapor pressures below $10^{-5}$ torr range | | |
| Phenylmercuric acetate | $9.0 \times 10^{-6}$ | P092 |
| Dimethoate | $8.5 \times 10^{-6}$ | P044 |
| Octachlorocamphene | $8.1 \times 10^{-6}$ | P123 |
| Silvex | $7.1 \times 10^{-6}$ | U233 |
| Aldrin | $6.0 \times 10^{-6}$ | P004 |
| 4,4'-Methylene bis(2-chloro-aniline) | $6.0 \times 10^{-6}$ | U158 |
| 3-Methylcholanthrene | $3.8 \times 10^{-6}$ | U157 |
| Hexachlorocyclohexane | $3.3 \times 10^{-6}$ | U129 |
| Cyclophosphamide | $2.8 \times 10^{-6}$ | U058 |
| Ethyl,4,4'-Dichlorobenzilate | $2.2 \times 10^{-6}$ | U038 |
| Dieldrin | $7.8 \times 10^{-7}$ | P037 |
| α-Naphthylthiourea | $5.1 \times 10^{-7}$ | P072 |
| 3,3-Dimethylbenzidine | $\underline{\ \ }.9 \times 10^{-7}$ | U095 |
| Beryllium dust | $2.7 \times 10^{-7}$ | P015 |
| 2-Acetyl-1-aminofluorene | $2.1 \times 10^{-7}$ | U005 |
| Endrin | $2.0 \times 10^{-7}$ | P051 |
| 3,3'-Dimethoxybenzidine | $1.9 \times 10^{-7}$ | U091 |
| DDD | $1.5 \times 10^{-7}$ | U060 |
| DDT | $1.5 \times 10^{-7}$ | U061 |
| Chlorambucil | $1.4 \times 10^{-7}$ | U035 |
| 3,3'-Dichlorobenzidine | $1.1 \times 10^{-7}$ | U073 |
| 2,4-Dichlorophenoxyacetic acid | $1.0 \times 10^{-7}$ | U240 |
| Arsenic (III) oxide | $8.3 \times 10^{-8}$ | P012 |

Appendix B.  (Continued)

| Waste Name | Ambient Vapor Presssure (torr) @ 25°C | Waste Code |
|---|---|---|
| **Nonvolatile Wastes -** (continued) | | |
| Di-n-octylphthalate | $6.8 \times 10^{-8}$ | U107 |
| Mytomycin C | $4.1 \times 10^{-8}$ | U010 |
| Melphalan | $4.0 \times 10^{-8}$ | U150 |
| 3,4-Benzacridine | $8.0 \times 10^{-10}$ | U016 |
| Benzo[a]pyrene | $1.8 \times 10^{-10}$ | U022 |
| 1,2-Benzanthracene | $1.0 \times 10^{-10}$ | U018 |
| Dibenz[a,h]anthracene | $5.2 \times 10^{-11}$ | U063 |
| Strychnine | $1.2 \times 10^{-11}$ | P108 |
| Warfarin | $1.0 \times 10^{-11}$ | P001 |
| 7,12-Diemthylbenz[a,h]anthracene | $3.9 \times 10^{-12}$ | U094 |
| Diethylstilbesterol | $2.1 \times 10^{-12}$ | U089 |
| Hexachlorophene | $1.0 \times 10^{-12}$ | U064 |
| Brucine | $1.0 \times 10^{-12}$ | P018 |
| Chrysene | $1.0 \times 10^{-13}$ | U050 |
| Daunomycin | $9.4 \times 10^{-15}$ | U059 |
| 4-Bromophenylphenylether | $4.1 \times 10^{-20}$ | U030 |
| α,α-Dimethylbenzyl-hydroperoxide | $1.1 \times 10^{-24}$ | U096 |

**Compounds with Negligible (but undetermined) Vapor Pressures**

| | Waste Code | Evidence |
|---|---|---|
| Thallium carbonate | U215 | Melting point 273°C, decomposes to $Tl_2O$, the product boils at 1080°C. |
| Thallium sulfate | P115 | Melts at 632°C. |
| Lead phosphate | U145 | Melting point at 1014°C. |
| Lead subacetate | U146 | Melting point at 475°C. |
| Thallium nitrate | U217 | Decomposes at 800°C. |
| Strontium sulfide | P107 | Melting point at 2000°C. |
| Zinc phosphide | P122 | Melting point at 1100°C. |
| Thallium (I) chloride | U216 | Vapor pressure of 10 torr at 507°C. |

Appendix B.  (Continued)

| Waste Name | Waste Code | Evidence |
|---|---|---|

Compounds with Negligible (but undetermined) Vapor Pressures - (continued)

| Waste Name | Waste Code | Evidence |
|---|---|---|
| Arsenic | D004 | Vapor pressure of 1 torr at 372°C |
| Asbestos | U013 | Melting point above 1000°C. |
| Thallium selenite | P114 | Melting point over 400°C. |
| Lead acetate | U144 | Decomposes to the oxide above 280°C. |
| Calcium cyanide | U032 | Decomposes to nonvolatile oxides at 200°C. |
| Sodium cyanide | P106 | Vapor pressure 150 torr at 800°C. |
| Aluminum phosphide | P006 | Melting point at 1350°C. |
| Copper cyanides | P029 | Melting point at 473°C. |
| Zinc cyanide | P121 | Melting point at 800°C. |
| Potassium cyanide | P098 | Melting point at 634°C. |
| Vanadium pentoxide | P120 | Decomposes at 1750°C, melting point at 690°C. |
| Thallic oxide | P113 | Melting point at 717°C |
| Ammonium vanadate | P119 | Melting point at 200°C |
| Silver | D011 | Boiling point at 2000°C |
| Barium | D005 | Boiling point at 1600°C |
| Cadmium | D006 | Boiling point at 765°C |
| Chromium | D007 | Boiling point at 2672°C |
| Lead | D008 | Boiling point at 1740°C |
| Selenium | D010 | Boiling point at 685°C |

# Appendix C: RCRA Waste Catagorization Based on Aqueous Volatility (Henry's Constant)

| Waste Name | Henry's Constant (atm-m$^3$/mole) | Waste Code |
|---|---|---|
| **Highly Volatile Wastes** - Values of H above 10$^{-3}$ | | |
| Bis(2-ethyl hexyl phthalate) | 26.6 | U028 |
| Cyanogen | 9.91 | P031 |
| Reserpine | 4.28 | U200 |
| Nickel carbonyl | 0.5 | P073 |
| Dichlorodifluoromethane | 0.415 | U075 |
| Chloromethane | 0.38 | U045 |
| Chloroethene | 0.199 | U043 |
| Phosphine | 0.19 | P096 |
| Cyclohexane | 0.18 | U056 |
| 2-Nitropropane | 0.12 | U171 |
| Trichloromonofluoromethane | 5.8 x 10$^{-2}$ | U121 |
| 2,3,4,6-Tetrachlorophenol | 4.5 x 10$^{-2}$ | U212 |
| 1,3-Pentadiene | 4.2 x 10$^{-2}$ | U186 |
| Pentachloronitrobenzene | 2.9 x 10$^{-2}$ | U185 |
| Tetrachloroethylene | 2.87 x 10$^{-2}$ | U210 |
| Hexachloropropene | 2.5 x 10$^{-2}$ | U243 |
| Tetrachloromethane | 2.13 x 10$^{-2}$ | U211 |
| Hexachlorocyclopentadiene | 1.60 x 10$^{-2}$ | U130 |
| 1,1-Dichloroethylene | 1.50 x 10$^{-2}$ | U078 |
| Cumene | 1.40 x 10$^{-2}$ | U055 |
| DDD | 1.26 x 10$^{-2}$ | U060 |
| Carbon disulfide | 1.2 x 10$^{-2}$ | P022 |
| Mercury | 1.14 x 10$^{-2}$ | D009 |
| Hexachloroethane | 9.85 x 10$^{-3}$ | U131 |
| Hexachlorobutadiene | 9.14 x 10$^{-3}$ | U128 |
| Trichloroethylene | 8.92 x 10$^{-3}$ | U228 |
| 3-Methylcholanthrene | 7.7 x 10$^{-3}$ | U157 |
| 1,2-Dichloroethylene (CIS) | 6.6 x 10$^{-3}$ | U079 |
| Toluene | 6.64 x 10$^{-3}$ | U220 |
| Furan | 5.7 x 10$^{-3}$ | U124 |
| Benzene | 5.55 x 10$^{-3}$ | U019 |
| 1,1-Dichloroethane | 5.45 x 10$^{-3}$ | U076 |
| 1,2-Dichloroethylene (trans) | 5.32 x 10$^{-3}$ | U079 |
| o-Xylene | 5.27 x 10$^{-3}$ | U239 |
| Bromomethane | 5.26 x 10$^{-3}$ | U029 |
| Methyl iodide | 5.0 x 10$^{-3}$ | U138 |
| 1,1,1-Trichloroethane | 4.92 x 10$^{-3}$ | U226 |
| Toxaphene | 4.89 x 10$^{-3}$ | U224 |
| Methanethiol | 4.0 x 10$^{-3}$ | U153 |
| Chlorobenzene | 3.93 x 10$^{-3}$ | U037 |

Appendix C.   (Continued)

| Waste Name | Henry's Constant $(atm-m^3/mole)$ | Waste Code |
|---|---|---|

**Highly Volatile Wastes** - (continued)

| Waste Name | Henry's Constant | Waste Code |
|---|---|---|
| Chloroform | $3.39 \times 10^{-3}$ | U044 |
| Cyanogen chloride | $3.2 \times 10^{-3}$ | P033 |
| Methylene chloride | $3.19 \times 10^{-3}$ | U080 |
| 1,2-Dichloropropane | $2.8 \times 10^{-3}$ | U083 |
| 1,1,1,2-Tetrachloroethane | $2.76 \times 10^{-3}$ | U208 |
| 4-Bromopropylphenylether | $2.74 \times 10^{-3}$ | U030 |
| m-Dichlorobenzene | $2.63 \times 10^{-3}$ | U071 |
| m-Xylene | $2.55 \times 10^{-3}$ | U239 |
| p-Xylene | $2.51 \times 10^{-3}$ | U239 |
| Hexachlorohexahydro-exo,exo-<br>dimethanonaphthalene | $2.49 \times 10^{-3}$ | P060 |
| p-Dichlorobenzene | $2.37 \times 10^{-3}$ | U072 |
| Benzene, 1,3,5-trinitro | $2.3 \times 10^{-3}$ | U234 |
| Pentachloroethane | $2.17 \times 10^{-3}$ | U184 |
| Octachlorocamphene | $2.01 \times 10^{-3}$ | P123 |
| o-Dichlorobenzene | $1.94 \times 10^{-3}$ | U070 |
| Dimethylcarbamoylchloride | $1.8 \times 10^{-3}$ | P097 |
| 1,3-Dichloropropane | $1.77 \times 10^{-3}$ | U084 |
| Hexachlorobenzene | $1.7 \times 10^{-3}$ | U127 |
| Heptachlor | $1.48 \times 10^{-3}$ | P059 |
| Pentachlorobenzene | $1.3 \times 10^{-3}$ | U183 |
| 1,1,2-Trichloroethane | $1.18 \times 10^{-3}$ | U227 |
| 1,2-Dichloroethane | $1.10 \times 10^{-3}$ | U077 |

**Moderately Volatile Wastes** - Values of H below $10^{-3}$ to $10^{-5}$

| Waste Name | Henry's Constant | Waste Code |
|---|---|---|
| Ethyl ether | $8.69 \times 10^{-4}$ | U117 |
| 2-Cyclohexyl,4,6-dinitrophenol | $8.37 \times 10^{-4}$ | P034 |
| N-Nitroso-di-n-butylamine | $7.9 \times 10^{-4}$ | U172 |
| 2,6-Dinitrotoluene | $7.42 \times 10^{-4}$ | U106 |
| 2-Chloroethyl vinyl ether | $7.35 \times 10^{-4}$ | U042 |
| Ethylene dibromide | $6.25 \times 10^{-4}$ | U067 |
| Bromoform | $5.32 \times 10^{-4}$ | U225 |
| Aldrin | $4.96 \times 10^{-4}$ | P004 |
| Naphthalene | $4.8 \times 10^{-4}$ | U165 |
| 1,1,2,2-Tetrachloroethane | $4.7 \times 10^{-4}$ | U209 |
| N-nitroso-N-methyl urethane | $4.17 \times 10^{-4}$ | U178 |
| Dipropylamine | $3.32 \times 10^{-4}$ | U110 |
| Methylene bromide | $3.16 \times 10^{-4}$ | U068 |
| β-Chloronapthalene | $3.15 \times 10^{-4}$ | U047 |
| Methyl methacrylate | $3.11 \times 10^{-4}$ | U162 |

Appendix C.  (Continued)

| Waste Name | Henry's Constant (atm-m$^3$/mole) | Waste Code |
|---|---|---|

Moderately Volatile Wastes - (continued)

| Waste Name | Henry's Constant (atm-m$^3$/mole) | Waste Code |
|---|---|---|
| Benzenethiol | $3.10 \times 10^{-4}$ | P014 |
| Formaldehyde | $2.92 \times 10^{-4}$ | U122 |
| Ethyl acrylate | $2.71 \times 10^{-4}$ | U113 |
| Bischloromethylether | $2.50 \times 10^{-4}$ | P016 |
| Benzylchloride | $2.36 \times 10^{-4}$ | P028 |
| Dihydrosafrole | $2.30 \times 10^{-4}$ | U090 |
| Diallate | $1.99 \times 10^{-4}$ | U062 |
| Dinoseb | $1.82 \times 10^{-4}$ | P020 |
| Benzal chloride | $1.70 \times 10^{-4}$ | U017 |
| 5-Nitro-o-toluidine | $1.67 \times 10^{-4}$ | U181 |
| 1,2-Dibromo-3-Chloropropane | $1.59 \times 10^{-4}$ | U066 |
| Ethyl Methacrylate | $1.49 \times 10^{-4}$ | U118 |
| Tris(2,3-dibromopropyl)phosphate | $1.46 \times 10^{-4}$ | U235 |
| 4-Methyl-2-pentanone | $1.32 \times 10^{-4}$ | U161 |
| Ethyl acrylate | $1.20 \times 10^{-4}$ | U112 |
| Benzotrichloride | $1.12 \times 10^{-4}$ | U023 |
| Tetrahydrofuran | $1.08 \times 10^{-4}$ | U213 |
| Trichloromethylmercaptan | $1.04 \times 10^{-4}$ | P118 |
| Bis-2-chloroisopropyl ether | $1.03 \times 10^{-4}$ | U027 |
| 1,2,4,5-Tetrachlorobenzene | $1.0 \times 10^{-4}$ | U207 |
| N-Nitrosopiperidine | $9.78 \times 10^{-5}$ | U179 |
| Acrylonitrile | $9.2 \times 10^{-5}$ | U009 |
| Methapyriline | $7.6 \times 10^{-5}$ | U135 |
| Acrolein | $6.79 \times 10^{-5}$ | P003 |
| 1,4-Dichloro-2-butene | $6.78 \times 10^{-5}$ | U074 |
| Trichloroacetaldehyde | $6.77 \times 10^{-4}$ | U034 |
| Osmium tetroxide | $5.86 \times 10^{-5}$ | P087 |
| N-Nitrosomethylvinylamine | $5.65 \times 10^{-5}$ | P084 |
| Kepone | $5.6 \times 10^{-5}$ | U142 |
| DDT | $5.2 \times 10^{-5}$ | U061 |
| Nitroglycerin | $5.18 \times 10^{-5}$ | P081 |
| Chlordane | $4.8 \times 10^{-5}$ | U036 |
| Diethylphthalate | $4.75 \times 10^{-5}$ | U088 |
| 1-Chloro-2,3-epoxypropane | $3.8 \times 10^{-5}$ | U041 |
| Paraldehyde | $3.66 \times 10^{-5}$ | U182 |
| Ethylene oxide | $3.63 \times 10^{-5}$ | U115 |
| Dichloroethylether | $2.58 \times 10^{-5}$ | U025 |
| Cyclohexanone | $2.56 \times 10^{-5}$ | U057 |
| Endosulfan | $2.5 \times 10^{-5}$ | P050 |
| Propanenitrile | $2.4 \times 10^{-5}$ | P101 |
| 2-Butanone | $2.4 \times 10^{-5}$ | U159 |

Appendix C.  (Continued)

| Waste Name | Henry's Constant (atm-m$^3$/mole) | Waste Code |
|---|---|---|
| **Moderately Volatile Wastes** - (continued) | | |
| 2-Picoline | $2.4 \times 10^{-5}$ | U191 |
| Nitrobenzene | $2.4 \times 10^{-5}$ | U169 |
| Methyl aziridine | $2.22 \times 10^{-5}$ | P067 |
| Hydrofluoric acid | $2.0 \times 10^{-5}$ | U134 |
| 1-Propanamine | $2.0 \times 10^{-5}$ | U194 |
| 2,6-Dichlorophenol | $2.0 \times 10^{-5}$ | U082 |
| α,α-Dimethylphenethylamine | $1.86 \times 10^{-5}$ | P046 |
| Acetophenone | $1.41 \times 10^{-5}$ | U004 |
| 2,4-Dimethylphenol | $1.18 \times 10^{-5}$ | U101 |
| Crotonaldehyde | $1.13 \times 10^{-5}$ | U053 |
| N-Nitrosopyrrolidine | $1.13 \times 10^{-5}$ | U180 |
| Dieldrin | $1.1 \times 10^{-5}$ | P037 |
| Ethylenimine | $1.1 \times 10^{-5}$ | P054 |
| Safrole | $1.08 \times 10^{-5}$ | U203 |
| Isobutanol | $1.03 \times 10^{-5}$ | U140 |
| 4-Chloro-o-toluidine | $1.02 \times 10^{-5}$ | U049 |
| **Slightly Volatile Wastes** -  Values of H from $10^{-5}$ to $10^{-7}$ | | |
| Chloromethyl methylether | $9.12 \times 10^{-6}$ | U046 |
| Diisopropylfluorophosphate | $9.1 \times 10^{-6}$ | P043 |
| Pronamide | $9.0 \times 10^{-6}$ | U192 |
| oo-Diethyl-o-pyrazinyl-phosphorothioate | $8.58 \times 10^{-6}$ | P040 |
| α-Toluidine hydrochloride | $7.55 \times 10^{-6}$ | U222 |
| Ammoniun Picrate | $7.4 \times 10^{-6}$ | P009 |
| Di-n-propylnitrosamine | $7.2 \times 10^{-6}$ | U111 |
| n-Butanol | $7.0 \times 10^{-6}$ | U031 |
| Acetone | $6.8 \times 10^{-6}$ | U002 |
| Tetranitromethane | $6.33 \times 10^{-6}$ | P112 |
| 2,4,5-Trichlorophenol | $6.0 \times 10^{-6}$ | U230 |
| 2,4-Dichlorophenol | $5.62 \times 10^{-6}$ | U081 |
| Phorate | $5.47 \times 10^{-6}$ | P094 |
| N-Nitroso-N-ethylurea | $5.4 \times 10^{-6}$ | U176 |
| 2-Naphthalamine | $5.4 \times 10^{-6}$ | U168 |
| 2,4,6-Trichlorophenol | $4.82 \times 10^{-6}$ | U231 |
| Chloroacetaldehyde | $4.7 \times 10^{-6}$ | P023 |
| o-Chlorophenol | $4.7 \times 10^{-6}$ | U048 |
| 4-Pyridinamine | $4.4 \times 10^{-6}$ | P008 |
| Endrin | $4.2 \times 10^{-6}$ | P051 |

Appendix C.  (Continued)

| Waste Name | Henry's Constant $(atm-m^3/mole)$ | Waste Code |
|---|---|---|
| **Slightly Volatile Wastes -**  (continued) | | |
| 1-Naphthalamine | $4.1 \times 10^{-6}$ | U167 |
| Isosafrole | $4.08 \times 10^{-6}$ | U141 |
| Furfural | $3.6 \times 10^{-6}$ | U125 |
| 1,4-Naphthalenedione | $3.6 \times 10^{-6}$ | U166 |
| 3-Chloropropionitrile | $3.5 \times 10^{-6}$ | P027 |
| Allyl alcohol | $3.47 \times 10^{-6}$ | P005 |
| Dimethyl sulfate | $3.37 \times 10^{-6}$ | U103 |
| Aniline | $3.07 \times 10^{-6}$ | U012 |
| p-Chloroaniline | $3.0 \times 10^{-6}$ | P024 |
| Methyl hydrazine | $3.0 \times 10^{-6}$ | P068 |
| Disulfoton | $2.59 \times 10^{-6}$ | P039 |
| Acetonitrile | $2.47 \times 10^{-6}$ | U003 |
| N-Nitroso-N-methylurea | $2.20 \times 10^{-6}$ | U177 |
| Cresols | $2.0 \times 10^{-6}$ | U052 |
| Phenacetin | $1.4 \times 10^{-6}$ | U187 |
| 4,6-Dinitro-o-cresol | $1.4 \times 10^{-6}$ | P047 |
| Phenol | $1.3 \times 10^{-6}$ | U188 |
| Parathion | $1.21 \times 10^{-6}$ | P089 |
| Methanol | $1.1 \times 10^{-6}$ | U154 |
| Dibutylphthalate | $1.09 \times 10^{-6}$ | U069 |
| p-Nitroaniline | $1.0 \times 10^{-6}$ | P077 |
| Bromoacetone | $9.9 \times 10^{-7}$ | P017 |
| Indeno [1,2,3,-cd] pyrene | $7.2 \times 10^{-7}$ | U137 |
| 1,4-Diethylenedioxide | $7.14 \times 10^{-7}$ | U108 |
| 1,4-Dioxane | $7.0 \times 10^{-7}$ | U152 |
| N-Nitrosodiethylamine | $6.0 \times 10^{-7}$ | U174 |
| Hydrazine | $6.0 \times 10^{-7}$ | U133 |
| Selenium dioxide | $6.0 \times 10^{-7}$ | U204 |
| Ethyl-4,4'-dichlorobenzilate | $5.89 \times 10^{-7}$ | U038 |
| Glycidylaldeyde | $5.80 \times 10^{-7}$ | U126 |
| p-Benzoquinone | $5.0 \times 10^{-7}$ | U197 |
| Lindane | $4.93 \times 10^{-7}$ | D013 |
| Pentachlorophenol | $4.8 \times 10^{-7}$ | U242 |
| Hydrocyanic acid | $4.65 \times 10^{-7}$ | P063 |
| Formic acid | $4.4 \times 10^{-7}$ | U123 |
| Acrylic acid | $4.0 \times 10^{-7}$ | U008 |
| α-Hexachlorocyclohexane | $3.16 \times 10^{-7}$ | U129 |
| Di-n-octylphthalate | $3.0 \times 10^{-7}$ | U107 |
| 4-Chloro-m-cresol | $2.83 \times 10^{-7}$ | U039 |
| Bis-2-chloromethoxymethane | $2.77 \times 10^{-7}$ | U024 |
| Dimethylphthalate | $2.10 \times 10^{-7}$ | U102 |

Appendix C.   (Continued)

| Waste Name | Henry's Constant (atm-m$^3$/mole) | Waste Code |
|---|---|---|

**Slightly Volatile Wastes -** (continued)

| | | |
|---|---|---|
| Methyl parathion | $1.97 \times 10^{-7}$ | P071 |
| Saccharin | $1.90 \times 10^{-7}$ | U202 |
| Methylthiouracil | $1.80 \times 10^{-7}$ | U164 |
| Nicotine | $1.6 \times 10^{-7}$ | P075 |
| N-Phenylthiourea | $1.47 \times 10^{-7}$ | P093 |
| 4,4'-Methylenebis(2-chloroaniline) | $1.40 \times 10^{-7}$ | U158 |
| Bromoacetone | $1.17 \times 10^{-7}$ | P017 |
| Malononitrile | $1.0 \times 10^{-7}$ | U149 |

**Nonvolatile Wastes -** Values of H below $10^{-7}$

| | | |
|---|---|---|
| 2,4-Dinitrotoluene | $7.6 \times 10^{-8}$ | U105 |
| 2,4-D | $7.5 \times 10^{-8}$ | P035 |
| Dimethylamine | $5.9 \times 10^{-8}$ | U092 |
| 2,4,5-T | $3.44 \times 10^{-8}$ | U232 |
| Ethylmethanesulfonate | $3.14 \times 10^{-8}$ | U119 |
| Dimethylnitrosamine | $3.0 \times 10^{-8}$ | P082 |
| Chlornaphazine | $2.8 \times 10^{-8}$ | U026 |
| 1-(α-Chlorophenylthiourea) | $2.51 \times 10^{-8}$ | P026 |
| Ethyl carbamate | $2.0 \times 10^{-8}$ | U238 |
| 1,2,7,8-Dibenzopyrene | $2.0 \times 10^{-8}$ | U064 |
| 2-Propyn-1-ol | $2.0 \times 10^{-8}$ | P102 |
| Benzidine | $1.91 \times 10^{-8}$ | U021 |
| Silvex | $1.80 \times 10^{-8}$ | U233 |
| Acrylamide | $1.49 \times 10^{-8}$ | U007 |
| 1,2:3,4-Diepoxybutene | $1.02 \times 10^{-8}$ | U085 |
| n-Nitrosodiethanolamine | $8.0 \times 10^{-9}$ | U173 |
| 3,4-Benzacridine | $7.0 \times 10^{-9}$ | U016 |
| Pyridine | $7.0 \times 10^{-9}$ | U196 |
| 1,2-Benzanthracene | $3.4 \times 10^{-9}$ | U018 |
| Diethyl-p-nitrophenylphosphate | $2.83 \times 10^{-9}$ | P041 |
| Fluoroacetamide | $2.33 \times 10^{-9}$ | P057 |
| Toluenediamine | $2.30 \times 10^{-9}$ | U221 |
| 3,3'-Dimethylbenzidine | $1.75 \times 10^{-9}$ | U095 |
| Benzo[a]pyrene | $1.38 \times 10^{-9}$ | U022 |
| 7,12-Dimethylbenz[A]anthracene | $1.03 \times 10^{-9}$ | U094 |
| 3,3'-Dichlorobenzidine | $1.0 \times 10^{-9}$ | U073 |
| Phenylmercuricacetate | $1.0 \times 10^{-9}$ | P092 |
| Thiofanox | $9.37 \times 10^{-10}$ | P045 |
| 2-Acetylaminofluorene | $4.4 \times 10^{-10}$ | U005 |
| Mitomycin C | $2.5 \times 10^{-10}$ | U101 |
| Cyclophosphamide | $2.37 \times 10^{-10}$ | U058 |

Appendix C.   (Continued)

| Waste Name | Henry's Constant (atm-m$^3$/mole) | Waste Code |
|---|---|---|

**Nonvolatile Wastes -** (continued)

| Waste Name | Henry's Constant | Waste Code |
|---|---|---|
| α-naphthylthiourea | $2.26 \times 10^{-10}$ | P072 |
| Urcil,5[Bis-2-chloromethylamino] | $1.0 \times 10^{-10}$ | U237 |
| Aldicarb | $1.0 \times 10^{-10}$ | P070 |
| Dibenz[AH]anthracene | $1.0 \times 10^{-10}$ | U063 |
| Methomyl | $1.0 \times 10^{-10}$ | P066 |
| Dimethoate | $1.0 \times 10^{-10}$ | P044 |
| Etylenebis(dithiocarbamic acid) | $1.0 \times 10^{-10}$ | U114 |
| 2-Methylacetonitrile | $1.0 \times 10^{-10}$ | P069 |
| 4-Nitrophenol | $1.0 \times 10^{-10}$ | U170 |
| Maleic anhydride | $1.0 \times 10^{-10}$ | U147 |
| Phthalic anhydride | $1.0 \times 10^{-10}$ | U190 |
| Diethylstilbesterol | $5.1 \times 10^{-11}$ | U089 |
| 2,4-D salts and esters | $3.6 \times 10^{-11}$ | U240 |
| 3,3'-Dimethoxybenzidine | $1.0 \times 10^{-11}$ | U091 |
| 1,3-Propane sultone | $1.0 \times 10^{-11}$ | U193 |
| 1,2-Diphenylhydrazine | $1.0 \times 10^{-11}$ | U109 |
| Streptozotocin | $1.0 \times 10^{-11}$ | U206 |
| Melphalan | $1.0 \times 10^{-11}$ | U150 |
| Chlorambucil | $2.1 \times 10^{-12}$ | U035 |
| Arsenic III oxide | $1.7 \times 10^{-12}$ | P012 |
| Chrysene | $1.0 \times 10^{-12}$ | U050 |
| 5-(Aminomethyl)-3-isoxazolol | $1.0 \times 10^{-13}$ | P007 |
| Resorcinol | $1.0 \times 10^{-13}$ | U201 |
| Strychnine | $1.0 \times 10^{-14}$ | P108 |
| Warfarin | $1.0 \times 10^{-15}$ | P001 |
| Brucine | $1.0 \times 10^{-18}$ | P018 |
| Hexachlorophene | $1.0 \times 10^{-18}$ | U132 |
| Daunomycin | $2.2 \times 10^{-19}$ | U059 |

**Wastes with no Henry's Constant Value Available**

| Waste Name | Waste Code |
|---|---|
| Chromium | D007 |
| Arsenic | D004 |
| Lead | D008 |
| Cadmium | D006 |
| Barium | D004 |
| Acetaldehyde | U001 |
| Selenium | D010 |
| Silver | D011 |
| Arsenic acid | P010 |
| Sodium cyanide | P106 |

Appendix C.   (Continued)

| Waste Name | Henry's Constant (atm-m$^3$/mole) | Waste Code |
|---|---|---|

Wastes with no Henry's Constant Value Available - (continued)

| | |
|---|---|
| 2,4-Dinitrophenol | P048 |
| Benzenesulfonyl chloride | U020 |
| 2,4,5TP Silvex | D017 |
| Lindane | D013 |
| Aluminum phosphide | P006 |
| Copper cyanide | P029 |
| Dichlorophenyl arsine | P036 |
| Zinc cyanide | P121 |
| Nickel cyanide | P074 |
| Potassium cyanide | P098 |
| Nitrogen dioxide | P078 |
| Tetraethyl lead | P110 |
| Tetraethyl pyrophosphate | P111 |
| Bromine cyanide | U246 |
| Bis(dimethylthiocarbamoyldisulfide) | U244 |
| Silver cyanide | P104 |
| Nitric oxide | P076 |
| Calcium chromate | U032 |
| Calcium cyanide | P021 |
| Endothall | P088 |
| Acetyl chloride | U006 |
| Dimethylaminoazobenzene | U093 |
| Auramine | U014 |
| Octamethylpyrophosphoramide | P085 |
| N,N-Diethylhydrazine | U086 |
| α,α-Diethyl-S-methyl-dithiophosphate | U087 |
| Azaserine | U015 |
| Epinephrine | P042 |
| α,α-Dimethylbenzeylhydroperoxide | U096 |
| Tetraethyldithiopyrophosphate | P109 |
| Arsenic V oxide | P011 |
| Carbonyl fluoride | U033 |
| Mercuric fulminate | P065 |
| Strontiun sulfide | P107 |
| Potassium silver cyanide | P099 |
| Hexaethyl tetraphosphate | P062 |
| Diethyl arsine | P038 |
| Thallium selenate | P114 |
| Selenourea | P103 |
| 1-Acetyl-2-Thiourea | P002 |
| Toluenediisocyanate | U223 |
| Hydrogen sulfide | U135 |

Appendix C.  (Continued)

| Waste Name | Henry's Constant (atm-m$^3$/mole) | Waste Code |
|------------|-----------------------------------|------------|

Wastes with no Henry's Constant Value Available - (continued)

| Waste Name | Waste Code |
|------------|------------|
| Trypan blue | U236 |
| Zinc phosphide | P122 |
| 1,2-Dimethylhydrazine | U099 |
| Methyl isocyanate | P064 |
| Amitrole | U011 |
| 1,1-Dimethylhydrazine | U098 |
| Iron dextran | U139 |
| Thallium I acetate | U214 |
| Thiosemicarbazide | P116 |
| Thallium I carbonate | U215 |
| Thalloas sulfate | P115 |
| Ethylene thiourea | U116 |
| Dimethylcarbamoylchloride | U097 |
| Lasiocarpine | U143 |
| N-Methyl-N'-nitro-N-nitrosoguanidine | U163 |
| Ammoniun vanadate | P119 |
| Sodium azide | P105 |
| Phosgene | P095 |
| Sulfur selenide | U205 |
| Hydroxydimethylarsineoxide | U136 |
| Phosphorus sulfide | U189 |
| Fluorine | P056 |
| Barium cyanide | P013 |
| Sodium fluoroacetate | P058 |
| Beryllium dust | P015 |
| Vanadium pentoxide | P120 |
| Methylchlorocarbonate | U156 |
| Lead phosphate | U145 |
| Maleic hydrazide | U148 |
| Lead subacetate | U146 |
| Thallium I nitrate | U217 |
| Thallium I chloride | U216 |
| Thioacetamide | U218 |
| Asbestos | U013 |
| Methylethyl ketone peroxide | U160 |
| Thiourea | U219 |
| Lead acetate | U144 |

# Appendix D: RCRA Waste Categorization Based on Relative Soil Volatility

| Waste Name | Waste Code | Relative Soil Volatility |
|---|---|---|
| Highly Volatile Wastes - Those with relative soil volatility greater than 1 | | |
| Formaldehyde | U122 | 809.4 |
| Cyanogen | P031 | 551.9 |
| Bromomethane | U029 | 544.0 |
| Dichlorodifluoromethane | U075 | 439.1 |
| Ethene, chloro | U043 | 336.5 |
| Methanethiol | U153 | 219.4 |
| Ethylene oxide | U115 | 195.1 |
| Hydrofluoric acid | U134 | 179.0 |
| Cyanogen chloride | P033 | 152.4 |
| Phosgene | P095 | 143.7 |
| Hydrocyanic acid | P063 | 140.5 |
| Acetaldehyde | U001 | 139.1 |
| Nitrogen dioxide | P078 | 118.0 |
| Furan | U124 | 76.9 |
| Trichloromonofluoromethane | U121 | 65.6 |
| 1,1-Dichloroethylene | U078 | 64.0 |
| Ethyl ether | U117 | 62.8 |
| Pentadiene | U186 | 50.2 |
| Methylene chloride | U080 | 46.4 |
| Carbon disulfide | P022 | 41.0 |
| 1-Propanamine | U194 | 36.0 |
| Chloroacetaldehyde | P023 | 35.8 |
| Acetyl chloride | U006 | 35.7 |
| Trans-1,2-dichloroethylene | U079 | 35.7 |
| Acrolein | P003 | 34.5 |
| Methyl iodide | U138 | 33.6 |
| Nickel carbonyl | P073 | 30.6 |
| Ethylenimine | P054 | 30.0 |
| N-nitrosopyrrolidine | U180 | 28.4 |
| Methyl isocyanate | P064 | 26.7 |
| Acetone | U002 | 26.3 |
| Chloromethyl methylether | U046 | 23.8 |
| Cis-1,2-dichloroethylene | U079 | 22.0 |
| 1,1-Dimethylhydrazine | U098 | 20.3 |
| Methanol | U154 | 20.1 |
| 1,1-Dichloroethane | U076 | 19.4 |
| Tetrahydrofuran | U213 | 17.6 |
| Chloroform | U044 | 15.8 |
| Acetonitrile | U003 | 15.6 |
| Acrylonitrile | U009 | 13.4 |
| 2-Methylaziridine | P067 | 12.2 |

402

Appendix D.    (Continued)

| Waste Name | Waste Code | Relative Soil Volatility |
|---|---|---|
| **Highly Volatile Wastes** - (continued) | | |
| Methylchlorocarbonate | U156 | 11.6 |
| Benzene | U019 | 11.4 |
| 2-Butanone | U159 | 10.6 |
| 1,1,1-Trichloroethane | U226 | 10.1 |
| Bromine cyanide | U246 | 9.71 |
| Tetrachloromethane | U211 | 9.29 |
| 1,2-Dimethylhydrazine | U099 | 8.78 |
| Ethyl acetate | U112 | 8.76 |
| 1,2-Dichloroethane | U077 | 7.61 |
| Methyl hydrazine | P068 | 7.31 |
| Trichloroethylene | U228 | 6.26 |
| N-methyl-N'-nitro-N-<br>  nitroso-isoguanidine | U163 | 5.70 |
| Propane nitrile | P101 | 5.39 |
| Glycidylaldehyde | U126 | 5.02 |
| Formic acid | U123 | 4.93 |
| 1,2-Dichloropropane | U083 | 4.70 |
| Dipropylamine | U110 | 4.33 |
| 1,4-Dioxane | U152 | 3.99 |
| 1,4-Diethylene dioxide | U108 | 3.94 |
| Allyl alcohol | P005 | 3.69 |
| Methyl methacrylate | U162 | 3.55 |
| Methylene bromide | U068 | 3.47 |
| Hydrogen sulfide | U135 | 3.43 |
| Ethyl acrylate | U113 | 3.40 |
| Bis(chloromethyl)ether | P016 | 3.39 |
| N-nitroso-N-methylurea | U177 | 3.30 |
| 2-Chloroethyl vinyl ether | U042 | 2.95 |
| Trichloroacetylaldehyde | U034 | 2.89 |
| Toluene | U220 | 2.79 |
| N-nitroso-n-methylurethane | U178 | 2.72 |
| 4-Pyridinamine | P008 | 2.66 |
| 1,3-Dichloropropane | U084 | 2.63 |
| Diethylarsine | P038 | 2.59 |
| Hydrazine | U133 | 2.54 |
| Crotonaldehyde | U053 | 2.27 |
| Pyridine | U196 | 2.25 |
| Paraldehyde | U182 | 2.20 |
| 1-Chloro-2,3-epoxypropane | U041 | 2.00 |
| 4-Methyl-2-pentanone | U161 | 1.90 |
| Ethylmethacrylate | U118 | 1.78 |
| 1,1,2-Trichloroethane | U227 | 1.64 |
| 2-Nitropropane | U171 | 1.60 |

Appendix D.   (Continued)

| Waste Name | Waste Code | Relative Soil Volatility |
|---|---|---|
| **Highly Volatile Wastes** - (continued) | | |
| N-nitroso-N-ethylurea | U176 | 1.51 |
| 2-Propyn-1-ol | P102 | 1.28 |
| N-Nitrosodiethanolamine | U173 | 1.28 |
| Isobutanol | U140 | 1.16 |
| Chlorobenzene | U037 | 1.11 |
| 2-Picoline | U191 | 1.04 |
| **Moderately Volatile Wastes** - Those with relative soil volatility from 1 to $10^{-3}$ | | |
| Tetranitromethane | P112 | $9.3 \times 10^{-1}$ |
| 1,2,3,4-Diepoxybutane | U085 | $8.1 \times 10^{-1}$ |
| Ethylene dibromide | U067 | $8.1 \times 10^{-1}$ |
| 1,1,1,2-Tetrachloroethane | U208 | $7.7 \times 10^{-1}$ |
| n-Butanol | U031 | $7.6 \times 10^{-1}$ |
| Cyclohexane | U056 | $7.4 \times 10^{-1}$ |
| Chloromethane | U045 | $7.1 \times 10^{-1}$ |
| Osmiumtetraoxide | P087 | $6.3 \times 10^{-1}$ |
| Dimethylnitrosamine | P082 | $5.7 \times 10^{-1}$ |
| Acrylic acid | U008 | $5.0 \times 10^{-1}$ |
| Bromoacetone | P017 | $4.7 \times 10^{-1}$ |
| Trichloromethyl mercaptan | P118 | $4.7 \times 10^{-1}$ |
| Cyclohexanone | U057 | $4.6 \times 10^{-1}$ |
| Cumene | U055 | $4.1 \times 10^{-1}$ |
| 1,4-Dichloro-2-butene | U074 | $3.6 \times 10^{-1}$ |
| Bromoform | U225 | $3.5 \times 10^{-1}$ |
| N-nitroso-di-n-butylamine | U172 | $3.3 \times 10^{-1}$ |
| 1,1,2,2-Tetrachloroethane | U209 | $3.2 \times 10^{-1}$ |
| m-Xylene | U239 | $3.1 \times 10^{-1}$ |
| p-Xylene | U239 | $3.1 \times 10^{-1}$ |
| Dimethylamine | U092 | $3.0 \times 10^{-1}$ |
| o-Xylene | U239 | $2.7 \times 10^{-1}$ |
| Pentachloroethane | U184 | $2.5 \times 10^{-1}$ |
| Dimethyl carbonyl chloride | P097 | $2.5 \times 10^{-1}$ |
| Furfural | U125 | $2.4 \times 10^{-1}$ |
| Benzene, 1,3,5-trinitro | U234 | $2.0 \times 10^{-1}$ |
| m-Dichlorobenzene | U071 | $1.7 \times 10^{-1}$ |
| N-Nitrosodiethylamine | U174 | $1.7 \times 10^{-1}$ |
| Benzenethiol | P014 | $1.4 \times 10^{-1}$ |
| 3-Chloropropionitrile | P027 | $1.4 \times 10^{-1}$ |
| 2,6-Dinitrotoluene | U106 | $1.3 \times 10^{-1}$ |
| Dichloroethyl ether | U025 | $1.2 \times 10^{-1}$ |
| o-Dichlorobenzene | U070 | $1.2 \times 10^{-1}$ |
| Metapyriline | U155 | $1.2 \times 10^{-1}$ |

Appendix D.  (Continued)

| Waste Name | Waste Code | Relative Soil Volatility |
|---|---|---|
| **Moderately Volatile Wastes** - (continued) | | |
| Benzylchloride | P028 | $1.1 \times 10^{-1}$ |
| Fluoroacetimide | P057 | $1.0 \times 10^{-1}$ |
| 2,3,4,6-Tetrachlorophenol | U212 | $1.0 \times 10^{-1}$ |
| Aniline | U012 | $8.9 \times 10^{-2}$ |
| Toxaphene (technical chlorinated camphene) | D015 | $8.9 \times 10^{-2}$ |
| Bis(2-chloroethoxy)methane | U024 | $7.9 \times 10^{-2}$ |
| o-Chlorophenol | U048 | $6.9 \times 10^{-2}$ |
| 1,2-Benzanthracene | U018 | $6.6 \times 10^{-2}$ |
| Bis(2-chloroisopropyl)ether | U027 | $6.5 \times 10^{-2}$ |
| Phenol | U188 | $6.4 \times 10^{-2}$ |
| Dimethyl sulfate | U103 | $5.1 \times 10^{-2}$ |
| p-Dichlorobenzene | U072 | $5.5 \times 10^{-2}$ |
| Acetophenone | U004 | $4.4 \times 10^{-2}$ |
| Diisopropylfluorophosphate | P043 | $4.3 \times 10^{-2}$ |
| o-Cresols | U052 | $4.2 \times 10^{-2}$ |
| Hexachloroethane | U131 | $3.9 \times 10^{-2}$ |
| Ethyl carbamate | U238 | $3.8 \times 10^{-2}$ |
| Di-N-propylnitrosamine | U111 | $3.7 \times 10^{-2}$ |
| 1,2-Dibromo-3-chloropropane | U066 | $3.3 \times 10^{-2}$ |
| Ethylmethane sulfonate | U119 | $3.0 \times 10^{-2}$ |
| Hexachloropropene | U243 | $2.9 \times 10^{-2}$ |
| Benzalchloride | U017 | $2.6 \times 10^{-2}$ |
| Naphthalene | U165 | $2.6 \times 10^{-2}$ |
| N-Nitrosopiperidine | U179 | $2.3 \times 10^{-2}$ |
| 5-Nitro-o-toluidine | U181 | $2.3 \times 10^{-2}$ |
| α,α-Dimethylphenethylen-amine | P046 | $2.3 \times 10^{-2}$ |
| Dihydrosafrole | U090 | $2.1 \times 10^{-2}$ |
| Nitrobenzene | U169 | $1.9 \times 10^{-2}$ |
| Malononitrile | U149 | $1.8 \times 10^{-2}$ |
| 1,2,3-Propanetriol, trinitroate | P081 | $1.7 \times 10^{-2}$ |
| m-Cresols | U052 | $1.7 \times 10^{-2}$ |
| Hexachlorobutadiene | U128 | $1.7 \times 10^{-2}$ |
| 1,4-Napthalenedione | U166 | $1.6 \times 10^{-2}$ |
| p-Cresols | U052 | $1.5 \times 10^{-2}$ |
| Ammonium picrate | P009 | $1.5 \times 10^{-2}$ |
| Selenium dioxide | U204 | $1.5 \times 10^{-2}$ |
| 1-Acetyl-2-thiourea | P002 | $1.3 \times 10^{-2}$ |
| Benzotrichloride | U023 | $1.1 \times 10^{-2}$ |
| 1,4-Dimethylphenol | U101 | $1.1 \times 10^{-2}$ |
| Uracil, 5[bis(2-chloromethyl)-amino] | U237 | $1.1 \times 10^{-2}$ |

Appendix D.  (Continued)

| Waste Name | Waste Code | Relative Soil Volatility |
|---|---|---|
| **Moderately Volatile Wastes** - (continued) | | |
| Tetraethyl lead | P110 | $1.1 \times 10^{-2}$ |
| 2,4,-Dichlorophenol | U081 | $9.2 \times 10^{-3}$ |
| 4-Chloro-o-toluidene | U049 | $8.8 \times 10^{-3}$ |
| p-Benzoquinone | U197 | $8.7 \times 10^{-3}$ |
| Safrole | U203 | $5.6 \times 10^{-3}$ |
| Hexachlorocyclopentadiene | U130 | $4.8 \times 10^{-3}$ |
| 1-Napthalamine | U167 | $4.7 \times 10^{-3}$ |
| 2-Napthalamine | U168 | $4.7 \times 10^{-3}$ |
| Nicotine | P075 | $4.4 \times 10^{-3}$ |
| Acrylamide | U007 | $3.9 \times 10^{-3}$ |
| Dinosob (2,4-dinitro-6- secbutyl-phenol) | P020 | $3.7 \times 10^{-3}$ |
| 2,4,5-Trichlorophenol | U230 | $3.5 \times 10^{-3}$ |
| Benzene sulfonyl chloride | U020 | $3.0 \times 10^{-3}$ |
| 2-Methylacetonitrile | P069 | $2.6 \times 10^{-3}$ |
| 4-Bromophenyl phenyl ether | U030 | $2.6 \times 10^{-3}$ |
| Isosafrole | U141 | $2.0 \times 10^{-3}$ |
| Toluene diisocyanate | U223 | $1.7 \times 10^{-3}$ |
| Methyl parathion | P071 | $1.6 \times 10^{-3}$ |
| Toluene diamine | U221 | $1.6 \times 10^{-3}$ |
| p-Chloroaniline | P024 | $1.3 \times 10^{-3}$ |
| N-phenylthiourea | P093 | $1.3 \times 10^{-3}$ |
| 2,6-Dichlorophenyl | U082 | $1.3 \times 10^{-3}$ |
| 4,6-Dinitro-o-cresol (and salts) | P047 | $1.3 \times 10^{-3}$ |
| 2,4,6-Trichlorophenol | U231 | $1.1 \times 10^{-3}$ |
| **Slightly Volatile Wastes** - Those with relative soil volatility below $10^{-3}$ to $10^{-6}$ | | |
| Tetraethyldithiopyrophosphate | P109 | $9.9 \times 10^{-4}$ |
| α,α-Dimethylbenzo- hydroperoxide | U096 | $8.9 \times 10^{-4}$ |
| Diethyl phthalate | U088 | $5.4 \times 10^{-4}$ |
| 4-Chloro-m-cresol | U039 | $4.9 \times 10^{-4}$ |
| Diallate | U062 | $4.8 \times 10^{-4}$ |
| β-Chloronapthalene | U047 | $4.1 \times 10^{-4}$ |
| p-Nitroaniline | P077 | $3.4 \times 10^{-4}$ |
| Phenacetin | U187 | $2.6 \times 10^{-4}$ |
| Dimethylphthalate | U102 | $2.6 \times 10^{-4}$ |
| Saccharin and salts | U202 | $2.0 \times 10^{-4}$ |
| o,o-Diethyl-o-pyrazinylphos- phorothioate | P040 | $1.9 \times 10^{-4}$ |

Appendix D.  (Continued)

| Waste Name | Waste Code | Relative Soil Volatility |
|---|---|---|

**Slightly Volatile Wastes** - (continued)

| Waste Name | Waste Code | Relative Soil Volatility |
|---|---|---|
| 2-Cyclohexyl-4,6-dinitrophenol | P034 | $1.8 \times 10^{-4}$ |
| 1,2,4,5-Tetrachlorobenzene | U207 | $1.5 \times 10^{-4}$ |
| Pentachloronitrobenzene | U185 | $1.4 \times 10^{-4}$ |
| 5-(Aminomethyl)-3-isoxazolol | P007 | $1.3 \times 10^{-4}$ |
| o-Toluidine hydrochloride | U222 | $1.0 \times 10^{-4}$ |
| Mercury | U151 | $9.2 \times 10^{-5}$ |
| Dibutyl phthalate | U069 | $6.0 \times 10^{-5}$ |
| Octamethylpyrophosphoramide | P085 | $5.9 \times 10^{-5}$ |
| 1,3-Propane sultone | U193 | $5.8 \times 10^{-5}$ |
| Phorate | P094 | $5.0 \times 10^{-5}$ |
| Reserpine | U200 | $>4.1 \times 10^{-5}$ |
| Tris(2,3-dibromopropyl) phosphate | U235 | $<3.8 \times 10^{-5}$ |
| 1-(o-Chlorophenyl)thiourea | P026 | $3.0 \times 10^{-5}$ |
| Tetraethylpyrophospate | P111 | $2.8 \times 10^{-5}$ |
| 1,2-Benzenediol, 4-[hydroxy-2-(methylamino)ethyl) | P042 | $2.8 \times 10^{-5}$ |
| Pentachlorobenzene | U183 | $2.6 \times 10^{-5}$ |
| Ethylenebis(diethiocarbonic acid) | U114 | $2.5 \times 10^{-5}$ |
| Promamide | U192 | $2.5 \times 10^{-5}$ |
| Phosphine | P026 | $2.4 \times 10^{-5}$ |
| Heptachlor | P059 | $1.6 \times 10^{-5}$ |
| Thiofanox | P045 | $1.2 \times 10^{-5}$ |
| Phthalic anhydride | U190 | $1.2 \times 10^{-5}$ |
| Disulfoton | P039 | $1.1 \times 10^{-5}$ |
| Resorcinol | U201 | $1.1 \times 10^{-5}$ |
| 2,4-D | P035 | $1.1 \times 10^{-5}$ |
| Azaserine | U015 | $9.5 \times 10^{-6}$ |
| Streptozotocin | U206 | $7.8 \times 10^{-6}$ |
| Hexahclorohexahydro,exo,exo dimethanonaphthalene | P060 | $7.3 \times 10^{-6}$ |
| Aldicarb | P070 | $7.3 \times 10^{-6}$ |
| Pentachlorophenol | U242 | $6.7 \times 10^{-6}$ |
| 2,4-Dinitrotoluene | U105 | $6.3 \times 10^{-6}$ |
| Maleic anhydride | U147 | $5.1 \times 10^{-6}$ |
| Diethyl-p-nitrophenyl phosphate | P041 | $4.7 \times 10^{-6}$ |
| Methylthiouracil | U164 | $4.0 \times 10^{-6}$ |
| 1,2-Diphenylhydrazine | U109 | $3.9 \times 10^{-6}$ |

Appendix D.   (Continued)

| Waste Name | Waste Code | Relative Soil Volatility |
|---|---|---|
| **Slightly Volatile Wastes - (continued)** | | |
| Methomyl | P066 | $3.9 \times 10^{-6}$ |
| 4-Nitrophenol | U170 | $3.3 \times 10^{-6}$ |
| Parathion | P089 | $2.2 \times 10^{-6}$ |
| 2,4,5'-T | U232 | $1.5 \times 10^{-6}$ |
| **Nonvolatile Wastes** - Those with relative soil volatility below $10^{-6}$ | | |
| Benzidine | U021 | $7.4 \times 10^{-7}$ |
| Bis-(2-ethylhexyl)phthalate | U028 | $7.0 \times 10^{-7}$ |
| Hexachlorobenzene | U127 | $6.5 \times 10^{-7}$ |
| Dimethoate | P044 | $5.6 \times 10^{-7}$ |
| Endosulfan | P050 | $5.0 \times 10^{-7}$ |
| Phenyl mercuric acetate | P092 | $4.9 \times 10^{-7}$ |
| Chlorodane, tech | U036 | $4.9 \times 10^{-7}$ |
| Silvex | U233 | $4.3 \times 10^{-7}$ |
| Octachlorocamphene | P123 | $3.9 \times 10^{-7}$ |
| 4,4'-Methylenebis (2-chloroaniline) | U158 | $3.7 \times 10^{-7}$ |
| Aldrin | P004 | $3.1 \times 10^{-7}$ |
| Chlornaphazine | U026 | $2.7 \times 10^{-7}$ |
| 3-Methcholanthrene | U157 | $2.3 \times 10^{-7}$ |
| α-hexachlorocyclohexane | U129 | $1.9 \times 10^{-7}$ |
| Cyclophosphamide | U058 | $1.7 \times 10^{-7}$ |
| Ethyl 4,4'-dichlorobenzilate | U038 | $1.2 \times 10^{-7}$ |
| Beryllium dust | P015 | $9.0 \times 10^{-8}$ |
| Dieldrin | P037 | $4.0 \times 10^{-8}$ |
| α-Naphthylthiourea | P072 | $3.6 \times 10^{-8}$ |
| 3,3-Dimethylbenzidine | U095 | $2.0 \times 10^{-8}$ |
| 2-Acetylaminoflourene | U005 | $1.4 \times 10^{-8}$ |
| 3,3'-Dimethoxybenzidine | U091 | $1.2 \times 10^{-8}$ |
| Endrin | P051 | $1.0 \times 10^{-8}$ |
| DDD | U060 | $8.4 \times 10^{-9}$ |
| Chloroambucil | U035 | $8.0 \times 10^{-9}$ |
| DDT | U061 | $8.0 \times 10^{-9}$ |
| 3,3'-Dichlorobenzidine | U073 | $7.0 \times 10^{-9}$ |
| Arsenic (III) oxide | P012 | $5.9 \times 10^{-9}$ |
| Di-n-octylphthalate | U107 | $3.5 \times 10^{-9}$ |
| Melphalan | U150 | $2.3 \times 10^{-9}$ |
| Mitomycin C | U010 | $2.2 \times 10^{-9}$ |

Appendix D.   (Continued)

| Waste Name | Waste Code | Relative Soil Volatility |
|---|---|---|
| **Nonvolatile Wastes** - (continued) | | |
| 3,4-Benzacridine | U016 | $5.3 \times 10^{-11}$ |
| Benzo(a)pyrene | U022 | $1.2 \times 10^{-11}$ |
| 1,2-Benzanthracene | U108 | $6.6 \times 10^{-12}$ |
| Ideno[1,2,3-CD]pyrene | U137 | $6.0 \times 10^{-12}$ |
| Dibenzo[a,h]anthracene | U063 | $3.1 \times 10^{-12}$ |
| Strychnine and salts | P018 | $6.6 \times 10^{-13}$ |
| Warfarin | P001 | $5.7 \times 10^{-13}$ |
| 7,12-Dimethylbenz[a]anthracene | U094 | $2.4 \times 10^{-13}$ |
| Diethylstilbestrol | U089 | $1.3 \times 10^{-13}$ |
| 1,2:7,8-Dibenzopyrene | U064 | $5.8 \times 10^{-14}$ |
| Brucine | P018 | $5.0 \times 10^{-14}$ |
| Hexachlorophene | U132 | $5.0 \times 10^{-14}$ |
| Chrysene | U050 | $6.6 \times 10^{-15}$ |
| Daunomycin | U059 | $4.1 \times 10^{-16}$ |

# Appendix E: Diffusion Coefficients in Air and Water for RCRA Wastes Identified as Highly Volatile from Water

| Waste Name | Waste Code | Henry's Law Constant (atm-m³/mol) | Diffusion Coefficient in Air* (cm²/sec) | Diffusion Coefficient in Water** (cm²/sec) | Liquid-Phase Mass Transfer Coefficient (cm/hr) |
|---|---|---|---|---|---|
| Bis(2-ethylhexyl phthalate) | U028 | 26.6 | 0.0378 | $3.78 \times 10^{-6}$ | 0.138 |
| Cyanogen | P031 | 9.91 | 0.1144(2) | $1.14 \times 10^{-5}$ | 18.40 |
| Reserpine | U200 | 4.28 | 0.0339 | $3.37 \times 10^{-6}$ | 0.111 |
| Dichlorodifluoromethane | U075 | 2.75 | 0.0944(2) | $1.13 \times 10^{-5}$ | 0.248 |
| Nickel Carbonyl | P073 | 0.5 | ††† | ††† | 0.217 |
| Chloromethane | U045 | 0.38 | 0.1085 | $1.18 \times 10^{-5}$ | 18.76 |
| Phosphine | P096 | 0.19 | 0.1570 | $1.11 \times 10^{-5}$ | 22.75 |
| Cyclohexane | U056 | 0.18 | 0.0039 | $9.10 \times 10^{-6}$ | 0.298 |
| 2-Nitropropane | U171 | 0.12 | 0.0084 | $9.90 \times 10^{-6}$ | 0.289 |
| Pentachloroethane | U184 | 0.10 | 0.0117 | $8.20 \times 10^{-6}$ | 0.192 |
| Hexachlorobutadiene | U128 | $9.14 \times 10^{-2}$ | 0.0614 | $6.78 \times 10^{-6}$ | 0.169 |
| Trichlorofluoromethane | U121 | $5.8 \times 10^{-2}$ | 0.0862 | $1.02 \times 10^{-5}$ | 0.233 |
| Hexachloroethane | U131 | $1.3 \times 10^{-2}$ | 0.0674 | $7.64 \times 10^{-6}$ | 0.177 |
| 2,3,4,6-tetrachlorophenol | U212 | $4.5 \times 10^{-2}$ | 0.0624 | $6.88 \times 10^{-6}$ | 0.179 |
| 1,3-Pentadiene | U186 | $4.2 \times 10^{-2}$ | 0.0912 | $9.84 \times 10^{-6}$ | 0.331 |
| Tetrachloromethane | U211 | $3.0 \times 10^{-2}$ | 0.0028(1) | $1.00 \times 10^{-5}$ | 0.220 |
| Pentachloronitrobenzene | U185 | $2.9 \times 10^{-2}$ | 0.0533 | $6.17 \times 10^{-6}$ | 0.159 |
| Hexachloropropene | U243 | $2.5 \times 10^{-2}$ | 0.0636 | $7.09 \times 10^{-6}$ | 0.173 |
| 1,1,1-Trichloroethane | U226 | $3.42 \times 10^{-2}$ | 0.0794(1) | $9.30 \times 10^{-6}$ | 0.236 |
| Hexachlorocyclopentadiene | U130 | $1.60 \times 10^{-2}$ | 0.0621 | $6.49 \times 10^{-6}$ | 0.165 |
| 1,1-Dichloroethylene | U078 | $1.50 \times 10^{-2}$ | 0.1144(2) | $1.14 \times 10^{-5}$ | 0.277 |
| Cumene | U055 | $1.40 \times 10^{-2}$ | 0.0702 | $7.55 \times 10^{-6}$ | 0.249 |
| DDD | U060 | $1.26 \times 10^{-2}$ | 0.0494 | $5.19 \times 10^{-6}$ | 0.152 |
| Carbon disulfide | P022 | $1.2 \times 10^{-2}$ | 0.1045(1) | $1.28 \times 10^{-5}$ | 0.313 |
| Trichloroethylene | U228 | $8.92 \times 10^{-3}$ | 0.0875(1) | $1.03 \times 10^{-5}$†† | 0.238 |
| 3-Methylcholanthrene | U157 | $1.1 \times 10^{-3}$ | 0.0501 | $5.28 \times 10^{-6}$ | 0.167 |
| Toluene | U220 | $6.64 \times 10^{-3}$ | 0.0849(1) | $9.10 \times 10^{-6}$ | 0.284 |
| 1,2-Dichloroethylene | U079 | $6.6 \times 10^{-3}$ | 0.1144(2)† | $1.14 \times 10^{-5}$ | 0.277 |
| Furan | U124 | $5.7 \times 10^{-3}$ | 0.1070 | $1.24 \times 10^{-5}$ | 0.331 |
| Bromomethane | U029 | $5.3 \times 10^{-3}$ | 0.1141 | $1.46 \times 10^{-5}$ | 0.280 |
| Benzene | U019 | $5.5 \times 10^{-3}$ | 0.0932(1) | $1.03 \times 10^{-5}$ | 0.309 |
| Methyl iodide | U138 | $5.0 \times 10^{-3}$ | 0.1025 | $1.32 \times 10^{-5}$ | 0.229 |
| Toxaphene | U224 | $4.89 \times 10^{-3}$ | *** | *** | 0.170 |

Appendix E.  (Continued)

| Waste Name | Waste Code | Henry's Law Constant ($atm\text{-}m^3/mol$) | Diffusion Coefficient in Air* ($cm^2/sec$) | Diffusion Coefficient in Water** ($cm^2/sec$) | Liquid-Phase Mass Transfer Coefficient ($cm/hr$) |
|---|---|---|---|---|---|
| 1,1-Dichloroethane | U076 | $5.45 \times 10^{-3}$ | 0.0919(1) | $1.25 \times 10^{-5}$ | 0.274 |
| 1,3-Dichloropropane | U084 | $4.2 \times 10^{-3}$ | 0.0817 | $9.22 \times 10^{-6}$ | 0.257 |
| Methanethiol | U153 | $4.0 \times 10^{-3}$ | 0.1242 | $1.43 \times 10^{-5}$ | 18.26 |
| Chlorobenzene | U037 | $3.93 \times 10^{-3}$ | 0.0741(1) | $9.31 \times 10^{-6}$ | 0.257 |
| Chloroform | U044 | $3.39 \times 10^{-3}$ | 0.0888(1) | $1.19 \times 10^{-5}$ | 0.250 |
| Methylene chloride | U080 | $3.19 \times 10^{-3}$ | 0.1037(1) | $1.29 \times 10^{-5}$ | 0.296 |
| 1,1,2-Tetrachloroethane | U208 | $2.76 \times 10^{-3}$ | 0.0739 | $8.4 \times 10^{-6}$ | 0.210 |

*Estimated via the Fuller, Schettler, and Giddings Method in Lyman et al. (1982).

**Estimated via the Hayduk and Laudie Method in Lyman et al. (1982).

***Cannot calculate because chemical consists of a mixture of molecules.

(1) Experimental value obtained from Lugg (1968).

(2) Experimental value obtained from Barr and Watts (1972).

† Assumed to be the same value as the experimental value of it's isomer, 1,1-dichloroethylene.

†† Estimated by Wilke and Chang Method (Lyman 1981).

†††Cannot calculate diffusion coefficient because a value for the structural contribution of nickel is needed to compute molar volume.  A value for the structural contribution of nickel was not found in the information reviewed.

# Appendix F: Diffusion Coefficients in Air and Water for RCRA Wastes Identified as Highly Volatile from Soil

| Waste Name | Waste Code | Relative Soil Volatility | | Diffusion Coefficient in Air* (cm²/sec) | Diffusion Coefficient in Water** (cm²/sec) |
|---|---|---|---|---|---|
| | | Dry | Wet | | |
| Formaldehyde | U122 | 1906 | 809 | 0.1728 | 2.06 x 10⁻⁵ |
| Cyanogen | P031 | 1473 | 552 | 0.1185 | 1.33 x 10⁻⁵ |
| Bromomethane | U029 | 1700 | 544 | 0.1139 | 1.46 x 10⁻⁵ |
| Dichlorodifluoromethane | U075 | 1450 | 439 | 0.0944(2) | 1.13 x 10⁻⁵ |
| Chloroethene | U043 | 958 | 336 | 0.1225(2) | 1.29 x 10⁻⁵ |
| Methanethiol | U153 | 578 | 219 | 0.1242 | 1.43 x 10⁻⁵ |
| Ethylene oxide | U115 | 505 | 195 | 0.1329 | 1.54 x 10⁻⁵ |
| Dimethylamine | U092 | --- | 193 | 0.0567 | 1.07 x 10⁻⁵ |
| Hydrofluoric acid | U134 | 378 | 179 | 0.2553 | 3.44 x 10⁻⁵ |
| Cyanogen chloride | P033 | 428 | 152 | 0.1213 | 1.47 x 10⁻⁵ |
| Phosgene | P095 | 458 | 144 | 0.1010 | 1.23 x 10⁻⁵ |
| Hydrocyanic acid | P063 | 321 | 140 | 0.1677 | 1.87 x 10⁻⁵ |
| Acetaldehyde | U001 | 360 | 139 | 0.1415 | 1.71 x 10⁻⁵ |
| Methylethylketone peroxide | U160 | 302 | 100 | 0.0853 | 9.50 x 10⁻⁶ |
| Furan | U124 | 222 | 77 | 0.1070 | 1.24 x 10⁻⁵ |
| Trichlorofluoromethane | U121 | 223 | 66 | 0.0862 | 1.02 x 10⁻⁵ |
| 1,1-Dichloroethylene | U078 | 202 | 64 | 0.1144(2) | 1.14 x 10⁻⁵ |
| Ethyl ether | U117 | 184 | 63 | 0.0892 | 9.70 x 10⁻⁶ |
| 1,3-Pentadiene | U186 | 145 | 50 | 0.0912 | 9.84 x 10⁻⁶ |
| Methylene chloride | U080 | 141 | 46 | 0.1037(1) | 1.29 x 10⁻⁵ |
| Carbon disulfide | P022 | 121 | 41 | 0.1045(1) | 1.28 x 10⁻⁵ |
| Chloroacetaldehyde | P023 | 108 | 36 | 0.1032 | 1.21 x 10⁻⁵ |
| Acetyl chloride | U006 | 108 | 36 | 0.1050 | 1.24 x 10⁻⁵ |
| 1-Propanamine | U194 | 101 | 36 | 0.0996 | 1.08 x 10⁻⁵ |
| Acrolein | P003 | 96 | 34 | 0.1131 | 1.29 x 10⁻⁵ |
| Methyl iodide | U138 | 116 | 34 | 0.1025 | 1.32 x 10⁻⁵ |

(1) Experimental value obtained from Lugg (1968).
(2) Experimental value obtained from Barr and Watts (1972).
*Estimated via the Fuller, Schettler, and Giddings Method in Lyman et al. (1982).
**Estimated via the Hayduk and Laudie Method in Lyman et al. (1982).

*Other Noyes Publications*

# INNOVATIVE THERMAL HAZARDOUS ORGANIC WASTE TREATMENT PROCESSES

by

## Harry Freeman

U.S. Environmental Protection Agency

### Pollution Technology Review No. 125

This book contains discussions of 21 thermal processes identified by the U.S. Environmental Protection Agency as innovative processes for treating or destroying hazardous organic wastes. The intent of the information provided is to assist in the evaluation of the processes by researchers and others interested in alternative processes for treating and disposing of hazardous wastes.

Today's rapidly developing and changing technologies and industrial products and practices frequently carry with them the increased generation of solid and hazardous wastes. These materials, if improperly dealt with, can threaten both public health and the environment.

While the processes included in the book differ widely in many respects (i.e., waste streams for which they are designed and state of development), they are similar in that they offer innovative approaches to solving problems presented by the generation of hazardous wastes. Some of the included processes might be regarded as emerging technologies. Others are in commercial operation and are already well beyond any such categorization as emerging technology.

Information about the subject processes was provided voluntarily by the process developers. The criteria used for selection of a process included the innovativeness of the process when compared with conventional existing processes and the potential contribution the process could make to the evolving field of hazardous waste management technology.

## CONTENTS

1. SUMMARY

2. INTRODUCTION

3. OVERVIEW

4. WET OXIDATION
Wet Air Oxidation
*Zimpro, Inc.*
Catalyzed Wet Oxidation
*IT Corporation*
Supercritical Fluid Oxidation
*Modar, Inc.*
High Temperature Wet Oxidation
*Methods Engineering, Inc.*

5. CHEMICAL TRANSFORMATION
Aqueous Phase Alkaline Destruction
of Halogenated Organic Wastes
*Battelle, Northwest*
Catalytic Dehalogenation of
Hazardous Waste
*GARD Division, Chamberlain Mfg. Co.*

6. MOLTEN GLASS
Joule Heated Glass Melter
*Battelle, Northwest*
Electromelt Pyro-Converter®
*Penberthy Electromelt International*

7. FLUIDIZED BED INCINERATION
Multisolid Fluidized Bed
*Battelle*
Circulating Bed Waste Incineration
*GA Technologies*
Low Temperature Fluidized Bed
*Waste Tech Services, Inc.*

8. PYROLYSIS
Pyrolytic Decomposition
*Midland-Ross Corporation*
High Temperature Pyrolysis
with Oxygen
*Russell and Axon*

9. MOLTEN SALT
Molten Salt Destruction
*Rockwell*

10. ADVANCED INCINERATORS
CONSERTHERM Rotary Kiln Oxidizer
*Industronics, Inc.*
Fast Rotary Reactor
*PEDCo Technology Corporation*
"CYCLIN" Cyclone Incinerator
*IGT*

11. ELECTRIC REACTOR
High Temperature Fluid Wall
*Thagard Research Corporation*
Advanced Electric Reactor
*Huber*

12. PLASMA SYSTEMS
Pyroplasma
*Pyrolysis Systems, Inc.*
Plasma Temperature Incinerator
*Applied Energetics*

ISBN 0-8155-1049-7 (1985)

125 pages

*Other Noyes Publications*

# LINER MATERIALS
# FOR HAZARDOUS AND TOXIC WASTES
# AND MUNICIPAL SOLID WASTE LEACHATE

by

## Henry E. Haxo, Jr., Robert S. Haxo, Nancy A. Nelson, Paul D. Haxo, Richard M. White, Suren Dakessian, Michael A. Fong

Matrecon, Inc.

*Pollution Technology Review No. 124*

This book presents studies which assess the relative effectiveness and durability of a wide variety of liner materials for use with hazardous and toxic wastes or with municipal solid waste (MSW) leachate.

Studies of land disposal of wastes have clearly indicated the need for positive control and containment measures to prevent contamination of surface water and groundwater systems that might result from on-land storage and disposal of hazardous wastes. Further, in sanitary landfills, used for MSW disposal, particularly in areas subject to high humidity and rainfall, leachates generated by water percolating through the landfill can also seriously degrade surface and groundwaters.

Recognizing this serious potential for pollution, the USEPA sponsored a series of studies to determine durable materials of very low permeability which might provide a long-term solution to the problem. Lining of landfills with such materials could also allow the use of sites which were previously unacceptable as landfills. The results of these studies are detailed in the book.

Listed below is a condensed table of contents including **part titles and selected chapter titles.**

*I. EXPOSURE TO HAZARDOUS AND TOXIC WASTES*

1. SUMMARY AND CONCLUSIONS

2. TECHNICAL APPROACH AND RESEARCH PLAN

3. SELECTION AND CHARACTERIZATION OF WASTES AND SLUDGES

4. SELECTION AND PROPERTIES OF LINER MATERIALS

5. EXPOSURE IN PRIMARY CELLS

6. IMMERSION TESTING OF POLYMERIC MEMBRANE LINERS

7. OUTDOOR EXPOSURE TESTS OF POLYMERIC MEMBRANE LINERS

8. POUCH TEST FOR POLYMERIC MEMBRANE LINER MATERIALS

9. ANALYSIS OF PYROLYSIS OF POLYMERIC MEMBRANE LINERS

*II. EXPOSURE TO MUNICIPAL SOLID WASTE LEACHATE*

10. TECHNICAL APPROACH

11. DESIGN AND CONSTRUCTION OF LANDFILL SIMULATORS

12. SELECTION OF LINER MATERIALS FOR EXPOSURE

13. CHARACTERIZATION OF MSW AND FILLING OF SIMULATORS

14. EXPOSURE TESTING OF PRIMARY AND BURIED LINER SPECIMENS IN MSW LANDFILL SIMULATORS

15. EXPOSURE TESTING OF POLYMERIC MEMBRANES BY IMMERSION IN MSW LEACHATE AND IN WATER

16. PERMEABILITY OF POLYMERIC MEMBRANE LINERS

17. SEAMING OF POLYMERIC LINER SPECIMENS AND EXPOSURE TESTING OF SEAMS

18. RECOVERY AND TESTING OF SAMPLES OF A POLYVINYL CHLORIDE LINER FROM A DEMONSTRATION LANDFILL

19. METHANE PRODUCTION IN SIMULATORS

20. DISCUSSION OF RESULTS

21. COSTS OF LINERS FOR LANDFILLS

REFERENCES

APPENDIXES

ISBN 0-8155-1048-9 (1985)

435 pages

# COSTS FOR HAZARDOUS WASTE INCINERATION
## Capital, Operation and Maintenance, Retrofit

by

### R.J. McCormick, R.J. DeRosier, K. Lim, R. Larkin, H. Lips

Acurex Corporation

*Pollution Technology Review No. 123*

Relationships between capital, and operation and maintenance (O&M) costs for hazardous waste incineration and the various waste-specific, design-specific, and operational factors that affect these costs are discussed in this book. These cost relationships were to be designed so that total capital investment and unit (dollars per pound, etc.) O&M cost estimates could be calculated for a variety of waste compositions, different incineration system designs and configurations, and a wide range of system operating conditions and performance requirements.

Also covered is a methodology that can be used to estimate the costs of retrofitting/upgrading various components of existing hazardous waste incineration facilities, to comply with RCRA performance requirements.

Based on the preliminary design specifications, equipment costs are estimated using a series of empirical relationships between cost and capacity, materials of construction, or other relevant design features. Estimated costs for equipment installation, indirects, and contingency are then combined with the equipment costs to determine total capital investment.

A condensed table of contents listing **part titles, chapter titles, and selected subtitles** is given below.

*I. CAPITAL AND OPERATION AND MAINTENANCE COST REQUIREMENTS*

**1. INPUT DATA SPECIFICATIONS**
Waste Quantities and Characteristics
Facility Characteristics

**2. DESIGN ASSUMPTIONS AND ENGINEERING CALCULATIONS**
Land and Site Development Requirements
Front-End Storage and Handling Equipment Design and Operation
Combustion System Operation
Waste Heat Boiler Operation
Quench Operation
Scrubber System Operation
Flue Gas Handling Equipment (ID Fan) Operation

**3. CAPITAL COST ESTIMATION**

**4. ANNUAL COST ESTIMATION**

**5. UNIT DISPOSAL COST CALCULATION**

*II. RETROFIT COST RELATIONSHIPS*

**6. INCINERATOR SYSTEMS CONSIDERED**

**7. ENGINEERING ECONOMIC PREMISES**

**8. COMBUSTION SYSTEM RETROFIT**
Burner Replacement
Refractory Replacement
Combustion Chamber Replacement

**9. QUENCH/WASTE HEAT BOILER ADDITION**
Quench Addition
Waste Heat Boilers
Low-Temperature Quenches

**10. SCRUBBING SYSTEM ADDITION/ REPLACEMENT/MODIFICATION**
Complete System Addition
Particulate Scrubbing System Addition/Replacement
Acid Gas Absorption System Addition/Replacement/Modification

**11. FLUE GAS HANDLING EQUIPMENT**
Induced Draft Fan Addition/ Replacement
Stack Replacement

**12. TOTAL INCINERATION SYSTEM REPLACEMENT**

**13. DOWNTIME CONSIDERATIONS**

*III. IONIZING WET SCRUBBER COSTS*

**14. INPUT DATA REQUIREMENTS**

**15. CAPITAL COSTS**

**16. OPERATING REQUIREMENTS**

ISBN 0-8155-1047-0 (1985)

274 pages